名师讲坛——Java 微服务架构实战

（SpringBoot+SpringCloud+Docker+RabbitMQ）

李兴华 编著

清華大學出版社

北 京

内 容 简 介

Java 微服务架构是当下最为流行的软件架构设计方案，可以快速地进行代码编写与开发，维护起来也非常方便。利用微架构技术，可以轻松地实现高可用、分布式、高性能的项目结构开发，同时也更加安全。

本书一共 15 章，核心内容为 SpringBoot、SpringCloud、Docker、RabbitMQ 消息组件。其中，SpringBoot 是 SpringMVC 技术的延伸，使用它进行程序开发会更简单，服务整合也会更容易。SpringCloud 是当前微架构的核心技术方案，属于 SpringBoot 的技术延伸，它可以整合云服务，基于 RabbitMQ 和 GITHUB 进行微服务管理。除此以外，本书还重点分析了 OAuth 统一认证服务的应用。

本书适用于从事 Java 开发且有架构与项目重构需求的读者，也适用于相关技术爱好者，同时也可作为应用型高等院校及培训机构的学习教材。

图书在版编目（CIP）数据

Java 微服务架构实战：SpringBoot+SpringCloud+Docker+RabbitMQ/李兴华编著．—北京：清华大学出版社，2020.1（2023.4重印）

（名师讲坛）

ISBN 978-7-302-50607-2

I. ①J… II. ①李… III. ①JAVA 语言-程序设计 IV. ①TP312.8

中国版本图书馆 CIP 数据核字（2018）第 151303 号

责任编辑：贾小红
封面设计：魏润滋
版式设计：楠竹文化
责任校对：马军令
责任印制：曹婉颖

出版发行：清华大学出版社
网　　址：http://www.tup.com.cn，http://www.wqbook.com
地　　址：北京清华大学学研大厦 A 座　　**邮　　编：**100084
社 总 机：010-83470000　　**邮　　购：**010-62786544
投稿与读者服务：010-62776969，c-service@tup.tsinghua.edu.cn
质量反馈：010-62772015，zhiliang@tup.tsinghua.edu.cn
印 装 者：三河市铭诚印务有限公司
经　　销：全国新华书店
开　　本：185mm×260mm　　**印　　张：**19　　**字　　数：**475 千字
版　　次：2020 年 1 月第 1 版　　**印　　次：**2023 年 4 月第 5 次印刷
定　　价：69.80 元

产品编号：080068-01

前 言

我们在用心做事，做最好的教育，写最好的原创图书。

笔者是一名从事 Java 开发快二十年的技术爱好者，一位普通的培训班老师，喜欢和学生们一边开着玩笑，一边教会他们当下流行与实用的技术。很多时候我会跟学生说："信息产业是一个不断发展变化的行业，没有人可以精确预测这个行业的未来发展方向，更没有人可以在这个行业里拥有绝对的技术实力。同样，也没有永远不过时的技术。我们能做的只是努力地学习与提升，每一天都要在踩坑与填坑的路上不断爬行，磕磕碰碰习惯了，解决问题所花费的时间就越来越少了。想要在这个行业走得长远，一定要喜欢这个行业，喜欢钻研。"

遥想起 2003 年开源风在中国兴起时，SSH（Spring 1.x + Struts 1.x + Hibernate 2.x）整合开发框架是当时最大的技术亮点。作为开发者的我们，最大的感受是再也不需要去编写那些重复的代码了，利用开发框架我们几乎可以解决当时所有的问题。然而技术的经典是短暂的，随着时间的流逝，SSH 的光环也不再辉煌。后来，有了 SSH2（Spring 2.x + Struts 2.x + Hibernate 3.x），又有了 SSM（Spring + Shiro + MyBatis）。随着开发框架的不断增加，以及 Spring 对各类开发框架的不断支持，新的问题出现了——参与整合的配置文件过多，项目的集成化太高。大家转而开始寻找新的解决方案。就在所有人都认为 Pivotal 公司（Spring 项目所属公司）已经停滞不前的时候，其在 2016 年推出了一套完善的轻量级分布式解决方案，就是今天流行的微架构（或称微服务），之中的主要技术手段是 SpringBoot + SpringCloud。

微架构的出现，很好地适应了这个时代对快速发展变化的要求。它不再提倡一体化的项目设计，而是对项目进行有效的"业务区"（可以简单理解为不同的子系统）划分，并利用合理的技术对业务性能做出提升和改善，同时又极大地简化了配置文件的使用与 profile 配置。总而言之，微架构是开发之中看起来非常简单的一种实现技术，但简单的背后考究的却是开发者对于开源技术的熟练程度。

SpringBoot 作为一种 Web 整合开发框架，很好地解决了 Web 程序的编写困难，可以更简单、高效地实现 MVC 设计模式。更为重要的是，它可以轻松地整合当前各类主流的开发项目，如消息组件、SQL 数据库、NoSQL 数据库、邮件服务等，因此能极大地缩短项目的开发周期，更快地响应客户的需求变更。SpringCloud 作为 SpringBoot 的延续，可以基于 Restful 流行架构实现 RPC 业务中心的搭建，可以基于消息组件实现远程配置动态的抓取，还可以与 Docker 相结合，采用虚拟化手段实现便捷的云服务管理。可以说，微架构的出现与云时代是密不可分的。

本书是笔者多年开发经验的总结，写作时力求能一针见血地分析透 Java 微服务的设计架构（见下图）与各类技术实现。全书围绕着当前的主流方案（高性能+高可用+分布式）进行展开，不仅讲解了所有微架构中的内容，还给出了真实有效的学习案例；不仅可以与虚拟化 Docker 整合开发，还可以实现大型企业分布式授权 OAuth 解决方案。可以说，本书就像 Java 微服务实现架构的一个技术宝典，读者学习后完全可以直接在实际项目之中进行应用。另外，由于微架构涉及到的技术非常广泛，对于某些技术还不十分清楚的读者，可以登录魔乐科技网站（www.mldn.cn）进行视频学习。

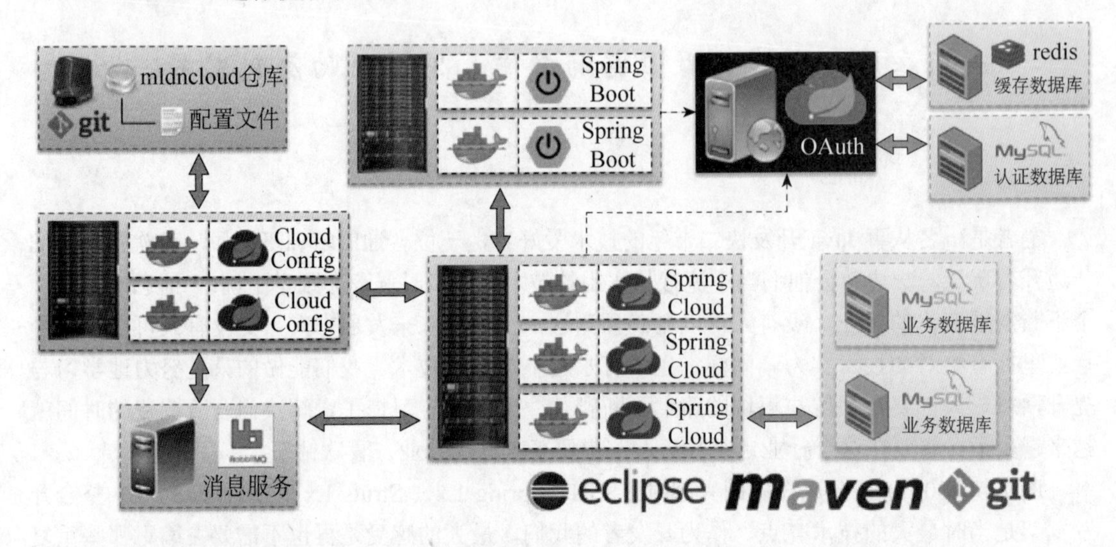

笔者崇尚原创，所出版的图书也均为原创。笔者将“技术实现优先”这一原则贯穿于全书，采用步骤分解的模式详细讲解每一步的开发，希望读者可以通过本书学习到微服务的技术精髓。另外，由于技术更新迭代过快，加之本人水平有限，书中难免有表达不到位或不明确的地方，欢迎读者批评指正，万分感谢。

创作不易。感谢我最爱的妻子和我的家人，是你们的付出与支持才让我可以安心创作，同时也祝福我年幼的儿子可以健康快乐地成长。

本书特色

☑ 资深 Java 讲师进行技术剖析，全面把握学习命脉，问题分析一针见血。
☑ 140 个课程案例，完美演示微服务的方方面面。
☑ 基于 Maven 实现项目管理，与真实项目完美衔接。
☑ 丰富的架构图示说明，轻松掌握微架构设计方案。
☑ 手把手步骤学习法，轻松掌握微架构开发。
☑ OAuth 使用分析与代码实现，掌握企业级 RPC 认证与授权解决方案。
☑ 微服务与 Docker 虚拟化技术结合使用，轻松实现云服务。

本书章节安排

全书涉及到的技术包括：SpringBoot、Thymeleaf、Jetty、Redis 整合、C3P0 整合、Druid 整合、MyBatis 整合、ActiveMQ 整合、RabbitMQ 整合、Kafka 整合、Shiro 整合、SpringDataJPA

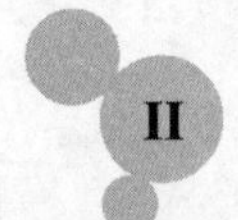

整合、Mail 整合、Actuator 监控、Restful、RestTemplate、Eureka、Ribbon、Feign、Hystrix、Turbine、Zuul、SpringCloudConfig、SpringCloudBus、SpringCloudStream、SpringCloudSleuth、Zipkin、OAuth、RabbitMQ 和 Docker。

考虑到学习层次，本书共分为 3 个组成部分：SpringBoot 篇、SpringCloud 篇和微服务辅助篇。

第一部分：**SpringBoot** 篇

☑ 第 1 章 **SpringBoot 编程起步**：本章将为读者讲解 SpringBoot 的发展背景与 SpringBoot 编程起步。

☑ 第 2 章 **SpringBoot 程序开发**：本章将为读者详细讲解 SpringBoot 开发常用的各项技术，包括代码测试、Jetty 配置、资源加载、访问路径、profile 配置、项目打包等。

☑ 第 3 章 **Thymeleaf 模板渲染**：Thymeleaf 是模板技术，也是当下 Web 开发中使用最多的一项技术，在 SpringBoot 中默认支持有此模板使用，本章将为读者讲解 Thyemeleaf 之中的使用语法以及与 JSP 语法的关联。

☑ 第 4 章 **SpringBoot 与 Web 应用**：主要讲解 https 协议整合、Tomcat 发布、全局异常处理、文件上传等。

☑ 第 5 章 **SpringBoot 服务整合**：主要讲解 C3P0、Druid、MyBatis、SpringDataJPA、ActiveMQ、RabbitMQ、Kafka、Redis、Shiro、Mail、Actuator 监控等组件的整合应用。

第二部分：**SpringCloud** 篇

☑ 第 6 章 **SpringCloud 简介**：主要讲解 RPC 技术的主要作用及 SpringCloud 技术实现架构。

☑ 第 7 章 **SpringCloud 与 Restful**：主要讲解 Restful 架构的基础实现方案、RestTemplate 调用微服务以及 SpringSecurity 基础认证处理。

☑ 第 8 章 **Eureka 注册服务**：主要讲解 Eureka 的作用、Eureka 微服务创建、Eureka 集群搭建、打包部署等。

☑ 第 9 章 **SpringCloud 服务组件**：主要讲解 Ribbon 负载均衡、Feign 接口转换、Hystrix 熔断机制以及 Zuul 代理机制。

☑ 第 10 章 **SpringCloudConfig**：与 GitHub 结合实现分布式配置文件管理、加密处理、SpringCloudBus 更新服务。

☑ 第 11 章 **SpringCloudStream**：讲解微服务中信息采集的搭建，主要与 RabbitMQ 整合。

☑ 第 12 章 **SpringCloudSleuth**：讲解微服务调用监控跟踪、Zipkin、数据采集。

☑ 第 13 章 **OAuth 认证管理**：分析 SpringSecurity 实现方案缺陷、OAuth 与 RPC 结合流程，并基于 SQL 数据库与 Redis 数据库实现 OAuth 认证与授权管理。

第三部分：微服务辅助篇

☑ 第 14 章 **RabbitMQ 消息组件**：RabbitMQ 与 Spring 微服务有着密不可分的关联，本章将为读者讲解 RabbitMQ 的安装、管理、Java 开发与集群使用。

☑ **第 15 章 Docker 虚拟化容器：**虚拟化与云开发是流行话题，本章主要讲解 Docker 虚拟化容器管理技术，同时讲解微服务与 Docker 的整合开发以及 DockerCompose 组件的使用。

寄语读者

本书全篇由笔者根据实践项目与教学经验总结而来，虽经过再三斟酌和审校，仍难免存在技术理解上的偏差和解释不到位的地方，欢迎读者批评指正。您的宝贵建议将帮助我们修正此书，大家一起努力，将传道、授业、解惑贯彻到底。

本书用到的程序源代码，读者可扫描图书封底的“文泉云盘”二维码获取其下载方式，也可登录清华大学出版社网站（www.tup.com.cn）进行下载。技术学习部分，读者可登录魔乐科技官网（http://www.mldn.cn）及沐言优拓官网（http://www.yootk.com）进行学习，也可登录笔者的新浪微博进行留言交流。

最后，希望本书成为您的良师益友。祝您读书快乐！

李兴华

目录

Contents

第一部分 SpringBoot 篇

第二部分 SpringCloud 篇

第三部分　微服务辅助篇

第一部分

SpringBoot 篇

- SpringBoot 与 Restful 标准
- SpringBoot 微服务创建
- Thymeleaf 语法标准
- SpringBoot 与服务整合

第 1 章

SpringBoot 编程起步

通过本章学习，可以达到以下目标：

1. 理解基于 Maven 的传统项目开发问题。
2. 理解 SpringBoot 开发框架的主要作用。
3. 编写第一个 SpringBoot 程序。

SpringBoot 是当下最为流行的 Java Web 端开发框架，该框架由 Spring（Pivotal 公司）开源组织提供，使用该框架可以解决传统项目之中混乱的 Maven 依赖管理问题，同时可以基于 Maven 快速进行项目的打包与发布。

1.1 传统开发中痛的领悟

在 Java 项目开发中，MVC 已经成为了一种深入人心的设计模式，几乎所有正规的项目之中都会使用到 MVC 设计模式。采用 MVC 设计模式可以有效地实现显示层、控制层、业务层、数据层的结构分离，如图 1-1 所示。

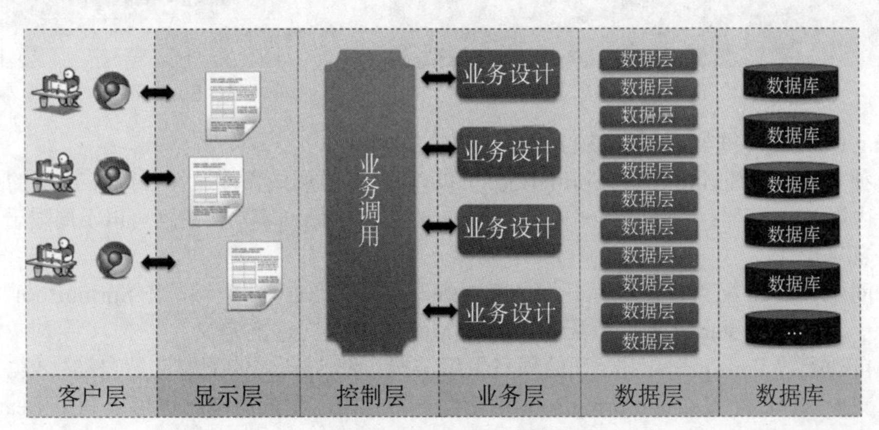

图 1-1 MVC 设计模式

虽然 MVC 开发具有良好的可扩展性，但是在实际的开发过程中，许多开发者依然会感受到如下的问题。

☑ 采用原生 Java 程序实现 MVC 设计模式时，一旦整体项目设计不到位，就会存在大量的重复代码，并且项目维护困难。

☑ 为了简化 MVC 各个层的开发，可以引用大量的第三方开发框架，如 Spring、Hibernate、MyBatis、Shiro、JPA、SpringSecurity 等，但这些框架都需要在 Spring 中实现整合，其结果就是会存在大量的配置文件。
☑ 当使用一些第三方的服务组件（如 RabbitMQ、Kafka、JavaMail 等）时，需要编写大量重复的配置文件，而且还需要根据环境定义不同的 profile（如 dev、beta、product）。
☑ 使用 Maven 作为构建工具时，需要配置大量的依赖关系，且程序需要被打包为*.war 文件并部署到应用服务器上才可以执行。
☑ Restful 作为接口技术应用得越来越广泛，但如果使用 Spring 来搭建 Restful 服务，则需要引入大量的 Maven 依赖库，并且需要编写许多的配置文件。

基于以上种种因素，很多人开始寻求更加简便的开发方式，而遗憾的是，这种简便的开发没有被官方的 JavaEE 所支持。JavaEE 官方支持的技术标准依然只提供最原始的技术支持。

1.2 SpringBoot 简介

SpringBoot 是 Spring 开发框架提供的一种扩展支持，其主要目的是希望通过简单的配置实现开发框架的整合，使开发者的注意力可以完全放在程序业务功能的实现上，其核心在于通过“零配置”的方式来实现快速且简单的开发。图 1-2 显示了 Spring 官方网站中 SpringBoot 项目，图 1-3 显示了 SpringBoot 当前的开发版本。

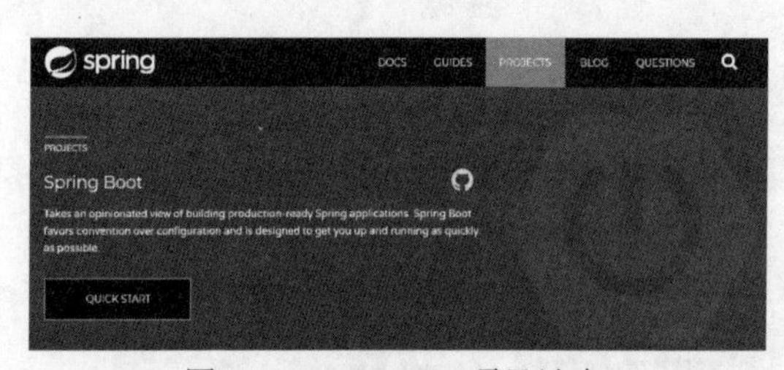

图 1-2 SpringBoot 项目站点

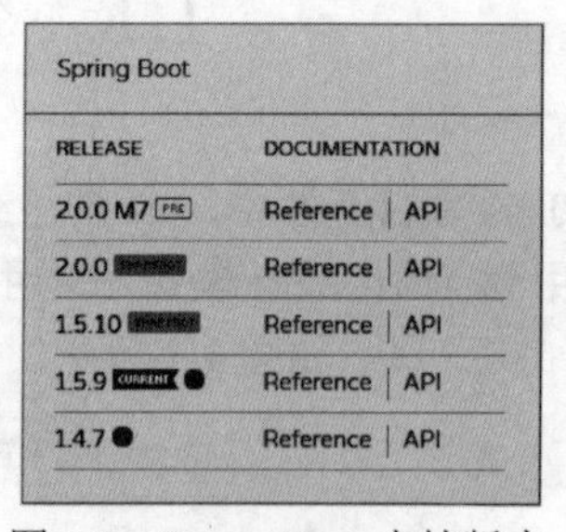

图 1-3 SpringBoot 支持版本

Spring Boot 开发框架有如下核心功能。
☑ 独立运行的 Spring 项目：SpringBoot 可以以 jar 包的形式直接运行在拥有 JDK 的主机上。
☑ 内嵌 Web 容器：SpringBoot 内嵌了 Tomcat 容器与 Jetty 容器，这样可以不局限于 war 包的部署形式。
☑ 简化 Maven 配置：在实际开发中需要编写大量的 Maven 依赖，在 SpringBoot 中会提供一系列使用 starter 的依赖配置来简化 Maven 配置文件的定义。
☑ 自动配置 Spring：采用合理的项目组织结构，使 Spring 的配置注解自动生效。
☑ 减少 XML 配置：在 SpringBoot 中依然支持 XML 配置，同时也可以利用 Bean 和自动配置机制减少 XML 配置文件的定义。

1.3 SpringBoot 编程起步

SpringBoot 编程需要依赖于 Maven 或 Gradle 构建工具完成，这里将直接使用 Maven 进行开

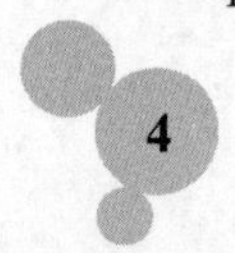

发，同时利用 Eclipse 来建立 Maven 项目。

1. 在 Eclipse 中创建一个新的 Maven 项目，项目类型为 quickstart，如图 1-4 所示。

Catalog: All Catalogs　Configure...

Filter:

Group Id	Artifact Id	Version
org.apache.maven.archetypes	maven-archetype-portlet	1.0.1
org.apache.maven.archetypes	maven-archetype-profiles	1.0-alpha-4
org.apache.maven.archetypes	maven-archetype-quickstart	1.1
org.apache.maven.archetypes	maven-archetype-site	1.1
org.apache.maven.archetypes	maven-archetype-site-simple	1.1

图 1-4　Eclipse 创建 Maven 项目

2. 设置 Maven 项目的信息（Group Id、Artifact Id、Version 等），本例建立的项目名称为 bootstart，如图 1-5 所示。

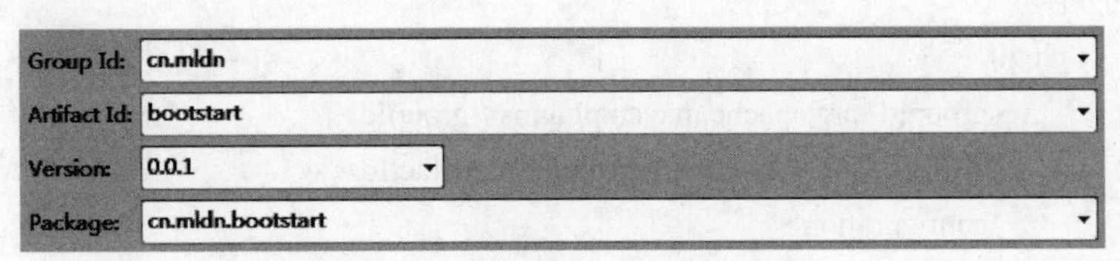

图 1-5　设置 Maven 的配置信息

3. 修改 pom.xml 配置文件，追加 SpringBoot 的依赖配置与相关插件。

```
<project xmlns="http://maven.apache.org/POM/4.0.0"
xmlns:xsi="http://www.w3.org/2001/XMLSchema-instance"
    xsi:schemaLocation="http://maven.apache.org/POM/4.0.0
http://maven.apache.org/xsd/maven-4.0.0.xsd">
    <modelVersion>4.0.0</modelVersion>
    <parent>                                                    <!-- 引入SpringBoot支持 -->
        <groupId>org.springframework.boot</groupId>
        <artifactId>spring-boot-starter parent</artifactId>
        <version>1.5.9.RELEASE</version>
    </parent>
    <groupId>cn.mldn</groupId>
    <artifactId>bootstart</artifactId>
    <version>0.0.1</version>
    <packaging>jar</packaging>
    <name>bootstart</name>
    <url>http://maven.apache.org</url>
    <properties>
        <jdk.version>1.8</jdk.version>
        <project.build.sourceEncoding>UTF-8</project.build.sourceEncoding>
    </properties>
    <dependencies>
```

```
        <dependency>
            <groupId>junit</groupId>
            <artifactId>junit</artifactId>
            <scope>test</scope>
        </dependency>
        <dependency>                                                <!-- 定义依赖配置 -->
            <groupId>org.springframework.boot</groupId>
            <artifactId>spring-boot-starter-web</artifactId>
        </dependency>
    </dependencies>
    <build>
        <finalName>bootstart</finalName>
        <plugins>
            <plugin>                                                <!-- 配置编译插件 -->
                <groupId>org.apache.maven.plugins</groupId>
                <artifactId>maven-compiler-plugin</artifactId>
                <configuration>
                    <source>${jdk.version}</source>               <!-- 源代码使用的开发版本 -->
                    <target>${jdk.version}</target>               <!-- 需要生成的目标class文件的
                                                                       编译版本 -->
                    <encode>${project.build.sourceEncoding}</encode>
                </configuration>
            </plugin>
        </plugins>
    </build>
</project>
```

本程序采用官方文档给出的配置方式实现了 SpringBoot 项目的创建。这里，spring-boot-starter-parent 就是官方给出的快速构建 SpringBoot 项目的公共父 pom.xml 配置文件支持。

注意：配置完成后要更新项目。

本例的项目开发是基于 Eclipse 完成的，开发者修改完 pom.xml 配置文件之后，一定要更新项目（快捷键为 Alt + F5），如图 1-6 所示。

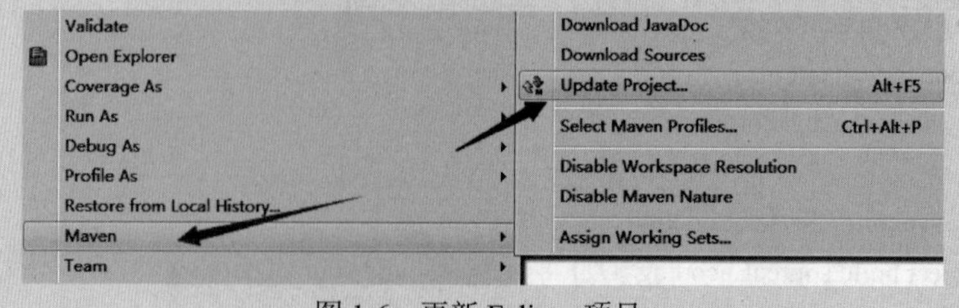

图 1-6　更新 Eclipse 项目

4. 编写第一个 SpringBoot 程序。

```
package cn.mldn.bootstart;
import org.springframework.boot.SpringApplication;
import org.springframework.boot.autoconfigure.EnableAutoConfiguration;
import org.springframework.stereotype.Controller;
import org.springframework.web.bind.annotation.RequestMapping;
import org.springframework.web.bind.annotation.ResponseBody;
@Controller                                                    // 控制器注解
@EnableAutoConfiguration                                       // 启用自动配置
public class SampleController {
    @RequestMapping("/")                                       // 访问映射路径
    @ResponseBody                                              // Restful返回形式
    public String home() {                                     // 控制器方法
        return "www.mldn.cn";                                  // 返回信息
    }
    public static void main(String[] args) throws Exception {
        SpringApplication.run(SampleController.class, args);  // 启动SpringBoot程序
    }
}
```

5. 这里使用了 Eclipse-STS（Spring Source Tool）插件，所以可以直接运行，如图 1-7 所示。

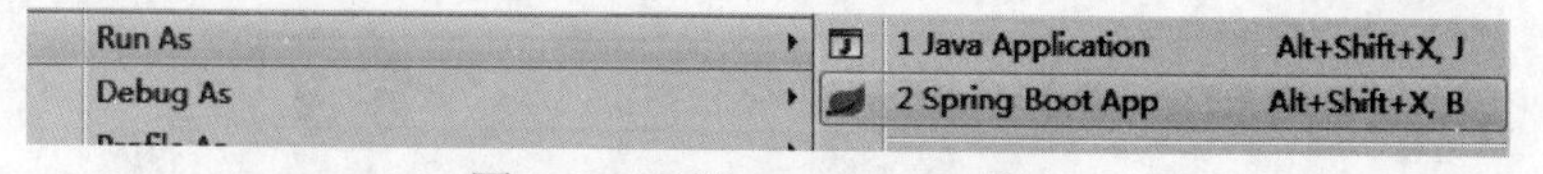

图 1-7　运行 SpringBoot 项目

提示：采用 Maven 工具直接运行。

如果 Eclipse 工具中没有安装 STS 开发插件，也可以通过 Maven 的方式运行。直接输入 spring-boot:run，就可以启动 SpringBoot 项目了，如图 1-8 所示。

图 1-8　使用 Maven 运行 SpringBoot 程序

考虑到代码编写的方便，本书强烈建议读者安装 STS 开发插件。本书中所有的项目也都是通过 STS 插件的方式开发的。同时需要提醒读者的是，当使用 Eclipse-STS 建立了 SpringBoot 项目时，会在项目后面标记“[boot]”提示信息。

6. SpringBoot 项目启动之后，开发者可以直接通过控制台看到如图 1-9 所示的信息提示，完成后的项目结构如图 1-10 所示。

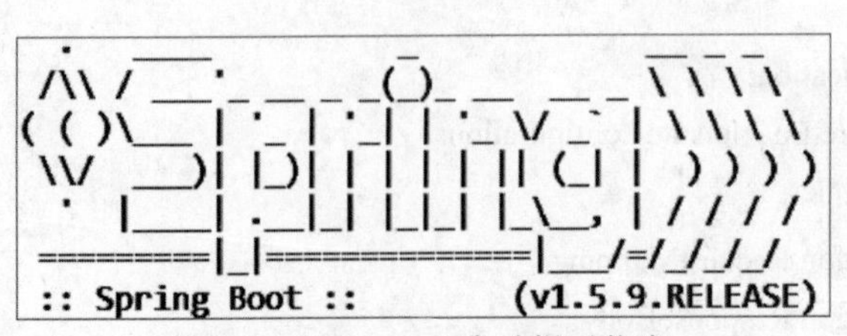

图 1-9　SpringBoot 启动提示信息

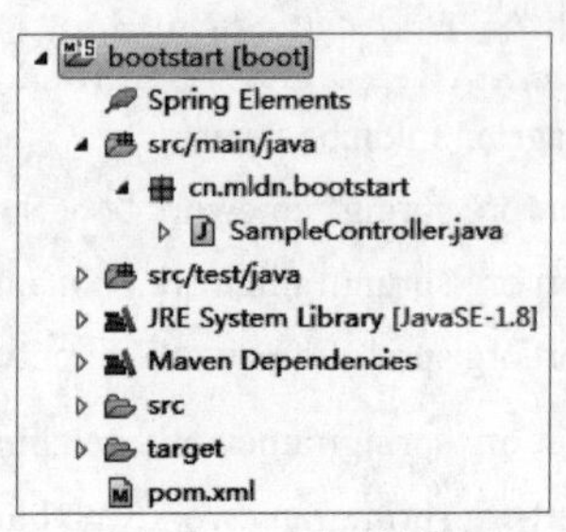

图 1-10　项目结构

同时也可以在控制台中看到如下的日志信息：

```
s.b.c.e.t.TomcatEmbeddedServletContainer    : Tomcat initialized with port(s): 8080 (http)
o.apache.catalina.core.StandardService      : Starting service [Tomcat]
org.apache.catalina.core.StandardEngine     : Starting Servlet Engine: Apache Tomcat/8.5.23
o.a.c.c.C.[Tomcat].[localhost].[/]          : Initializing Spring embedded WebApplicationContext
o.s.web.context.ContextLoader               : Root WebApplicationContext: initialization completed in 1388 ms
o.s.b.w.servlet.FilterRegistrationBean      : Mapping filter: 'characterEncodingFilter' to: [/*]
o.s.j.e.a.AnnotationMBeanExporter           : Registering beans for JMX exposure on startup
s.b.c.e.t.TomcatEmbeddedServletContainer    : Tomcat started on port(s): 8080 (http)
cn.mldn.bootstart.SampleController          : Started SampleController in 2.328 seconds (JVM running for 3.601)
```

由于 SpringBoot 自带 Tomcat 容器，所以项目启动后通过浏览器输入 http://localhost:8080，就可以直接访问控制器返回的信息，如图 1-11 所示。

图 1-11　SpringBoot 运行界面

1.4　本章小结

1. SpringBoot 提倡的是一种简洁的开发模式，可保证用户不被大量的配置文件和依赖关系所困扰。

2. SpringBoot 开发需要 Maven 或 Gradle 构建工具支持。

3. SpringBoot 使用一系列的注解来简化开发过程。

第 2 章

SpringBoot 程序开发

通过本章学习，可以达到以下目标：

1. 掌握 SpringBoot 开发标准。
2. 掌握 SpringBoot 中的常用注解。
3. 掌握内置对象在 SpringBoot 中的使用。
4. 掌握 SpringBoot 项目的打包与发布处理。
5. 掌握 Spring 多开发环境（profile）的配置。
6. 掌握 Spring 错误处理以及错误跳转处理。
7. 掌握 Spring 与 Tomcat 的结合使用。

SpringBoot 项目的实现主要依赖于构建工具，在使用构建工具定义的过程中往往需要提供一个标准的父 pom 定义。同时，一个良好的 SpringBoot 程序中会使用大量的注解来代替配置文件，以轻松实现项目的打包与部署。本章将为读者进一步详细讲解与 SpringBoot 有关的开发支持。

2.1　建立统一父 pom 管理

在项目中使用 SpringBoot，往往会需要引入一个标准的父 pom 配置。

```
<parent>                                                    <!-- 引入SpringBoot支持 -->
    <groupId>org.springframework.boot</groupId>
    <artifactId>spring-boot-starter-parent</artifactId>
    <version>1.5.9.RELEASE</version>
</parent>
```

利用这个父 pom 文件，可以方便地进行核心依赖库的导入，并且由父 pom 统一管理所有的开发版本。但在实际的 Maven 项目开发中，开发团队往往会根据自己的需要来自定义属于自己的父 pom，这样就会造成冲突。为了解决这样的问题，在 SpringBoot 里面，用户也可以直接以依赖管理的形式使用 SpringBoot。

1.【mldnboot 项目】建立一个用于管理父 pom 的 Maven 项目 mldnboot，如图 2-1 所示。

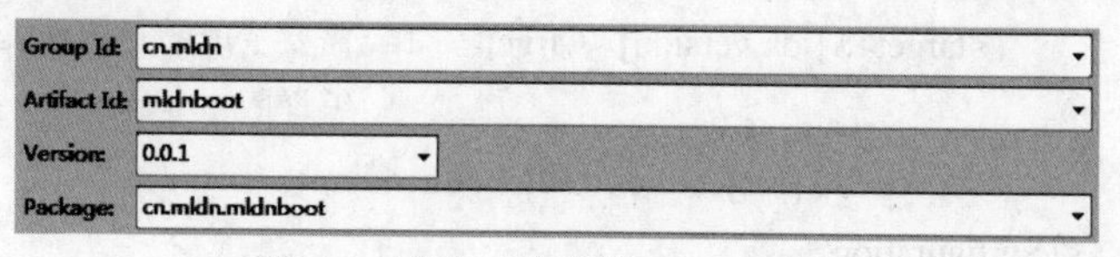

图 2-1　建立一个新的 Maven 项目

2.【mldnboot 项目】修改 pom.xml 配置文件，追加相关依赖配置项。

```
<project xmlns="http://maven.apache.org/POM/4.0.0"
xmlns:xsi="http://www.w3.org/2001/XMLSchema-instance"
    xsi:schemaLocation="http://maven.apache.org/POM/4.0.0
http://maven.apache.org/xsd/maven-4.0.0.xsd">
    <modelVersion>4.0.0</modelVersion>
    <groupId>cn.mldn</groupId>
    <artifactId>mldnboot</artifactId>
    <version>0.0.1</version>
    <packaging>pom</packaging>                              <!-- 定义为pom类型 -->
    <name>mldnboot</name>
    <url>http://maven.apache.org</url>
    <properties>
        <jdk.version>1.8</jdk.version>
    <spring-boot-dependencies.version>1.5.9.RELEASE</spring-boot-dependencies.version>
        <project.build.sourceEncoding>UTF-8</project.build.sourceEncoding>
    </properties>
    <dependencyManagement>
        <dependencies>
            <dependency>                                    <!-- 定义SpringBoot依赖管理 -->
                <groupId>org.springframework.boot</groupId>
                <artifactId>spring-boot-dependencies</artifactId>
                <version>${spring-boot-dependencies.version}</version>
                <type>pom</type>
                <scope>import</scope>
            </dependency>
        </dependencies>
    </dependencyManagement>
    <build>
        <finalName>mldnboot</finalName>
        <plugins>
            <plugin>                                        <!-- 配置编译插件 -->
                <groupId>org.apache.maven.plugins</groupId>
                <artifactId>maven-compiler-plugin</artifactId>
                <configuration>
                    <source>${jdk.version}</source><!-- 源代码使用的开发版本 -->
                    <target>${jdk.version}</target> <!-- 需要生成的目标class文件的
                                                        编译版本 -->
                    <encode>${project.build.sourceEncoding}</encode>
                </configuration>
```

```
        </plugin>
    </plugins>
  </build>
</project>
```

3.【mldnboot 项目】在 mldnboot 父项目之中建立一个新的 Maven 模块 mldnboot-base，如图 2-2 所示。

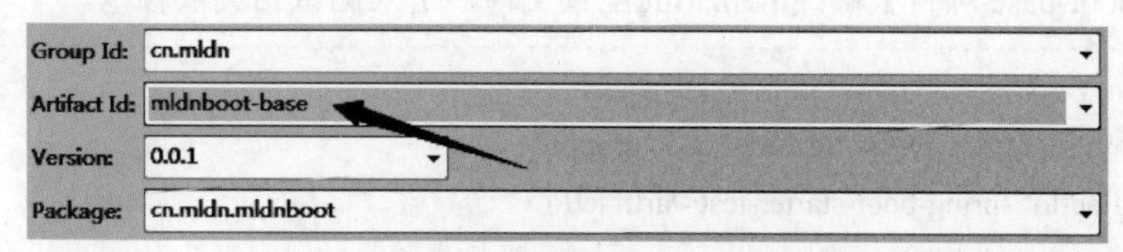

图 2-2　建立 mldnboot-base 子模块

4.【mldnboot-base 项目】修改 pom.xml 配置文件，追加要引入的 SpringBoot 依赖配置。

```
<dependency>
    <groupId>org.springframework.boot</groupId>
    <artifactId>spring-boot-starter-web</artifactId>
</dependency>
```

5.【mldnboot-base 项目】编写程序，实现 SpringBoot 基础开发。

```
package cn.mldn.mldnboot;
import org.springframework.boot.SpringApplication;
import org.springframework.boot.autoconfigure.EnableAutoConfiguration;
import org.springframework.stereotype.Controller;
import org.springframework.web.bind.annotation.RequestMapping;
import org.springframework.web.bind.annotation.ResponseBody;
@Controller                                              // 控制器注解
@EnableAutoConfiguration                                 // 启用自动配置
public class SampleController {
    @RequestMapping("/")                                 // 访问映射路径
    @ResponseBody                                        // Restful返回形式
    public String home() {                               // 控制器方法
        return "www.mldn.cn";                            // 返回信息
    }
    public static void main(String[] args) throws Exception {
        SpringApplication.run(SampleController.class, args);   // 启动SpringBoot程序
    }
}
```

本程序与第 1 章中的 SpringBoot 程序功能相同，正常启动 SpringBoot 后就可以通过浏览器得到相应的结果了。

2.2 SpringBoot 程序测试

SpringBoot 程序开发完成之后，需要对程序的功能进行测试，这时需要启动 Spring 容器。开发者可以直接利用 SpringBoot 提供的依赖包来实现控制层方法测试。

1.【mldnboot-base 项目】修改 pom.xml 配置文件，引入测试相关依赖包。

```
<dependency>
    <groupId>org.springframework.boot</groupId>
    <artifactId>spring-boot-starter-test</artifactId>
    <scope>test</scope>
</dependency>
<dependency>
    <groupId>junit</groupId>
    <artifactId>junit</artifactId>
    <scope>test</scope>
</dependency>
```

2.【mldnboot-base 项目】编写一个测试程序类。

```
package cn.mldn.mldnboot.test;
import org.junit.Test;
import org.junit.runner.RunWith;
import org.springframework.beans.factory.annotation.Autowired;
import org.springframework.boot.test.context.SpringBootTest;
import org.springframework.test.context.junit4.SpringJUnit4ClassRunner;
import org.springframework.test.context.web.WebAppConfiguration;
import cn.mldn.mldnboot.SampleController;
import junit.framework.TestCase;
@SpringBootTest(classes = SampleController.class)          // 定义要测试的SpringBoot类
@RunWith(SpringJUnit4ClassRunner.class)                    // 使用JUnit进行测试
@WebAppConfiguration                                       // 进行Web应用配置
public class TestSampleController {
    @Autowired
    private SampleController sampleController;             // 注入控制器对象
    @Test
    public void testHome() {                               // 使用JUnit测试
        TestCase.assertEquals(this.sampleController.home(), "www.mldn.cn");
    }
}
```

3.【mldnboot-base】测试程序编写完成之后，就可以启动测试了。如果测试通过，则返回如图 2-3 所示的界面。

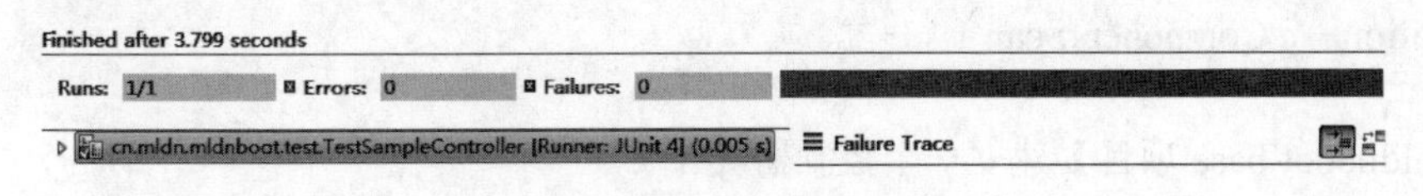

图 2-3　SpringBoot 测试成功

2.3　SpringBoot 注解分析

通过前面的学习可以发现，在整个 SpringBoot 程序里面使用了许多注解，这些注解的作用如表 2-1 所示。

表 2-1　SpringBoot 注解

No.	注解	说明
1	@Controller	进行控制器的配置注解，这个注解所在的类就是控制器类
2	@EnableAutoConfiguration	开启自动配置处理
3	@RequestMapping("/")	表示访问的映射路径，此时的路径为“/”，访问地址为 http://localhost:8080/
4	@ResponseBody	在 Restful 架构之中，该注解表示直接将返回的数据以字符串或 JSON 的形式获得

在给定的几个注解中，@EnableAutoConfiguration 为整个 SpringBoot 的启动注解配置，也就是说，这个注解应该随着程序的主类一起进行定义。但是该注解有一个前提，就是只能够扫描在同一程序类包中的配置程序，很明显其功能是不足的。

对于控制器程序类，由于在项目中有许多的控制器，那么最好将这些类统一保存在一个包中（如将所有的控制器程序类保存在 cn.mldn.mldnboot.controller 中，这是 cn.mldn.mldnboot 子包），在这样的环境下建议开发者使用@SpringBootApplication 注解实现启动配置。

注意：请严格遵守 SpringBoot 的自动配置约束。

在 SpringBoot 开发过程中，为了简化开发配置，往往会在 SpringBoot 启动类下创建若干个子包，这样子包中的注解就都可以自动扫描到（@EnableAutoConfiguration 注解不支持此功能），并且可以实现依赖关系的自动配置。以本程序为例，如果要进行标准开发，则程序的开发包结构如图 2-4 所示。

```
cn.mldn.mldnboot（父包）
        |- SpringBootStartApplicationMain（程序启动类）
        |- cn.mldn.mldnboot.controller（子包）
                |- XxxController（控制器程序类）
        |- cn.mldn.mldnboot.service（子包）
                |- IXxxService（业务接口）
```

图 2-4　SpringBoot 项目结构

此时保存在 cn.mldn.mldnboot 下的所有子包中配置的注解都可以被 Spring 容器自动扫描。如果不在指定的子包中，程序启动类就需要配置@ComponentScan 注解设置扫描包。这样的配置会显得整个项目非常啰嗦，如果不是必须的情况下，不建议这样配置。

可以简单地将@SpringBootApplication 理解为：@SpringBootApplication=@EnableAutoConfiguration +@ComponentScan（扫描父包）。

1.【mldnboot-base 项目】建立一个控制器类。

```
package cn.mldn.mldnboot.controller;
import org.springframework.stereotype.Controller;
import org.springframework.web.bind.annotation.RequestMapping;
import org.springframework.web.bind.annotation.ResponseBody;
@Controller                                         // 建立控制器
public class MessageController {
    @RequestMapping("/")                            // 访问映射路径
    @ResponseBody                                   // Restful返回形式
    public String home() {                          // 控制器方法
        return "www.mldn.cn";                       // 返回信息
    }
}
```

2.【mldnboot-base 项目】编写程序启动类（SpringBootStartApplication），使用@SpringBootApplication 进行注解。

```
package cn.mldn.mldnboot;
import org.springframework.boot.SpringApplication;
import org.springframework.boot.autoconfigure.SpringBootApplication;
@SpringBootApplication    // 启动SpringBoot程序，而后自带子包扫描
public class SpringBootStartApplication {
    public static void main(String[] args) throws Exception {
        SpringApplication.run(SpringBootStartApplication.class, args);     // 启动SpringBoot程序
    }
}
```

由于启动程序类保存在 cn.mldn.mldnboot 父包下，所以该包中所有的子包都将被自动扫描，而后自动实现配置。

2.4 配置访问路径

在实际的项目开发中，控制器的路径可能会有许多个。在进行控制器编写的时候，也会有以下两种运行模式。

☑ 控制器跳转模式：可以使用@Controller 注解定义，如果要实现 Restful 显示，也可以联合@ResponseBody 注解一起使用。

☑ Restful 显示：可以使用@RestController 注解，里面所有路径访问的信息都以 Restful 形式展示。

1.【mldnboot-base 项目】定义 MessageController 控制器程序类，使用 Restful 风格显示。

```
package cn.mldn.mldnboot.controller;
import org.springframework.web.bind.annotation.RequestMapping;
import org.springframework.web.bind.annotation.RestController;
@RestController                         // 建立控制器，所有路径以Restful形式运行
public class MessageController {
    @RequestMapping("/")                // 访问映射路径
    public String home() {              // 控制器方法
        return "www.mldn.cn";           // 返回信息
    }
}
```

此时的控制器程序类上使用了@RestController 注解，这样就可以避免在方法上使用@ResponseBody 注解。此时，MessageController 类中的所有映射路径都会以 Restful 形式展示。

提示：Restful 是 SpringCloud 技术的实现核心。

在控制器里面一旦使用了@RestController 注解，则意味着所有方法都将以 Restful 风格展示。这种做法未必适合于 SpringBoot 项目，因为在很多时候需要通过控制器跳转到显示层页面，而 Restful 是 SpringCloud 技术的实现关键。

2.【mldnboot-base 项目】前面定义的控制器类只能进行简单的信息返回，实际上也可以进行参数的接收处理。传递参数到控制器中最简单的做法是使用地址重写传递“访问路径?参数名称=内容”，在 MessageController 控制器程序类之中扩充一个新的 echo()方法。

```
    @GetMapping("/echo")                    // 只支持GET请求模式
    public String echo(String msg) {        // 接收msg参数
        return "【ECHO】" + msg ;           // 信息处理后返回
    }
```

此时如果要进行该路径的访问，则可以直接通过地址栏传递参数（http://localhost:8080/echo?msg=www.mldn.cn），并且参数的名称应该默认使用 msg，程序执行后的界面如图 2-5 所示。

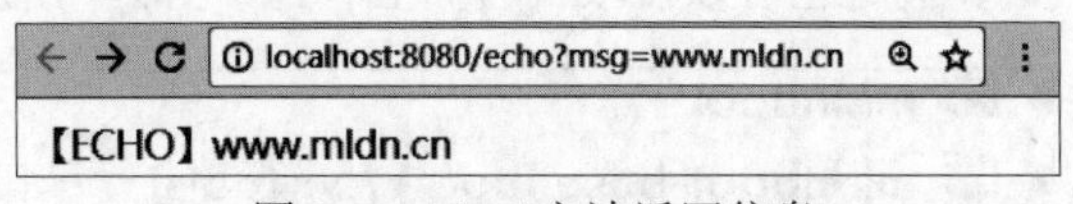

图 2-5　echo()方法返回信息

3.【mldnboot-base 项目】由于 SpringBoot 支持 Restful 风格处理，所以参数的接收可以采用路径参数的形式完成，但是需要在控制器方法的参数声明上使用@PathVariable 注解与访问路径的参数进行关联。

```
    @GetMapping("/echo/{message}")                          // 只支持GET请求模式
    public String echo(@PathVariable("message") String msg) {   // 接收msg参数
```

```
        return "【ECHO】" + msg ;                    // 信息处理后返回
    }
```

本程序需要通过地址传递参数，地址设置为 http://localhost:8080/echo/www.mldn.cn，程序显示结果如图 2-6 所示。

图 2-6　通过地址传递参数

提示：关于传递参数的选择。

在 Restful 架构中请求路径受多类语法支持，开发者可以结合 HTTP 请求模式（GET、POST、PUT、DELETE 等）与路径，实现多种组合，以处理不同类型的用户请求。参数的传递模式可以由开发团队自行定义。

2.5　SpringBoot 调试

在项目开发的过程中经常需要对代码进行反复修改，这样就会导致 SpringBoot 运行容器反复启动。为了解决这种频繁重启问题，SpringBoot 提供了自动加载配置的依赖库，以实现代码的动态加载。

范例：【mldnboot-base 项目】修改 pom.xml 配置文件，追加自动加载依赖库配置。

```
<dependency>
    <groupId>org.springframework</groupId>
    <artifactId>springloaded</artifactId>
</dependency>
<dependency>
    <groupId>org.springframework.boot</groupId>
    <artifactId>spring-boot-devtools</artifactId>
</dependency>
```

项目中配置了以上两个开发包之后，每当用户修改项目中程序类的时候都会由 SpringBoot 自动加载更新后的程序代码，同时也可以在项目名称上看到如图 2-7 所示的标记“[devtools]”。

图 2-7　SpringBoot 动态加载更新程序

2.6　使用内置对象

通过 SpringBoot 程序可以发现，SpringBoot 中控制器的形式和 SpringMVC 中是一样的，因

此在程序中使用 JSP 的内置对象也可以按照与 SpringMVC 同样的方式进行。

1.【mldnboot-base 项目】在 MessageController 控制器类之中追加新的方法，用于实现内置对象获取。

```
@GetMapping("/object")
public Object object(HttpServletRequest request,HttpServletResponse response) {
    Map<String,String> map = new HashMap<String,String>() ;
    map.put("客户端IP地址", request.getRemoteAddr()) ;
    map.put("客户端响应编码", response.getCharacterEncoding()) ;
    map.put("SessionID", request.getSession().getId()) ;
    map.put("项目真实路径", request.getServletContext().getRealPath("/")) ;
    return map ;                                // 以Restful风格返回
}
```

此时采用了与 SpringMVC 同样的方式来获取内置对象，并且将所有的信息保存在 Map 集合中，最后以 Restful 形式返回获取的信息（将 Map 集合自动变为 JSON 数据），程序运行界面如图 2-8 所示。

图 2-8 获取内置对象信息

2.【mldnboot-base 项目】除了在控制器的方法上使用参数来接收内置对象外，也可以利用 ServletRequestAttributes 形式来获取内置对象。

```
@GetMapping("/object")
public Object object() {
    HttpServletRequest request = ((ServletRequestAttributes)
RequestContextHolder.getRequestAttributes())
            .getRequest();              // 获取HttpServletRequest内置对象
    HttpServletResponse response = ((ServletRequestAttributes)
RequestContextHolder.getRequestAttributes())
            .getResponse();             // 获取HttpServletResponse内置对象
    Map<String,String> map = new HashMap<String,String>() ;
    map.put("客户端IP地址", request.getRemoteAddr()) ;
    map.put("客户端响应编码", response.getCharacterEncoding()) ;
    map.put("SessionID", request.getSession().getId()) ;
    map.put("项目真实路径", request.getServletContext().getRealPath("/"));
    return map ;                        // 以Restful风格返回
}
```

本程序实现了与上一程序完全相同的处理效果，唯一的区别是，控制器的方法不再需要明确地接收内置对象的参数，程序运行效果与图 2-8 相同。

2.7 使用 Jetty 容器

SpringBoot 在项目启动时默认情况下使用的是 Tomcat 容器，这一点可以通过日志直观看到。

```
Tomcat started on port(s): 8080 (http)
```

在实际的开发过程中，开发者往往会选择 Jetty 作为 Web 容器，由于 SpringBoot 也支持 Jetty 容器，所以开发者只需要修改 pom.xml 配置文件即可。

1.【mldnboot-base 项目】修改 pom.xml 文件，使用 Jetty 容器运行。

```
<dependency>
    <groupId>org.springframework.boot</groupId>
    <artifactId>spring-boot-starter-jetty</artifactId>
</dependency>
```

2.【mldnboot-base 项目】修改完 pom.xml 文件后，需要重新启动 SpringBoot 项目，此时就可以在日志中看到如下信息。

```
Jetty started on port(s) 8080 (http/1.1)
```

程序可以使用小巧的 Jetty 容器来运行 SpringBoot 项目，但是这种做法也仅仅是在开发过程中使用，在实际的生产环境下依然推荐使用 Tomcat 作为 Web 容器。

2.8 配置环境属性

SpringBoot 提倡的是一种“零配置”的设计框架，所以提供有许多默认的配置项。例如，SpringBoot 项目默认运行的 8080 端口就是一种默认配置。如果开发者需要修改 SpringBoot 的这种默认配置，可以在项目所在的 CLASSPATH 下添加 application.properties 配置文件。

1.【mldnboot-base 项目】建立一个新的源文件目录 src/main/resources。

2.【mldnboot-base 项目】在 src/main/resources 源文件目录中建立 application.properties 配置文件，目录结构如图 2-9 所示。

图 2-9　定义配置文件

注意：配置文件名称要相同。

SpringBoot 开发框架对一些结构（子包扫描）和配置文件（application.properties）做出了限定，这样开发者在使用框架开发的时候可以减少配置。如果开发者定义的配置文件名称不是 application.properties，那么 SpringBoot 将无法加载。

在本文件中进行 SpringBoot 项目默认端口的变更，将其修改为 80 端口运行。

```
# 设置Tomcat的运行服务所在端口
server.port=80
```

修改完成后重新启动 SpringBoot 项目（使用的是 Tomcat 容器），可以看到提示信息：Tomcat started on port(s): 80 (http)，表示当前的项目可以直接运行在 80 端口上。

3. 【mldnboot-base 项目】SpringBoot 项目默认情况下会将程序发布在根目录下，如果有需要，也可以配置上下文路径（ContextPath）。

```
# 设置Tomcat的运行服务所在端口
server.port=80
# 可以配置ContextPath访问路径，但是在实际开发中是不能够进行配置的
server.context-path=/mldnjava
```

本程序追加了一个 context-path 配置，所以项目的访问路径为 http://localhost/mldnjava/echo/www.mldn.cn（追加了/mldnjava 的路径前缀），页面运行效果如图 2-10 所示。

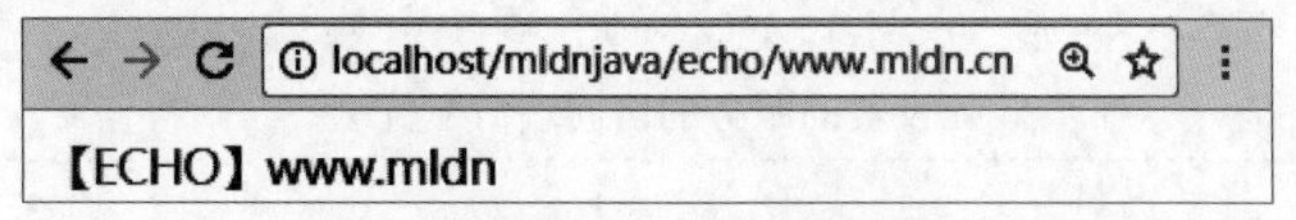

图 2-10　增加 ContextPath 配置

4. 【mldnboot-base 项目】在 SpringBoot 中可以使用的配置文件类型有两种：application.properties 和 application.yml，这两种配置文件都可以实现对 SpringBoot 环境的修改。下面将 application. properties 配置替换为 application.yml，内容如下：

```
server:
  port: 80                      # 此处设置的服务的访问端口配置
  context-path: /mldnjava       # 定义 ContextPath 路径
```

此时实现了与之前完全相同的配置，读者可以发现使用 application.yml 配置文件的结构要比使用 application.properties 更加清晰。

提示：关于 yml 配置文件说明。

yml 实际上是 YAML（Yet Another Markup Languange，一种标记语言）文件，这是一种结构化的数据文件，大量应用在各种开源项目之中，如 Apache Storm。

Spring 官方推荐使用 application.yml 来进行 SpringBoot 或 SpringCloud 框架的配置定义。如果项目中同时存在 application.yml 与 application.properties 配置文件并且配置冲突，将以 application.properties 文件中的配置为参考。在本书后面讲解的过程中，如无意外，将全部使用 application.yml 进行 SpringBoot 项目的配置。

2.9 读取资源文件

在实际的项目开发中，资源文件不可或缺，因为所有的提示文字信息都要在资源文件中进行定义，而且资源文件是实现国际化技术的主要手段。如果想在 SpringBoot 里面进行资源文件的配置，只需要做一些简单的 application.yml 配置即可，而且所有注入的资源文件都可以像最初的 Spring 处理那样，直接使用 MessageSource 进行读取。

1.【mldnboot-base 项目】在 src/main/resources 源文件夹下创建一个 i18n 的子目录（包）。

2.【mldnboot-base 项目】建立 src/main/resources/i18n/Messages.properties 文件，文件内容定义如下：

```
welcome.url=www.mldn.cn
welcome.msg=欢迎{0}光临！
```

3.【mldnboot-base 项目】修改 application.yml 配置文件，追加资源文件配置，项目结构如图 2-11 所示。

```
spring:
  messages:                    # 定义资源文件，多个资源文件使用“,”分割
    basename: i18n/Messages
server:
  port: 80                     # 此处设置的服务的访问端口配置
```

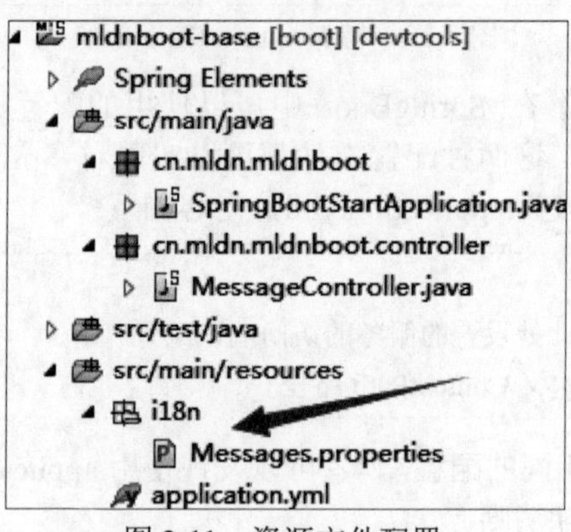

图 2-11 资源文件配置

4.【mldnboot-base 项目】在 MessageController 控制器中注入 org.springframework.context.MessageSource 接口对象，并且利用此对象实现资源文件读取。

```
@Autowired
private MessageSource messageSource; // 自动注入此资源对象
@GetMapping("/message")
public Object message() {
    Map<String,String> map = new HashMap<String,String>();
```

```
        map.put("welcome.url", this.messageSource.getMessage("welcome.url", null,Locale.getDefault()));
        map.put("welcome.msg", this.messageSource.getMessage("welcome.msg", new Object[] {"李兴华
"}, Locale.getDefault()));
        return map;
    }
```

当程序中配置了资源文件之后，就可以通过 MessageSource 接口中提供的 getMessage()方法进行资源的读取。本程序的运行效果如图 2-12 所示。

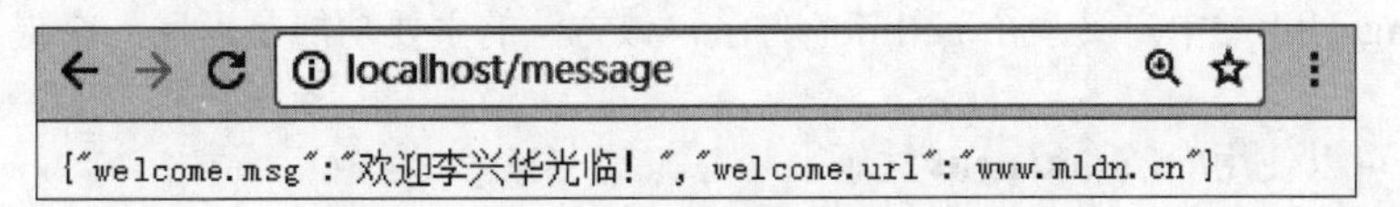

图 2-12　读取资源文件

提示：可以借用此机制实现国际化开发。

当程序可以实现资源文件读取的时候，就意味着可以实现国际化开发处理了。可以发现，MessageSource 接口中的 getMessage()方法里面需要接收一个 Locale 类的对象，此时就可以通过 Locale 类的设置来获取不同的资源文件。当然，也需要在项目中配置好不同语言的资源文件。例如，本程序在 src/main/resources/i18n 目录中又创建了 Messages_zh_CN.properties 和 Messages_en_US.properties（注意 baseName 的名称相同），如图 2-13 所示。

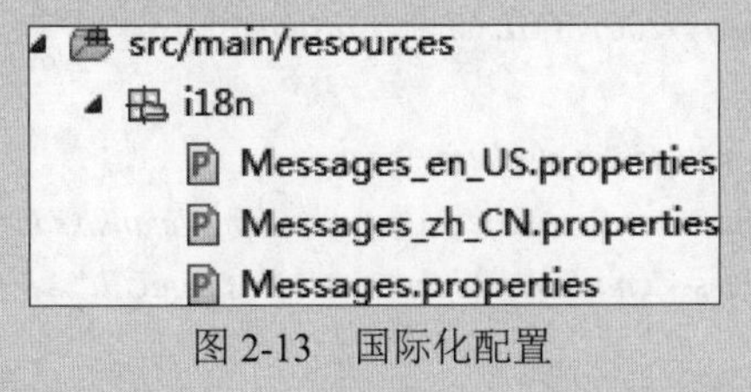

图 2-13　国际化配置

这样，当读取时可以采用不同的 Locale 对象实现指定语言的资源读取。例如，使用如下代码就可以实现 Messages_en_US.properties 资源文件的读取：

```
    map.put("welcome.msg", this.messageSource.getMessage("welcome.msg", new Object[] {"李兴华"},
new Locale("en","US"))) ;
```

需要提醒读者的是，即使提供了不同语言的资源文件，在 SpringBoot 中也依然需要提供 Messages.properties 配置文件，否则将无法实现资源文件的读取。

2.10　整合 Spring 配置

在进行 Spring 项目配置的时候，可以通过*.xml 文件配置，也可以通过 Bean（@Configuration 注解）配置。SpringBoot 延续了 Spring 这一特点，在 SpringBoot 项目中依然可以使用配置文件定义。

1.【mldnboot-base 项目】建立一个 MessageUtil 的工具类，该类的主要功能是进行配置的演示。

```
package cn.mldn.mldnboot.util;
public class MessageUtil {
    public String getInfo() {
        return "魔乐科技：www.mldn.cn" ;
    }
}
```

在 MessageUtil 类中定义一个 getInfo()方法，该方法的主要功能是返回一个提示信息。

提示：可以使用@Component 注解。

MessageUtil 类直接放在了程序启动类所在包的子包之中。在真实的开发中，开发者选择@Component 或@Repository 这样的注解是最方便的。本部分主要是为读者讲解配置文件与配置 Bean 的整合，所以没有采用注解配置。

2.【mldnboot-base 项目】在 src/main/resources 目录中创建 spring 的子目录，并且建立 spring-util.xml 配置文件。

```
<?xml version="1.0" encoding="UTF-8"?>
<beans xmlns="http://www.springframework.org/schema/beans"
    xmlns:xsi="http://www.w3.org/2001/XMLSchema-instance"
    xsi:schemaLocation="
        http://www.springframework.org/schema/beans
        http://www.springframework.org/schema/beans/spring-beans.xsd">
    <bean id="messageUtil" class="cn.mldn.mldnboot.util.MessageUtil" />      <!-- 定义Bean配置 -->
</beans>
```

3.【mldnboot-base 项目】在 MessageController 程序类中注入 MessageUtil 类对象，并且调用方法返回信息。

```
    @Autowired
    private MessageUtil messageUtil ;        // XML配置注入
    @GetMapping("/info")
    public Object info() {
        return this.messageUtil.getInfo() ; // 调用方法
    }
```

4.【mldnboot-base 项目】修改程序启动主类，定义要导入的 Spring 配置文件。

```
package cn.mldn.mldnboot;
import org.springframework.boot.SpringApplication;
import org.springframework.boot.autoconfigure.SpringBootApplication;
import org.springframework.context.annotation.ImportResource;
@SpringBootApplication    // 启动SpringBoot程序，而后自带子包扫描
```

```
@ImportResource(locations={"classpath:spring/spring-util.xml"})
public class SpringBootStartApplication {
    public static void main(String[] args) throws Exception {
        SpringApplication.run(SpringBootStartApplication.class, args);      // 启动SpringBoot程序
    }
}
```

本程序在定义启动主类时，利用@ImportResource 注解导入了所需要的 Spring 配置文件，而后会自动将配置文件中定义 bean 对象注入到 MessageController 类的属性中，程序运行结果如图 2-14 所示。

图 2-14　Spring 配置文件

5.【mldnboot-base 项目】SpringBoot 强调的就是“零配置”，虽然其本身支持配置文件定义，但很明显这样的处理形式不是最好的。如果确定要引入其他配置，强烈建议使用 Bean 的配置形式来完成。

```
package cn.mldn.mldnboot.config;                    // 放在主类所在子包
import org.springframework.context.annotation.Bean;
import org.springframework.context.annotation.Configuration;
import cn.mldn.mldnboot.util.MessageUtil;
@Configuration                                      // 本类为配置Bean
public class DefaultConfig {
    @Bean(name="messageUtil")                       // 定义Bean
    public MessageUtil getMessageUtil() {
        return new MessageUtil() ;                  // 配置Bean对象
    }
}
```

DefaultConfig 定义在程序主类所在的子包之中，这样就可以在 SpringBoot 程序启动时自动扫描配置并进行加载。对于程序的主类，也就没有必要使用@ImportResource 注解读取配置文件了。

提问：实际开发中使用配置文件还是使用 Bean 类配置？

在编写 SpringBoot 项目的过程之中，是采用*.xml 配置更好，还是利用 Bean 类配置会更好？

回答：崇尚“零配置”的 SpringBoot 项目建议使用 Bean 配置。

在 SpringBoot 项目中进行配置的时候，实际上有 3 种支持，按照优先选择顺序为：application.yml、Bean 配置和*.xml 配置文件。大部分的配置都可以在 application.yml（相当于传统项目中的 profile 配置作用）里面完成，但很多情况下会利用 Bean 类来进行扩展配置（本书主要使用此形式来作为扩展配置）。之所以提供*.xml 配置文件的支持，主要目的是帮助开发者用已有代码快速整合 SpringBoot 开发框架。

2.11　SpringBoot 项目打包发布

SpringBoot 作为微架构的主要实现技术，其发布项目的方式极为简单，只需要在项目中配置好插件，然后打包执行就可以了，并且这个执行不需要特别复杂的配置。

1.【mldnboot-base 项目】修改 pom.xml 配置文件，配置 SpringBoot 的打包插件。

```
<plugin>                              <!-- 该插件的主要功能是进行项目的打包发布处理 -->
    <groupId>org.springframework.boot</groupId>
    <artifactId>spring-boot-maven-plugin</artifactId>
    <configuration>                   <!-- 设置程序执行的主类 -->
        <mainClass>cn.mldn.mldnboot.SpringBootStartApplication</mainClass>
    </configuration>
    <executions>
        <execution>
            <goals>
                <goal>repackage</goal>
            </goals>
        </execution>
    </executions>
</plugin>
```

2.【mldnboot-base 项目】由于 Maven 增加了新的插件配置，所以需要对项目进行更新，如图 2-15 所示。更新时选择 mldnboot-base 项目，如图 2-16 所示。

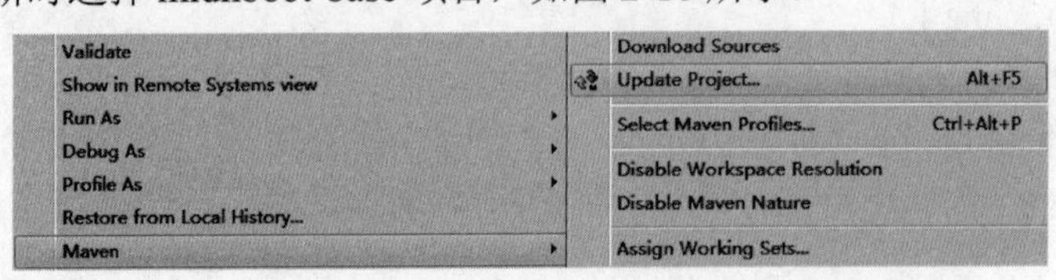

图 2-15　Maven 项目更新

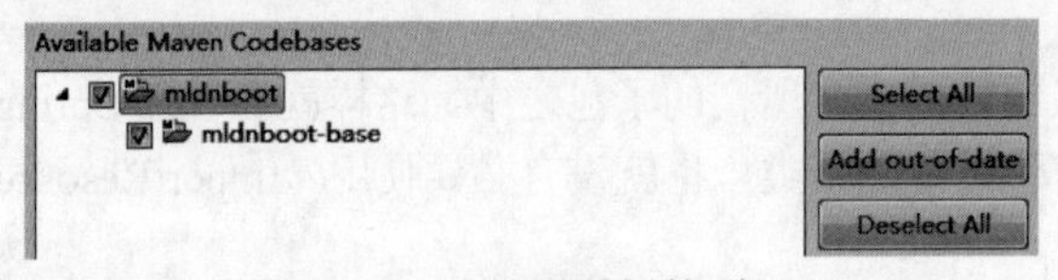

图 2-16　选择要更新的项目

3.【mldnboot-base 项目】将当前项目模块进行打包处理（clean package），如图 2-17 所示。打包完成后，会在项目的 target 目录下生成 mldnboot-base.jar 程序文件。

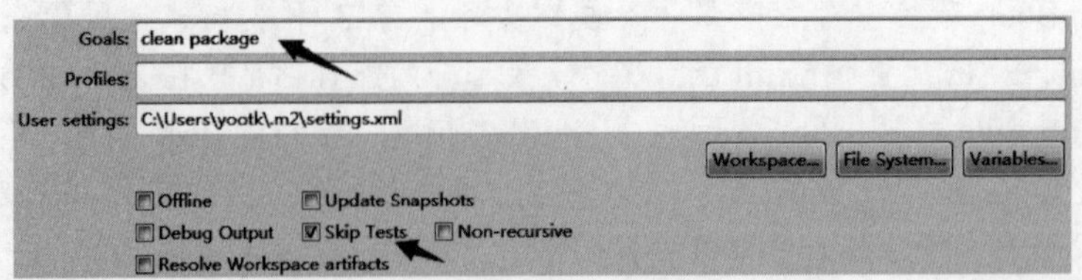

图 2-17　SpringBoot 项目打包

4.【mldnboot-base 项目】将 mldnboot-base.jar 文件复制到 D 盘根目录下，随后通过命令行方式执行此文件。

启动 SpringBoot 项目	java -jar mldnboot-base.jar

此时，SpringBoot 项目将以一个独立的*.jar 文件的方式执行。将此 jar 文件上传到任何配置有 JDK 的系统内，可以轻松实现项目的发布。

2.12　profile 配置

在项目开发过程中需要考虑不同的运行环境：开发环境（dev）、测试环境（beta）和生产环境（product）。在以往的开发过程中通常使用 Maven 构建工具进行控制，但却需要进行大量的配置。SpringBoot 考虑到此类问题，专门设计了 profile 支持。

1.【mldnboot-base 项目】修改 application.yml 配置文件，让其支持多 profile 配置。

```
spring:
  profiles:
    active: dev                     # 定义默认生效的环境
---
spring:
  profiles: dev
server:
  port: 7070                        # 此处设置开发服务的访问端口配置
---
spring:
  profiles: beta
server:
  port: 8080                        # 此处设置测试服务的访问端口配置
---
spring:
  profiles: product
server:
  port: 80                          # 此处设置生产服务的访问端口配置
```

在本配置文件中一共定义了 3 个环境（不同的 profile 之间使用“---”分割）。

☑　开发环境（profiles=dev、默认）：端口定义为 7070。

☑　测试环境（profiles=beta）：端口定义为 8080。

☑　生产环境（profiles=product）：端口定义为 80。

2.【mldnboot 项目】如果要正常进行打包，还需要修改 pom.xml 文件，追加 resource 配置。

```
<resources>
    <resource>
        <directory>src/main/resources</directory>
        <includes>
            <include>**/*.properties</include>
            <include>**/*.yml</include>
            <include>**/*.xml</include>
            <include>**/*.tld</include>
        </includes>
        <filtering>false</filtering>
    </resource>
    <resource>
        <directory>src/main/java</directory>
        <includes>
            <include>**/*.properties</include>
            <include>**/*.xml</include>
            <include>**/*.tld</include>
        </includes>
        <filtering>false</filtering>
    </resource>
</resources>
```

本程序主要的功能是进行源文件夹中内容的打包输出，配置完成后可以将配置文件打包到 *.jar 文件中。

3.【mldnboot-base 项目】为项目打包，这里直接通过 Eclipse 进行打包配置（此时无法设置 profile），如图 2-18 所示。

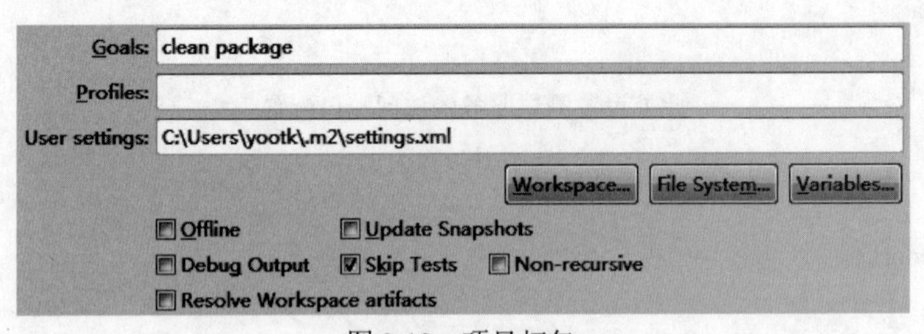

图 2-18　项目打包

4.【mldnboot-base 项目】项目打包完成后一定要运行程序，如果不做出任何的指派，那么默认配置的活跃 profile（dev）就将直接起作用（java -jar mldnboot-base.jar）。

5.【mldnboot-base 项目】如果要切换到不同的 profile 环境，可以在启动时动态配置。

```
java -jar mldnboot-base.jar --spring.profiles.active=product
```

此时在程序运行时将使用 product 作为运行环境配置。

 **提示：*.properties 与*.yml 配置不同。**

使用 application.yml 进行多 profile 配置的时候，只需要在一个配置文件中使用“---”分割不同的 profile 配置。但是此类模式不适合于 application.properties 配置，此时应该采用不同的 *.properties 保存不同的配置，才可以实现多 profile。

范例：定义 3 个针对不同运行环境的 application.properties 配置文件。

【开发】application-dev.properties	server.port=7070
【测试】application-beta.properties	server.port=8080
【生产】application-product.properties	server.port=80

随后还是需要有一个公共的 application.properties 配置文件，用于指派可以使用的 profile 配置。

范例：定义公共的 application.properties 配置文件。

```
spring.profiles.active=beta
```

随后的使用形式与 application.yml 配置相同。

2.13　本章小结

1. SpringBoot 的依赖管理除了可以作为项目的父 pom 引入之外，也可以采用依赖管理的形式进行配置。

2. SpringBoot 程序测试专门提供了 spring-boot-starter-test 依赖库，在测试时需要使用 @SpringBootTest 注解。

3. 在定义 SpringBoot 程序主类时使用@SpringBootApplication 注解，可以自动扫描子包中的配置项，实现自动配置。

4. @Controller 注解采用的是普通控制器的形式定义，而@RestController 注解可以直接以 Restful 方式运行。

5. SpringBoot 默认使用的是 Tomcat 容器，开发时也可以配置 spring-boot-starter-jetty 依赖库，使用 Jetty 容器。但是在实际部署时，建议使用 Tomcat 容器。

6. SpringBoot 支持*.properties 和*.yml 两类配置文件，在实际开发中建议通过 application.yml 实现环境配置。

7. SpringBoot 项目可以通过 spring-boot-maven-plugin 实现打包处理，这样就可以方便地通过 *.jar 文件来实现项目的发布。

第 3 章

Thymeleaf 模板渲染

通过本章学习，可以达到以下目标：

1. 掌握 Thymeleaf 模板的作用以及相关配置。
2. 掌握 Thymeleaf 中路径访问处理支持。
3. 掌握 Thymeleaf 页面处理语法。

SpringBoot 除了可以进行 Restful 运行之外，也可以像传统 MVC 设计模式那样，实现控制层与显示层的跳转处理。但 SpringBoot 所支持的显示层不再只是简单的 JSP 页面，而是 Thymeleaf 模板页面。此页面下利用模板语法，可以在 HTML 文件中实现 JSP 的相关逻辑。

3.1 Thymeleaf 简介

传统的 JSP 开发需要编写大量的 Scriptlet 程序代码，这样就使得页面非常混乱。虽然在 JSP 的后续发展中提供了标签编程与 JSTL 标签库，但其页面处理逻辑仍然是复杂的。

为了解决 JSP 代码过于臃肿的问题，在 SpringBoot 中默认引入了 Thymeleaf 模板程序。Thymeleaf 是 XML、XHTML、HTML5 模板引擎，可以用于 Web 与非 Web 应用。

Thymeleaf 提供了一种可以被浏览器正确显示、格式良好的模板创建方式，开发者可以通过它来创建经过验证的 XML 与 HTML 模板。相对于传统的逻辑程序代码，开发者只需将标签属性添加到模板中即可，而后这些标签属性就会在 DOM（文档对象模型）上执行预先制定好的逻辑，这样就极大地简化了显示层的程序逻辑代码。

1.【mldnboot 项目】本程序为了与之前的项目有所区分，将创建一个新的 mldnboot-thymeleaf 模块，并修改父 pom.xml 的定义，追加新模块配置。

```
<module>mldnboot-thymeleaf</module>
```

2.【mldnboot-thymeleaf 项目】修改 pom.xml 配置文件，追加 Thymeleaf 依赖库的配置。

```
<dependency>
        <groupId>org.springframework.boot</groupId>
        <artifactId>spring-boot-starter-thymeleaf</artifactId>
</dependency>
```

这样就可以在项目中使用 Thymeleaf 语法来实现显示层逻辑处理。

3.2　Thymeleaf 编程起步

Thymeleaf 需要按照传统 MVC 设计模式的方式来进行处理，所以在定义控制器的时候必须使用@Controller 注解来完成。通过控制器的 Model 类对象，可以传递相应属性到页面中显示。

1.【mldnboot-thymeleaf 项目】建立 ThymeleafController 程序类，该类将跳转到 Thyemelaf 模板页面。

```
package cn.mldn.mldnboot.controller;
import org.springframework.stereotype.Controller;
import org.springframework.ui.Model;
import org.springframework.web.bind.annotation.GetMapping;
@Controller
public class ThymeleafController {
    @GetMapping("/view")
    public String view(String mid, Model model) {
        model.addAttribute("url", "www.mldn.cn"); // request属性传递包装
        model.addAttribute("mid", mid);          // request属性传递包装
        return "message/message_show";           // 返回一个路径，该路径后缀默认是*.html
    }
}
```

2.【mldnboot-thymeleaf 项目】ThymeleafController 控制器会跳转到 message 目录下的 message_show.html 页面进行显示，而该页面一定要在 CLASSPATH 路径下配置。为了结构清晰，本程序将建立一个 src/main/view 的源文件，并且必须建立 templates 目录，随后在这个目录下创建所需要的子目录（本程序需要创建 message 子目录）。项目最终的目录结构如图 3-1 所示。

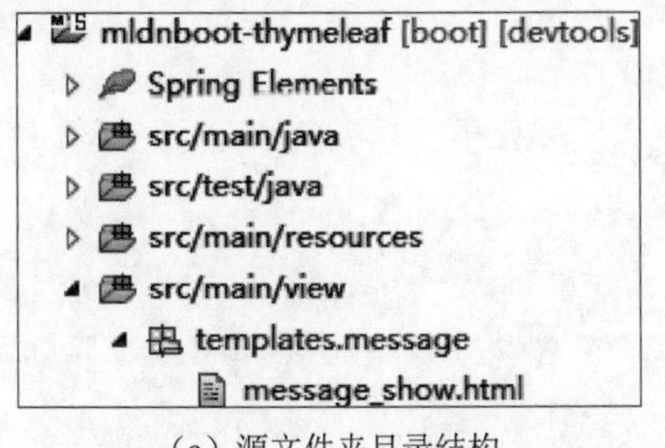

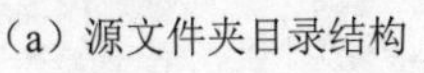

（a）源文件夹目录结构

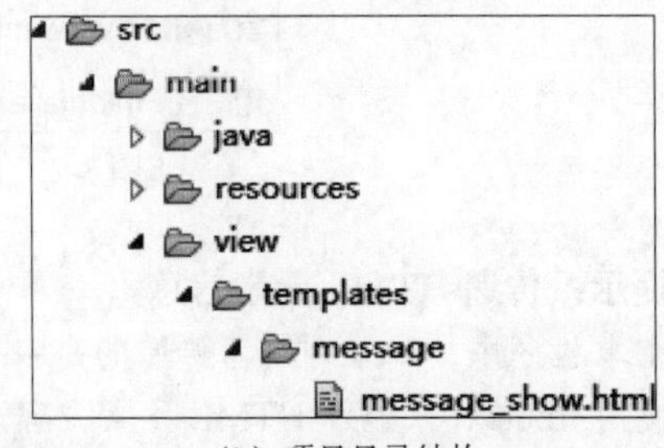

（b）项目目录结构

图 3-1　Thymeleaf 模板目录结构

提示：修改 Thymeleaf 的默认配置。

SpringBoot 项目中 Thymeleaf 的动态页面需要保存在 templates 目录中，页面的扩展名默认使用的是*.html，如果开发者觉得这样的设计不合理，也可以通过 application.yml 配置文件自行修改。

范例：修改 Thyemeleaf 的配置项。

```
spring:
  thymeleaf:
    suffix: .htm                          # 模板文件的后缀为*.html
    prefix: classpath:/mytemplates/       # 定义动态页面保存目录
```

虽然 SpringBoot 中可以修改 Thymeleaf 的默认配置项，但是在实际开发中不建议修改，还是遵从默认配置比较合理。

3.【mldnboot-thymeleaf 项目】编写 message_show.html 页面，实现控制层传递属性输出。

```
<!DOCTYPE HTML>
<html xmlns:th="http://www.thymeleaf.org">          <!-- 引入命名空间 -->
<head>
    <title>SpringBoot模板渲染</title>
    <meta http-equiv="Content-Type" content="text/html;charset=UTF-8"/>
</head>
<body>
    <p th:text="'官方网站：' + ${url}"/>              <!-- 输出url属性 -->
    <p th:text="'用户名：' + ${mid}"/>               <!-- 输出mid属性 -->
</body>
</html>
```

本程序使用<p>元素设定了要输出的内容。要想使用 Thymeleaf 的功能，必须以“th:属性”的形式处理，th:text 的主要作用是进行文本输出。而要想输出 request 属性中的内容，则需要采用“${属性名称}”的语法格式完成。随后启动程序，输入访问地址 http://localhost/view?mid=mldnjava，页面运行效果如图 3-2 所示。

图 3-2　Thymeleaf 模板运行

提示：传递 HTML 元素信息。

在本程序中，如果控制器传递的是一个 HTML 元素，那么对于 Thymeleaf 页面而言，就需要使用 th:utext 来显示 HTML 元素内容。

范例：【mldnboot-thymeleaf 项目】修改 ThymeleafController 控制器，传递 HTML 元素。

```
@Controller
public class ThymeleafController {
    @GetMapping("/view")
    public String view(String mid, Model model) {
        model.addAttribute("url",
            "<h1><span style='color:red'>www.mldn.cn</span></h1>");
        model.addAttribute("mid", mid);
```

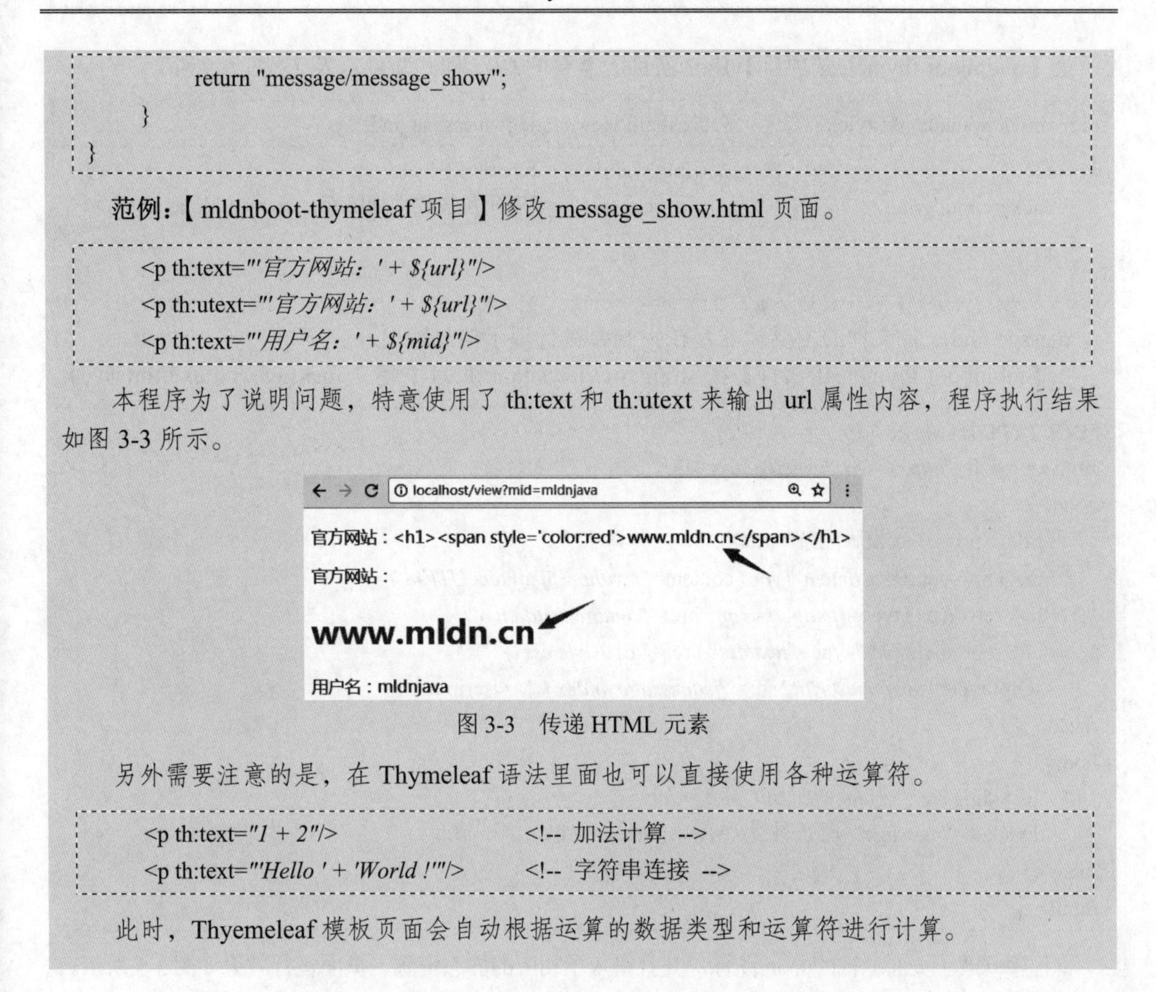

```
        return "message/message_show";
    }
}
```

范例：【mldnboot-thymeleaf 项目】修改 message_show.html 页面。

```
<p th:text="'官方网站：' + ${url}"/>
<p th:utext="'官方网站：' + ${url}"/>
<p th:text="'用户名：' + ${mid}"/>
```

本程序为了说明问题，特意使用了 th:text 和 th:utext 来输出 url 属性内容，程序执行结果如图 3-3 所示。

图 3-3　传递 HTML 元素

另外需要注意的是，在 Thymeleaf 语法里面也可以直接使用各种运算符。

```
<p th:text="1 + 2"/>                    <!-- 加法计算 -->
<p th:text="'Hello ' + 'World !'"/>     <!-- 字符串连接 -->
```

此时，Thyemeleaf 模板页面会自动根据运算的数据类型和运算符进行计算。

3.3　Thyemeleaf 静态资源

在进行 Web 信息显示的过程中，除了可以配置动态显示页面之外，也可以配置静态资源（如 *.html、*.css、*.js 等）。对于静态资源，要求其必须放在源文件夹的 static 目录中。本项目的页面结构如图 3-4 所示。

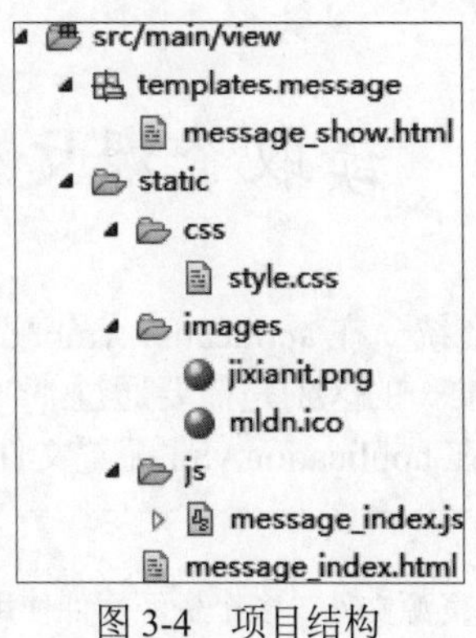

图 3-4　项目结构

1.【mldnboot-thymeleaf 项目】建立项目所需要的 CSS 样式文件以及 JS 脚本文件。

src/main/view/static/css/style.css

```
.message {
    background: gray;
    color: white ;
}
```

src/main/view/static/js/message_index.js

```
window.onload = function() {
    console.log("魔乐科技软件训练营：www.mldnjava.cn") ;
}
```

messge_index.js 文件的主要功能是在控制台进行提示信息输出。

2.【mldnboot-thymeleaf 项目】在 src/main/view/static 目录下建立 message_index.html 页面。

```
<!DOCTYPE HTML>
<html xmlns:th="http://www.thymeleaf.org">          <!-- 引入命名空间 -->
<head>
    <title>SpringBoot模版渲染</title>
    <meta http-equiv="Content-Type" content="text/html;charset=UTF-8"/>
    <link rel="icon" type="image/x-icon" href="/images/mldn.ico"/>
    <link rel="stylesheet" type="text/css" href="/css/style.css">
    <script type="text/javascript" src="/js/message_index.js"></script>
</head>
<body>
    <div><img src="/images/jixianit.png"></div>
    <div class="message">魔乐科技：www.mldn.cn</div>
</body>
</html>
```

本程序修改了页面运行的 icon 图标，并且引入了相应的静态资源，页面运行效果如图 3-5 所示。

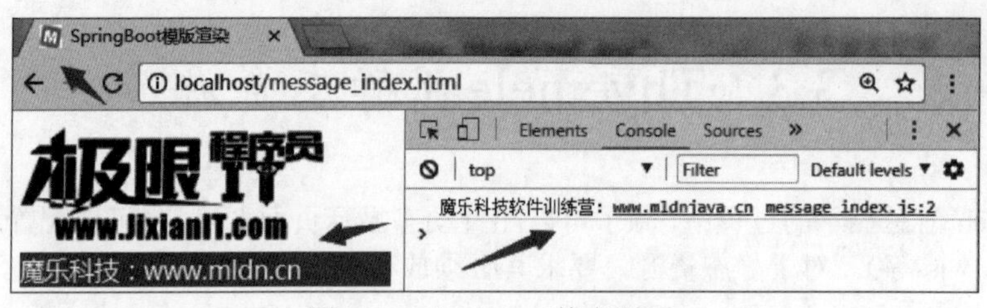

图 3-5　Thymeleaf 静态页面

3.4　读取资源文件

SpringBoot 项目中的资源文件会统一在 application.yml 配置文件中定义。当页面需要使用的时候，可以像输出属性一样完成，唯一的区别是需要通过“#{key}”的形式来获取资源内容。

1.【mldnboot-thymeleaf 项目】在 application.yml 配置文件中定义资源文件配置。

```
spring:
  messages:                    # 定义资源文件，多个资源文件使用“,”分割
```

```
    basename: i18n/Messages
```

2.【mldnboot-thymeleaf 项目】在 Messages.properties 配置文件里面定义资源内容。

```
welcome.url=www.mldn.cn
welcome.msg=欢迎{0}光临！
```

3.【mldnboot-thymeleaf 项目】要读取资源文件，需要动态页面的支持。在 Thyemeleaf Controller 控制器中建立一个新的方法，用于跳转到前端页面。

```
@GetMapping("/value")
public String value() {
    return "message/message_value";
}
```

4.【mldnboot-thymeleaf 项目】在 src/main/view/templates/message 目录中创建 message_value.html 页面，用于读取资源文件内容并进行显示。

```
<!DOCTYPE HTML>
<html xmlns:th="http://www.thymeleaf.org">          <!-- 引入命名空间 -->
<head>
    <title>SpringBoot模板渲染</title>
    <meta http-equiv="Content-Type" content="text/html;charset=UTF-8"/>
    <link rel="icon" type="image/x-icon" href="/images/mldn.ico"/>
</head>
<body>
    <h2 th:text="#{welcome.url}"/>                   <!-- 读取资源文件 -->
    <h2 th:text="#{welcome.msg('李兴华')}"/>      <!-- 读取资源文件 -->
</body>
</html>
```

在 Thyemeleaf 模板页面中直接使用资源的 key 获取资源信息，页面运行效果如图 3-6 所示。

图 3-6　Thyemeleaf 模板页面读取资源文件

3.5　路径处理

Web 开发过程中，路径的处理操作是最为麻烦的。如果要进行准确的路径定位，最好使用完整的路径，并明确写上用户的协议、主机名称、端口以及虚拟目录的名称。这些处理的难点在于 Thymeleaf 彻底消失了，因为其路径访问变得相当容易，只需要在动态页面中使用“@{路径}”即可访问。

提示：回顾原始实现。

在进行 Web 项目开发的过程中，相信不少开发者都编写过如下的类似代码：

```
<%
    String basePath = request.getScheme() + "://" +
        request.getServerName() + ":" + request.getServerPort() +
        request.getContextPath() + "/" ;
%>
```

而后再使用<base>元素（<base href="<%=basePath%>"/>）进行引用，可以解决路径操作问题。

1.【mldnboot-thymeleaf 项目】在 ThymeleafController 中创建一个新的方法，用于跳转。

```
@GetMapping("/path")
public String path() {
    return "message/message_path";
}
```

2.【mldnboot-thymeleaf 项目】建立 src/main/view/templates/message/message_path.html 页面。

```
<!DOCTYPE HTML>
<html xmlns:th="http://www.thymeleaf.org">         <!-- 引入命名空间 -->
<head>
    <title>SpringBoot模板渲染</title>
    <meta http-equiv="Content-Type" content="text/html;charset=UTF-8"/>
    <link rel="icon" type="image/x-icon" th:href="@{/images/mldn.ico}"/>
    <link rel="stylesheet" type="text/css" th:href="@{/css/style.css}"/>
    <script type="text/javascript" th:src="@{/js/message_index.js}"></script>
</head>
<body>
    <div><img th:src="@{/images/jixianit.png}"/></div>
    <div class="message">魔乐科技：www.mldn.cn</div>
</body>
</html>
```

在 message_path.html 页面中继续引用之前定义了的资源，而采用“@{路径}”的形式使得资源引用也十分简单，页面运行效果如图 3-7 所示。

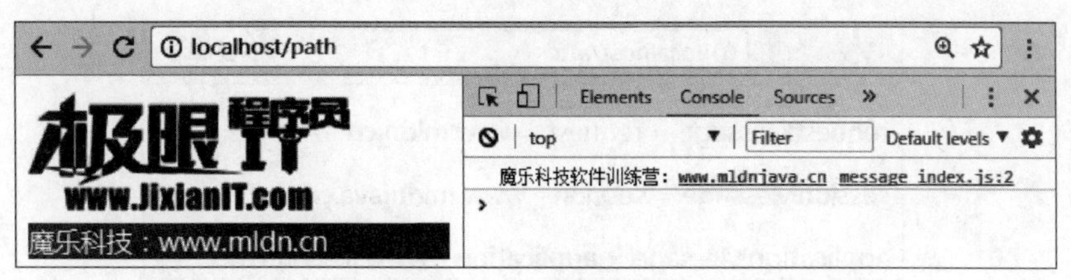

图 3-7　Web 资源引入

3.6　内置对象操作支持

在模板页面中，最为常用的功能就是输出控制器传递的属性。为了方便用户开发，Thymeleaf 支持内置对象的直接使用，也可以直接调用内置对象所提供的处理方法。

在通过控制器传递属性到 Thymeleaf 操作的时候，默认支持的属性获取范围为 request（${属性名称}）。如果要接收其他属性范围的内容，则需要指明范围，如 session 范围（${session.属性名称}）、application 范围（${application.属性名称}）。

1.【mldnboot-thymeleaf 项目】修改 ThymeleafController 控制器程序类，追加属性传递。本程序将传递 request、session 和 application 3 种属性范围的信息。

```
    @GetMapping("/attr")
    public String attr() {
        HttpServletRequest request = ((ServletRequestAttributes)
RequestContextHolder.getRequestAttributes())
                .getRequest();                    // 获取HttpServletRequest内置对象
        request.setAttribute("requestMessage", "request - www.mldn.cn");
        request.getSession().setAttribute("sessionMessage", "session - www.mldnjava.cn");
        request.getServletContext().setAttribute("applicationMessage", "application - www.jixianit.com");
        return "message/message_attr";
    }
```

2.【mldnboot-thymeleaf 项目】定义 src/main/view/templates/message/message_attr.html 页面，进行属性内容输出。

```
<p th:text="'requestMessage = ' + ${requestMessage}"/>
<p th:text="'sessionMessage = ' + ${session.sessionMessage}"/>
<p th:text="'applicationMessage = ' + ${application.applicationMessage}"/>
```

本程序在 Thymeleaf 模板页面中输出了控制器中传递的不同范围的属性内容。可以发现，只有 request 范围的属性可以直接通过表达式语法输出，而 session 与 application 范围的属性输出时，必须要有相应的范围标记，否则获取的内容就是 null。本程序的执行结果如图 3-8 所示。

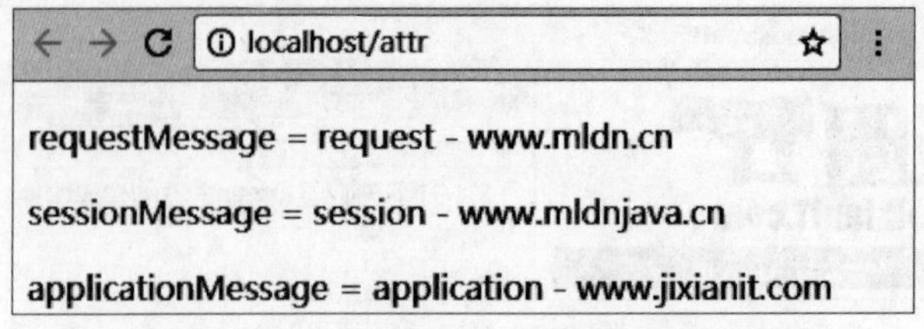

图 3-8　Thymeleaf 输出不同属性范围的信息

3.【mldnboot-thymeleaf 项目】在 Thymeleaf 中也支持对内置对象的直接处理。修改 message_attr.html 页面，增加内置对象的方法调用。

```
<p th:text="${#httpServletRequest.getRemoteAddr()}"/>
<p th:text="${#httpServletRequest.getAttribute('requestMessage')}"/>
<p th:text="${#httpSession.getId()}"/>
<p th:text="${#httpServletRequest.getServletContext().getRealPath('/')}"/>
```

本程序利用内置对象提供的方法获取了 IP 地址、request 属性、sessionId 以及项目真实路径，程序运行结果如图 3-9 所示。

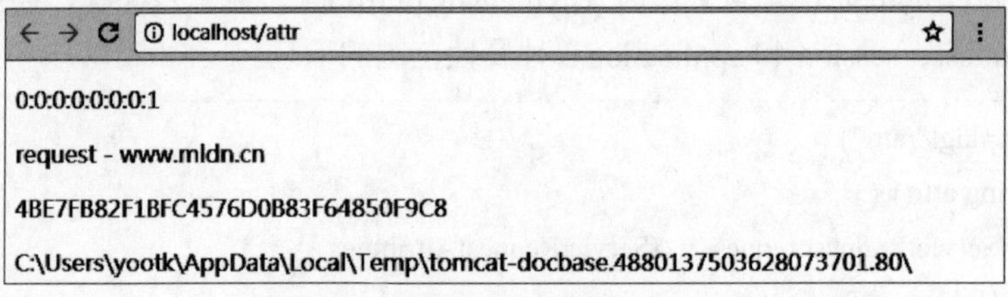

图 3-9　直接调用内置对象方法

3.7　对象输出

在实际页面中进行信息显示的时候，需要通过 VO 对象进行信息的传递。此时可以直接利用“${属性名称.成员属性}”的格式在页面中进行对象内容的输出。

1.【mldnboot-thymeleaf 项目】定义一个 VO 类 Member。

```
package cn.mldn.mldnboot.vo;
import java.io.Serializable;
import java.util.Date;
@SuppressWarnings("serial")
public class Member implements Serializable {
    private String mid ;                    // 用户编号
    private String name ;                   // 用户姓名
    private Integer age ;                   // 用户年龄
    private Double salary ;                 // 基本工资
    private Date birthday ;                 // 用户生日
```

```
    // setter、getter略
}
```

2.【mldnboot-thymeleaf 项目】修改 ThymeleafController 控制器程序，向页面传递对象信息。

```
@GetMapping("/member")
public String member(Model model) throws Exception {
    Member vo = new Member() ;
    vo.setMid("mldnjava");
    vo.setName("李兴华");
    vo.setSalary(3500.00);
    vo.setAge(18);
    vo.setBirthday(new SimpleDateFormat("yyyy-MM-dd").parse("1998-09-15"));
    model.addAttribute("member", vo) ;                // 传递页面属性
    return "message/message_member" ;
}
```

3.【mldnboot-thymeleaf 项目】建立 src/main/view/templates/message/message_member.html 页面，进行对象输出。

```
<p th:text="'用户编号：' + ${member.mid}"/>
<p th:text="'用户姓名：' + ${member.name}"/>
<p th:text="'用户年龄：' + ${member.age}"/>
<p th:text="'用户工资：' + ${member.salary}"/>
<p th:text="'出生日期：' + ${member.birthday}"/>
```

本程序在页面中使用“${属性名称.成员属性}”获取了 request 属性范围中传递的 member 对象的全部信息，页面执行结果如图 3-10 所示。

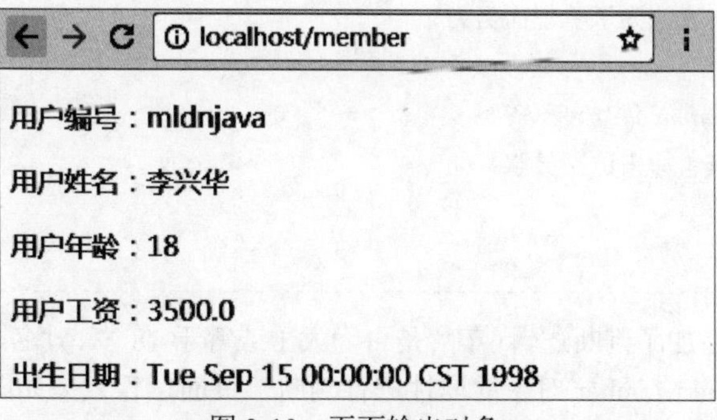

图 3-10　页面输出对象

提示：Thymeleaf 支持有简化的对象输出处理。

在 Thyemleaf 模板页面中可以发现，默认支持的对象成员获取语法需要频繁使用属性名称。为了简化输出，可以采用 th:object 标签处理。

范例：对象信息输出。

```
<div th:object="${member}">
    <p th:text="'用户编号：' + *{mid}"/>
    <p th:text="'用户姓名：' + *{name}"/>
    <p th:text="'用户年龄：' + *{age}"/>
    <p th:text="'用户工资：' + *{salary}"/>
    <p th:text="'出生日期：' + *{birthday}"/>
</div>
```

本程序使用了一个<div>元素，并在此元素中利用 th:object= *"${member}"*将需要输出的对象信息定义在父元素上，而后此元素的所有子元素就可以利用"*{成员属性}"获取对象中全部属性的内容。

另外需要提醒读者的是，$访问完整对象信息，*访问指定对象中的属性内容。如果访问的只是普通的内容（如传递字符串信息），两者在使用效果上没有区别。

3.8　页面逻辑处理

Thymeleaf 页面模板支持逻辑处理功能，如判断、循环处理等操作。开发者在页面中处理逻辑时，可以使用 and、or、关系比较（>、<、>=、<=、==、!=、lt、gt、le、ge、eq、ne）等运算符来完成。

1.【mldnboot-thymeleaf 项目】修改 src/main/view/templates/message/message_member.html 页面，追加逻辑判断。

```
<div th:object="${member}">
    <p th:if="*{age ge 18}">
        成年人应该为了梦想而努力！
    </p>
    <p th:if="*{name eq '李兴华'}">
        欢迎李老师来访，鼓掌~
    </p>
</div>
```

本程序在页面中追加了判断逻辑（年龄是否为大于或等于 18 岁、姓名是否为指定的字符串），这样会根据传递过来的 Member 对象的属性进行判断，页面运行效果如图 3-11 所示。

图 3-11　页面逻辑判断

2.【mldnboot-thymeleaf 项目】在 Thymeleaf 之中，如果使用 th:if 判断条件不满足时，也可以使用 th:unless 处理。

```
<p th:if="*{age ge 18}">
    成年人应该为了梦想而努力！
</p>
<p th:unless="*{age ge 18}">
    未成年人应该好好学习基础文化课程，好好锻炼身体！
</p>
```

3.【mldnboot-thymeleaf 项目】页面中可以使用 switch-case 来实现开关逻辑处理。

```
<div th:object="${member}">
    <p th:switch="*{mid}">
        <span th:case="mldnjava">欢迎“mldnjava”用户访问</span>
        <span th:case="jixianit">欢迎“jixianit”用户访问</span>
        <span th:case="*">没有匹配成功的数据，别匹配了！ </span>
    </p>
</div>
```

本程序使用 th:switch="*{mid}"语句对 mid 属性的内容进行 switch 判断。如果有匹配的信息，则进行内容输出；如果没有，则执行 th:case="*"的信息输出，页面运行效果如图 3-12 所示。

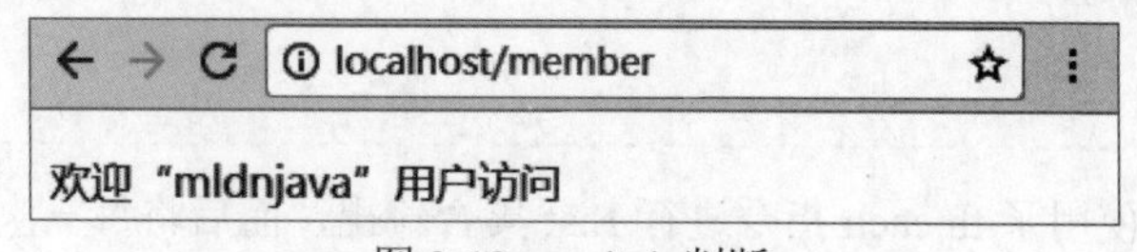

图 3-12　switch 判断

3.9　数据迭代处理

数据迭代显示是显示层的一个重要技术手段，在 Thymeleaf 模板中可以使用 th:each 指令实现 List 与 Map 集合的迭代输出。

1.【mldnboot-thymeleaf 项目】在 ThymeleafController 控制器中追加一个方法，该方法将创建一个 List 集合，随后通过 request 属性传递到页面。

```
@GetMapping("/list")
public String list(Model model) throws Exception {
    List<Member> allMembers = new ArrayList<Member>() ;
    for (int x = 0 ; x < 5 ; x ++) {
        Member vo = new Member() ;
        vo.setMid("mldnjava - " + x);
        vo.setName("李兴华 - " + x);
```

```
            vo.setSalary(3500.00);
            vo.setAge(16 + x);
            vo.setBirthday(new SimpleDateFormat("yyyy-MM-dd").parse("1998-09-15"));
            allMembers.add(vo) ;
        }
        model.addAttribute("allMembers", allMembers) ;          // 传递页面属性
        return "message/message_list" ;
    }
```

2.【mldnboot-thymeleaf 项目】建立 src/main/view/templates/message/message_list.html 页面。

```
<table border="1">
    <tr><td>No.</td><td>编号</td><td>姓名</td><td>年龄</td><td>工资</td><td>生日</td></tr>
    <tr th:each="member,memberStat:${allMembers}">
        <td th:text="${memberStat.index + 1}"/>
        <td th:text="${member.mid}"/>
        <td th:text="${member.name}"/>
        <td th:text="${member.age}"/>
        <td th:text="${member.salary}"/>
        <td th:text="${member.birthday}"/>
    </tr>
</table>
```

本程序在页面中使用了 th:each 指令进行 List 集合输出，而后将每一次迭代的结果都赋值给 member 对象，并将每一次迭代的基本信息都赋值给 memberStat 对象（可根据需要选择是否要使用此对象），这样就可以实现 List 集合显示，页面运行效果如图 3-13 所示。

localhost/list

No.	编号	姓名	年龄	工资	生日
1	mldnjava - 0	李兴华 - 0	16	3500.0	Tue Sep 15 00:00:00 CST 1998
2	mldnjava - 1	李兴华 - 1	17	3500.0	Tue Sep 15 00:00:00 CST 1998
3	mldnjava - 2	李兴华 - 2	18	3500.0	Tue Sep 15 00:00:00 CST 1998
4	mldnjava - 3	李兴华 - 3	19	3500.0	Tue Sep 15 00:00:00 CST 1998
5	mldnjava - 4	李兴华 - 4	20	3500.0	Tue Sep 15 00:00:00 CST 1998

图 3-13　迭代输出 List 集合

除了支持 List 集合输出之外，也可以采用同样的形式实现 Map 集合的输出。

3.【mldnboot-thymeleaf 项目】在 ThymeleafController 控制器中追加 Map 集合设置。

```
@GetMapping("/map")
public String map(Model model) throws Exception {
    Map<String,Member> allMembers = new HashMap<String,Member>();
```

```
    for (int x = 0 ; x < 5 ; x ++) {
        Member vo = new Member() ;
        vo.setMid("mldnjava - " + x);
        vo.setName("李兴华 - " + x);
        vo.setSalary(3500.00);
        vo.setAge(16 + x);
        vo.setBirthday(new SimpleDateFormat("yyyy-MM-dd").parse("1998-09-15"));
        allMembers.put("mldn-" + x, vo) ;
    }
    model.addAttribute("allMembers", allMembers) ;            // 传递页面属性
    return "message/message_map" ;
}
```

4.【mldnboot-thymeleaf 项目】建立 src/main/view/templates/message/message_map.html 页面。

```
<table border="1">
    <tr><td>No.</td><td>key</td><td>编号</td><td>姓名</td><td>年龄</td><td>工资</td><td>生日</td></tr>
    <tr th:each="memberEntry,memberStat:${allMembers}">
        <td th:text="${memberStat.index + 1}"/>
        <td th:text="${memberEntry.key}"/>
        <td th:text="${memberEntry.value.mid}"/>
        <td th:text="${memberEntry.value.name}"/>
        <td th:text="${memberEntry.value.age}"/>
        <td th:text="${memberEntry.value.salary}"/>
        <td th:text="${memberEntry.value.birthday}"/>
    </tr>
</table>
```

在进行 Map 集合输出时，每一次迭代所取出的对象类型都是 Map.Entry 接口实例，所以本程序使用 memberEntry 接收该接口对象，随后输出每一个 Map.Entry 对象中所保存的 key 与 value 信息，页面运行效果如图 3-14 所示。

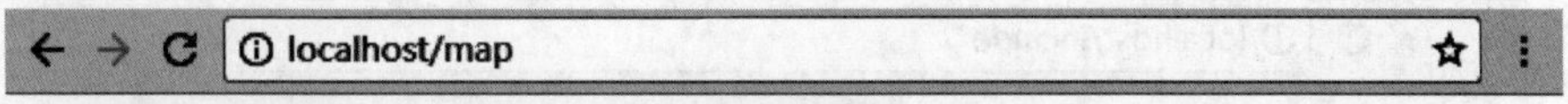

localhost/map

No.	key	编号	姓名	年龄	工资	生日
1	mldn-4	mldnjava - 4	李兴华 - 4	20	3500.0	Tue Sep 15 00:00:00 CST 1998
2	mldn-1	mldnjava - 1	李兴华 - 1	17	3500.0	Tue Sep 15 00:00:00 CST 1998
3	mldn-0	mldnjava - 0	李兴华 - 0	16	3500.0	Tue Sep 15 00:00:00 CST 1998
4	mldn-3	mldnjava - 3	李兴华 - 3	19	3500.0	Tue Sep 15 00:00:00 CST 1998
5	mldn-2	mldnjava - 2	李兴华 - 2	18	3500.0	Tue Sep 15 00:00:00 CST 1998

图 3-14　输出 Map 集合

3.10　包含指令

在页面开发中，包含是一个重要的指令，利用包含指令可以实现页面代码的重用处理。Thymeleaf 也同样支持数据的包含处理，而对于包含操作，在 Thymeleaf 模板中提供了两种支持语法。

☑　**th:replace：**使用标签进行替换，原始的宿主标签还在，但是包含标签不在。

☑　**th:include：**进行包含，原始的宿主标签消失，只保留包含的标签。

1.【mldnboot-thymeleaf 项目】建立 src/main/view/templates/commons/footer.html 页面。

```
<meta http-equiv="Content-Type" content="text/html;charset=UTF-8"/>
<foot th:fragment="companyInfo">
    <p><span th:text="${title}"/>（<span th:text="${url}"/>）</p>
</foot>
```

本程序设置了一个包含的名称信息为 companyInfo，同时还需要包含页面向本页面传递 title 与 url 两个参数信息。

2.【mldnboot-thymeleaf 项目】在 ThymeleafController 控制器类中追加一个新的方法，用于页面跳转。

```
@GetMapping("/include")
public String include(Model model) throws Exception {
    return "message/message_include" ;
}
```

3.【mldnboot-thymeleaf 项目】建立 src/main/view/templates/message/message_include.html 页面。

```
<div th:include="@{/commons/foot} :: companyInfo"
        th:with="title=魔乐科技,url=www.mldn.cn"/>
```

本程序使用 th:include 指令（替换掉父元素<div>）实现了页面的包含处理，同时利用 th:with 命令向被包含页面传递了两个参数。此时的页面运行效果如图 3-15 所示。

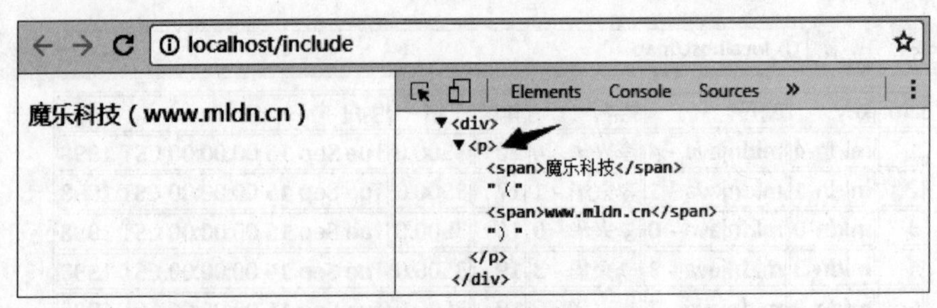

图 3-15　页面包含

3.11　Thymeleaf 数据处理

在 Thymeleaf 模板中还支持集合方法调用、字符串方法调用、日期格式化等操作。例如，在使用 List 集合的时候可以考虑采用 get()方法获取指定索引的数据，那么在使用 Set 集合的时候会考虑使用 contains()来判断某个数据是否存在，使用 Map 集合的时候也可以使用 containsKey()判断某个 key 是否存在，以及使用 get()根据 key 获取对应的 value。

1.【mldnboot-thymeleaf 项目】修改 src/main/view/templates/message/message_map.html 页面，调用 Map 方法。

```
<p th:text="${#maps.containsKey(allMembers,'mldn-3')}"/>
                                                    <!-- 判断key="mldn-3"是否存在 -->
<p th:text="${allMembers['mldn-3'].name}"/>         <!-- 根据 key 获取指定的成员属性 -->
```

2.【mldnboot-thymeleaf 项目】如果传递的是 Set 集合，也可以利用 Set 接口中的 contains()方法判断某个值是否存在。

☑　在 ThymeleafController 控制器中追加一个新的方法，利用 Set 传递属性。

```
@GetMapping("/set")
public String set(Model model) throws Exception {
    Set<String> all = new HashSet<String>();
    all.addAll(Arrays.asList("mldn","mldnjava","jixianit")) ;
    model.addAttribute("allInfos", all) ;          // 传递页面属性
    return "message/message_set" ;
}
```

☑　建立 src/main/view/templates/message/message_set.html 页面，判断某一个内容是否在集合中存在。

```
<p th:if="${#sets.contains(allInfos,'mldn')}">存在有"mldn"的信息！</p>
```

此时会判断在 allInfos 集合中是否存在 mldn 的内容，同时也可以利用 size()方法获取集合长度，页面运行效果如图 3-16 所示。

图 3-16　判断 Set 集合是否有指定数据

3.【mldnboot-thymeleaf 项目】在进行数据处理的时候，也可以直接使用字符串 String 类中定义的方法。

☑　在 ThymeleafController 控制器中追加一个新的方法，用于传递字符串属性。

```
@GetMapping("/string")
public String string(Model model) throws Exception {
        model.addAttribute("info", "www.mldn.cn") ;
        return "message/message_string" ;
}
```

☑ 建立 src/main/view/templates/message/message_string.html 页面，以处理字符串。

```
<p th:text="${'字符串替换：' + #strings.replace('www.mldn.cn','mldn.cn','jixianit.com')}"/>
<p th:text="${'字符串转大写：' + #strings.toUpperCase('www.mldn.cn')}"/>
<p th:text="${'字符串截取：' + #strings.substring(info,4)}"/>
```

在进行字符串数据处理时，可以直接使用控制器传递过来的属性，也可以直接定义具体的字符串内容。本程序处理后的结果如图 3-17 所示。

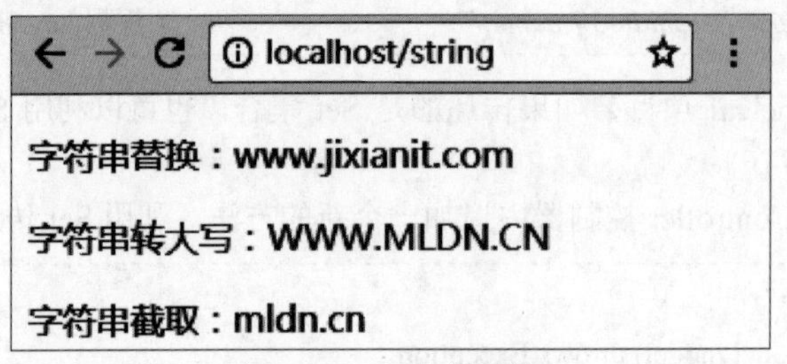

图 3-17　字符串处理

4.【mldnboot-thymeleaf 项目】在 Thyemeleaf 中还可以对输出的日期类型进行格式化处理。

☑ 在 ThymeleafController 控制器中追加一个新的方法，用于传递 Date 属性。

```
@GetMapping("/date")
public String date(Model model) throws Exception {
        model.addAttribute("mydate", new SimpleDateFormat("yyyy-MM-dd
HH:mm:ss").parse("2008-08-08 18:08:08"));
        return "message/message_date";
}
```

☑ 建立 src/main/view/templates/message/message_date.html 页面，以格式化日期显示。

```
<p th:text="${#dates.format(mydate,'yyyy-MM-dd')}"/>
<p th:text="${#dates.format(mydate,'yyyy-MM-dd HH:mm:ss.SSS')}"/>
```

本程序使用两种方式实现了日期时间的格式化处理，页面运行效果如图 3-18 所示。

图 3-18　格式化日期时间

3.12　本章小结

1. 在 SpringBoot 中如果要引入 Thymeleaf 模板，需要配置 spring-boot-starter-thymeleaf 依赖包。

2. Thymeleaf 可以有效地取代 JSP 页面，实现页面动态逻辑处理。

3. Thymeleaf 分为动态页面（templates）和静态资源（static）两类资源。

4. Thymeleaf 不仅仅可以实现控制器传递的属性输出，也可以实现资源文件的内容输出。

5. 在 Thymeleaf 中可以使用“@{路径}”的形式实现资源引入与跳转配置。

6. Thymeleaf 中可以使用判断、循环逻辑进行处理，也可以利用各种内置操作在页面上实现 List、Map、Set、字符串等数据处理，还可以使用日期格式化指令进行日期显示格式的处理。

第4章

SpringBoot与Web应用

通过本章学习，可以达到以下目标：

1. 掌握SpringBoot项目war包的生成与Tomcat发布。
2. 掌握SpringBoot基于https运行模式的配置。
3. 掌握SpringBoot错误处理。
4. 掌握SpringBoot与文件上传处理。
5. 掌握SpringBoot与拦截器的使用。

SpringBoot虽然极大地简化了Web项目开发与部署环节的配置，但是其依然属于Web项目，因此在实际工作中需要考虑与Tomcat的整合，当需要安全访问时还应该提供https支持。在本章中将为读者讲解SpringBoot项目与Tomcat的结合处理、错误处理机制、文件上传处理以及拦截器的使用。

4.1 配置Tomcat运行

在SpringBoot中默认支持Tomcat容器，所以当一个SpringBoot项目打包生成*.jar文件并且直接执行的时候就会自动启动内部的Tomcat容器。除了此种模式之外，也可以将Web项目打包为*.war文件，采用部署的形式通过Tomcat进行发布处理。

 **提示：Tomcat部署时的配置。**

在将SpringBoot打包为*.war文件的时候，如果想正常部署一定要注意以下两点：

1. 取消项目中的Jetty容器的配置。
2. 将所有的源文件夹目录设置输出资源，修改父pom.xml中的<resource>配置。

1. 【mldnboot-web项目】修改pom.xml配置文件，将程序的打包类型定义为*.war。

```
<packaging>war</packaging>                    <!-- 项目打包类型 -->
```

2. 【mldnboot-web项目】修改pom.xml配置文件，追加war文件打包插件。

```
<plugin>
    <groupId>org.apache.maven.plugins</groupId>
```

```
        <artifactId>maven-war-plugin</artifactId>
        <configuration>
            <warName>mldn</warName>                <!-- 打包后的war文件名称 -->
        </configuration>
    </plugin>
```

3.【mldnboot-web 项目】更新 Maven 项目，随后会提醒开发者当前项目中缺少 WEB-INF/web.xml 配置文件，此时需要开发者手工创建。创建完成的目录结构如图 4-1 所示。

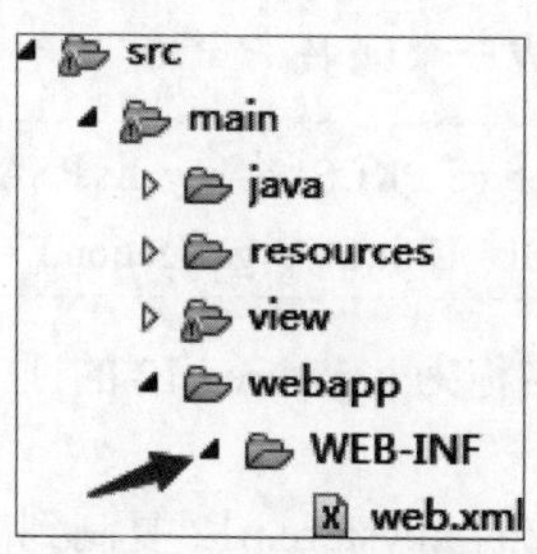

图 4-1　创建 Web 项目

4.【mldnboot-web 项目】如果现在项目要以 Tomcat 的形式运行，那么需要修改 SpringBoot 程序启动类定义，该类必须要继承 SpringBootServletInitializer 父类，同时还需要覆写 configure() 方法。

```
package cn.mldn.mldnboot;
import org.springframework.boot.SpringApplication;
import org.springframework.boot.autoconfigure.SpringBootApplication;
import org.springframework.boot.builder.SpringApplicationBuilder;
import org.springframework.boot.web.support.SpringBootServletInitializer;
@SpringBootApplication     // 启动SpringBoot程序，而后自带子包扫描
public class SpringBootStartApplication extends SpringBootServletInitializer {
    @Override
    protected SpringApplicationBuilder configure(
            SpringApplicationBuilder builder) {                    // 配置SpringBoot应用环境
        return builder.sources(SpringBootStartApplication.class) ;
    }
    public static void main(String[] args) throws Exception {
        SpringApplication.run(SpringBootStartApplication.class, args);     // 启动SpringBoot程序
    }
}
```

5.【mldnboot-web 项目】对项目进行打包部署（clean package），成功之后会在 target 目录中形成 mldn.war 程序文件，随后可以将此文件直接复制到 Tomcat 所在目录之中，而后启动 Tomcat 进行项目发布。

4.2 https安全访问

SpringBoot启动时默认采用http进行通信协议定义，考虑到安全，往往会使用https进行访问。正常来讲，https的访问是需要证书的，并且为了保证这个证书的安全，一定要在项目中使用CA进行认证。下面只是在本机做一个简单的模拟，利用Java提供的keytool命令实现证书的生成。

1.【操作系统】利用keytool生成一个证书。

```
keytool -genkey -alias mytomcat  -storetype PKCS12 -keyalg RSA -keysize 2048  -keystore keystore.p12 
-validity 3650 -dname "CN=Web Server,OU=Unit,O=Organization,L=City,S=State,C=US" -storepass mldnjava
```

该程序执行完成后会生成一个名称为keystore.p12的证书文件，该证书的别名为mytomcat，访问密码为mldnjava。

2.【mldnboot-web项目】将生成的keystore.p12复制到src/main/resources目录中，如图4-2所示。

图4-2　配置安全访问

3.【mldnboot-web项目】修改application.yml文件，配置ssl安全访问。

```
server:
  port: 443                                # https的端口设置为443
  ssl:
    key-store: classpath:keystore.p12    # keystore配置文件路径
    key-store-type: PKCS12               # keystore类型
    key-alias: mytomcat                  #  设置的别名
    key-password: mldnjava               #  访问密码
```

4.【mldnboot项目】资源目录中增加了*.p12文件，要想让其正常执行，还需要修改resource配置，追加输出文件类型配置。

```
<resource>
    <directory>src/main/resources</directory>
    <includes>
        <include>**/*.properties</include>
        <include>**/*.yml</include>
        <include>**/*.xml</include>
        <include>**/*.tld</include>
```

```
        <include>**/*.p12</include>
    </includes>
    <filtering>false</filtering>
</resource>
```

5.【mldnboot-web 项目】虽然现在程序配置了 https 支持，但考虑到用户访问时可能会使用 http 访问，所以需要做一个 Web 配置，使得通过 http 的 80 端口访问的请求直接映射到 https 的 443 端口上。

```
package cn.mldn.mldnboot.config;
import org.apache.catalina.connector.Connector;
import org.apache.tomcat.util.descriptor.web.SecurityCollection;
import org.apache.tomcat.util.descriptor.web.SecurityConstraint;
import org.springframework.boot.context.embedded.tomcat.TomcatEmbeddedServletContainerFactory;
import org.springframework.context.annotation.Bean;
import org.springframework.context.annotation.Configuration;
@Configuration
public class HttpConnectorConfig {                    // 此类专门负责http连接的相关配置
    public Connector initConnector() {
        Connector connector = new Connector("org.apache.coyote.http11.Http11NioProtocol") ;
        connector.setScheme("http");                  // 如果现在用户使用普通的http方式进行访问
        connector.setPort(80);                        // 用户访问的是80端口
        connector.setSecure(false);                   // 如果该连接为跳转，则表示不是一个新的连接对象
        connector.setRedirectPort(443);               // 设置转发操作端口
        return connector;
    }
    @Bean
    public TomcatEmbeddedServletContainerFactory servletContainerFactory() {
        TomcatEmbeddedServletContainerFactory factory = new
TomcatEmbeddedServletContainerFactory() {
            protected void postProcessContext(
                    org.apache.catalina.Context context) {   // 该方法主要进行请求处理的上下文配置
                                                             // 定义新的安全访问策略
                SecurityConstraint securityConstraint = new SecurityConstraint();
                securityConstraint.setUserConstraint("CONFIDENTIAL"); // 定义用户访问约束要求
                SecurityCollection collection = new SecurityCollection() ;
                collection.addPattern("/*");                          // 匹配所有的访问映射路径
                securityConstraint.addCollection(collection);         // 追加路径映射访问配置
                context.addConstraint(securityConstraint);
            };
        };
```

```
        factory.addAdditionalTomcatConnectors(this.initConnector());
        return factory;
    }
}
```

此时程序在通过 80 端口访问时，会自动跳转到 https 访问的 443 端口上。

4.3 数据验证

在进行 Web 开发过程中，用户提交数据的合法性是最基础的验证手段，在 SpringBoot 中可以直接使用 hibernate-vidator 组件包实现验证处理，而此组件包中支持的验证注解如表 4-1 所示。

表 4-1 hibernate-vidator 验证注解

No.	注解	描述
01	@Null	被注释的元素必须为 null
02	@NotNull	被注释的元素必须不为 null
03	@AssertTrue	被注释的元素必须为 true
04	@AssertFalse	被注释的元素必须为 false
05	@Min(value)	被注释的元素必须是一个数字，其值必须大于等于指定的最小值
06	@Max(value)	被注释的元素必须是一个数字，其值必须小于等于指定的最大值
07	@DecimalMin(value)	被注释的元素必须是一个数字，其值必须大于等于指定的最小值
08	@DecimalMax(value)	被注释的元素必须是一个数字，其值必须小于等于指定的最大值
09	@Size(max=, min=)	被注释的元素的大小必须在指定的范围内
10	@Digits (integer, fraction)	被注释的元素必须是一个数字，其值必须在可接受的范围内
11	@Past	被注释的元素必须是一个过去的日期
12	@Future	被注释的元素必须是一个将来的日期
13	@Pattern(regex=,flag=)	被注释的元素必须符合指定的正则表达式
14	@NotBlank(message =)	验证字符串非 null，且长度必须大于 0
15	@Email	被注释的元素必须是电子邮箱地址
16	@Length(min=,max=)	被注释的字符串的大小必须在指定的范围内
17	@NotEmpty	被注释的字符串的必须非空
18	@Range(min=,max=,message=)	被注释的元素必须在合适的范围内

1.【mldnboot-web 项目】在 src/main/resources 目录下创建 ValidationMessages.properties（文件名称为默认设置，不可更改）文件，该文件中要保留所有的错误提示信息。

```
member.mid.notnull.error=用户名不允许为空！
member.mid.email.error=用户名的注册必须输入正确的邮箱！
member.mid.length.error=用户名的格式错误！
```

```
member.age.notnull.error=年龄不允许为空！
member.age.digits.error=年龄必须输入合法数字！
member.salary.notnull.error=工资不允许为空！
member.salary.digits.error=工资必须是数字！
member.birthday.notnull.error=生日不允许为空！
```

2.【mldnboot-web 项目】建立一个 Member 程序类，并且在该类上使用验证注解。同时，验证出错时的错误信息引用之前 ValidationMessages.properties 文件中的定义。

```
public class Member implements Serializable {
    @NotNull(message="{member.mid.notnull.error}")
    @Email(message="{member.mid.email.error}")
    private String mid ;                // 用户编号，使用email作为用户名
    @NotNull(message="{member.age.notnull.error}")
    @Digits(integer=3,fraction=0,message="{member.age.digits.error}")
    private Integer age ;               // 用户年龄
    @NotNull(message="{member.salary.notnull.error}")
    @Digits(integer=20,fraction=2,message="{member.salary.digits.error}")
    private Double salary ;             // 基本工资
    @NotNull(message="{member.birthday.notnull.error}")
    private Date birthday ;             // 用户生日
    // setter、getter略
}
```

3.【mldnboot-web 项目】建立一个 MemberController 控制器程序类。

```
package cn.mldn.mldnboot.controller;
import java.text.SimpleDateFormat;
import java.util.Iterator;
import javax.validation.Valid;
import org.springframework.beans.propertyeditors.CustomDateEditor;
import org.springframework.stereotype.Controller;
import org.springframework.validation.BindingResult;
import org.springframework.validation.ObjectError;
import org.springframework.web.bind.WebDataBinder;
import org.springframework.web.bind.annotation.GetMapping;
import org.springframework.web.bind.annotation.InitBinder;
import org.springframework.web.bind.annotation.PostMapping;
import org.springframework.web.bind.annotation.ResponseBody;
import cn.mldn.mldnboot.vo.Member;
@Controller
public class MemberController {
```

```
    @GetMapping("/member_add_pre")
    public String addPre() {                                    // 增加前的准备操作路径
        return "member_add";
    }
    @PostMapping("/member_add")
    @ResponseBody
    public Object add(@Valid Member vo, BindingResult result) {    // 增加前的准备操作路径
        if (result.hasErrors()) {           // 执行的验证出现错误
            Iterator<ObjectError> iterator = result.getAllErrors().iterator();
                                                                // 获取全部错误
            while (iterator.hasNext()) {
                ObjectError error = iterator.next() ;           // 取出每一个错误
                System.out.println("【错误信息】code = " + error.getCode() + ", message = " +
error.getDefaultMessage());
            }
            return result.getAllErrors() ;
        } else {
            return vo;
        }
    }
    @InitBinder
    public void initBinder(WebDataBinder binder) {               //本程序需要对日期格式进行处理
        // 首先建立一个可以将字符串转换为日期的工具程序类
        SimpleDateFormat sdf = new SimpleDateFormat("yyyy-MM-dd") ;
        // 明确地描述此时需要注册一个日期格式的转化处理程序类
        binder.registerCustomEditor(java.util.Date.class, new CustomDateEditor(sdf, true));
    }
}
```

本程序为了方便读者理解，除了将错误提示信息以 Restful 方式返回之外，还直接在后台进行了错误信息的打印。如果用户输入的内容全部正确，则会返回用户输入的信息。

4.【mldnboot-web 项目】在 src/main/view 源文件夹中创建 templates/member_add.html 页面，定义用户信息增加表单。

```
<form th:action="@{member_add}" method="post">
    用户邮箱：<input type="text" name="mid" value="mldnjava@163.com"/><br/>
    用户年龄：<input type="text" name="age" value="16"/><br/>
    用户工资：<input type="text" name="salary" value="8587.23"/><br/>
    用户生日：<input type="text" name="birthday" value="2010-10-10"/><br/>
    <input type="submit" value="提交"/>
    <input type="reset" value="重置"/>
</form>
```

本程序由于存在 Member 数据验证逻辑，在用户信息输入正确时将返回如图 4-3 所示的界面。如果输入错误，则会返回如图 4-4 所示的界面。

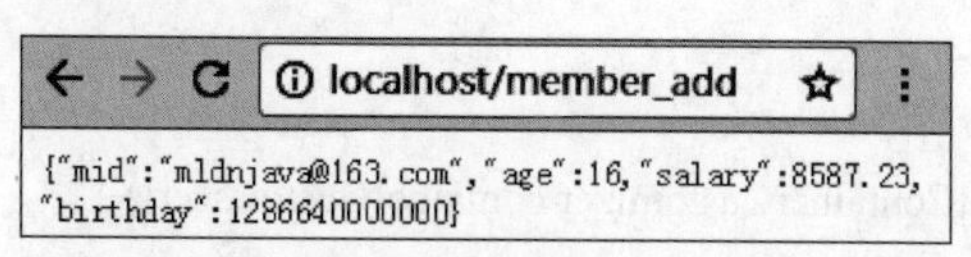

图 4-3　信息输入正确时返回 Member 对象

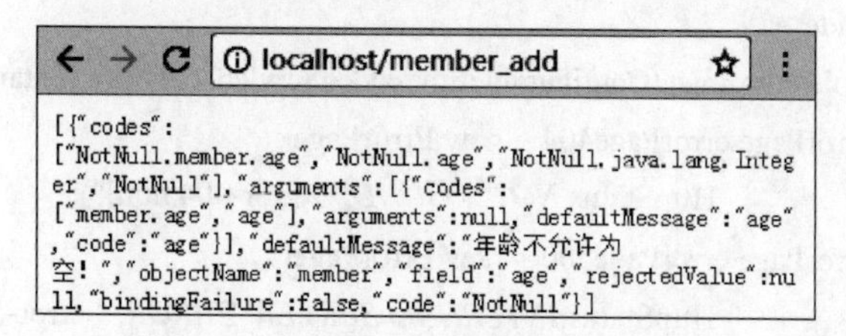

图 4-4　信息输入错误时返回错误信息

4.4　配置错误页

在 Web 项目开发过程中，错误信息提示页是一个重要的组成部分。无论多么合理的项目，也很难保证不出现类似于 404 或 500 的错误问题，而让用户直接看见满是异常信息的页面明显不是一个好的选择。这时，就需要有一个错误信息提示页。

1.【mldnboot-web 项目】错误页面一般都属于静态页面，这里在 src/main/view/static 目录下创建 error-404.html 和 error-500.html 两个页面，项目结构如图 4-5 所示。

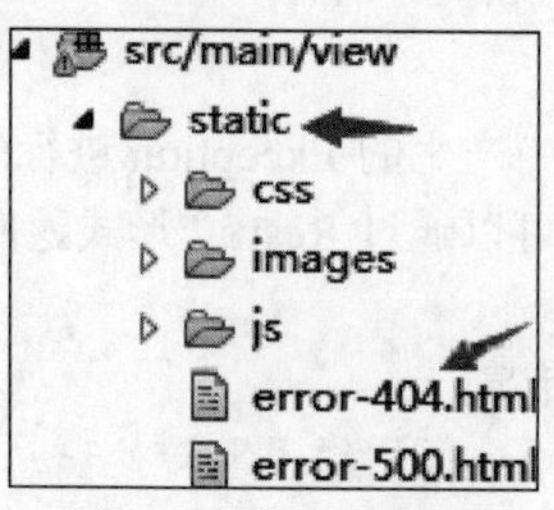

图 4-5　错误页结构

error-404.html 核心内容	error-500.html 核心内容
<h1>对不起，页面消失了（404）！</h1>	<h1>对不起，先让我静静的崩溃会儿（500）！</h1>

2.【mldnboot-web 项目】建立错误页配置。

```
package cn.mldn.mldnboot.config;
import org.springframework.boot.context.embedded.ConfigurableEmbeddedServletContainer;
import org.springframework.boot.context.embedded.EmbeddedServletContainerCustomizer;
import org.springframework.boot.web.servlet.ErrorPage;
import org.springframework.context.annotation.Bean;
import org.springframework.context.annotation.Configuration;
```

```
import org.springframework.http.HttpStatus;
@Configuration
public class ErrorPageConfig {
    @Bean
    public EmbeddedServletContainerCustomizer containerCustomizer() {
        return new EmbeddedServletContainerCustomizer() {
            @Override
            public void customize(ConfigurableEmbeddedServletContainer container) {
                ErrorPage errorPage404 = new ErrorPage(
                        HttpStatus.NOT_FOUND, "/error-404.html");          // 定义404错误页
                ErrorPage errorPage500 = new ErrorPage(
                        HttpStatus.INTERNAL_SERVER_ERROR, "/error-500.html");
                                                                         // 定义500错误页
                container.addErrorPages(errorPage404, errorPage500);     // 追加错误页
            }
        };
    }
}
```

配置完错误页之后，会根据用户请求时的 http 状态码跳转到不同的页面进行显示。

4.5 全局异常处理

全局异常处理指的是针对程序中产生的 Exception 进行的处理。产生了异常之后，可以统一跳转到一个页面进行错误提示，也可以通过 Restful 形式返回错误信息。

> **提示：关于全局错误与全局异常。**
>
> 全局错误指的是对 http 状态码进行的错误跳转处理，全局异常指的是发生某些异常（如果处理的是 Exception，则表示处理全部异常）之后的跳转页面。两者属于并行的概念，在项目开发中建议同时配置两者。
>
> 另外，如果想方便地观察本程序的执行结果，建议先将错误页的配置取消。

1.【mldnboot-web 项目】建立一个全局异常处理，该类可以处理所有的 Exception 异常。

```
package cn.mldn.mldnboot.advice;
import javax.servlet.http.HttpServletRequest;
import org.springframework.web.bind.annotation.ControllerAdvice;
import org.springframework.web.bind.annotation.ExceptionHandler;
import org.springframework.web.servlet.ModelAndView;
@ControllerAdvice                                    // 作为一个控制层的切面处理
```

```
public class GlobalExceptionAdvice {
    public static final String DEFAULT_ERROR_VIEW = "error";              // 错误显示页error.html
    @ExceptionHandler(Exception.class)                                    // 处理所有Exception类
    public ModelAndView defaultErrorHandler(HttpServletRequest request,
            Exception e) {                              // 出现异常之后会执行此方法
        ModelAndView mav = new ModelAndView(DEFAULT_ERROR_VIEW);   // 设置跳转路径
        mav.addObject("exception", e);                                    // 保存异常信息
        mav.addObject("url", request.getRequestURL());                    // 获得请求的路径
        return mav;
    }
}
```

2.【mldnboot-web 项目】建立 src/main/view/templates/error.html 页面，进行错误信息显示。

```
<p th:text="${'错误路径：' + url}"/>
<p th:text="${'错误信息：' + exception.message}"/>
```

3.【mldnboot-web 项目】建立一个控制器，主要功能是产生一个异常信息，以观察全局异常处理是否生效。

```
@RestController
public class ExceptionController {
    @GetMapping("/info")
    public String info() {
        int result = 10 / 0 ;                                             // 此处产生异常
        return "www.mldn.cn" ;
    }
}
```

在本程序中，只要访问/info 路径，就会产生异常，而产生异常之后将统一跳转到 error.html 页面。本程序运行结果如图 4-6 所示。

图 4-6　错误页跳转

提示：基于 Restful 错误信息提示。

本程序在发生异常之后采用跳转的形式来处理，而 SpringBoot 最大的特点是支持 Restful 处理，因此为了描述异常，也可以直接采用 Restful 的形式回应异常信息，即不再跳转到 HTML 页面进行显示。

范例：修改 GlobalExceptionAdvice 程序类，将其修改为 Restful 风格显示。

```
package cn.mldn.mldnboot.advice;
import javax.servlet.http.HttpServletRequest;
import org.springframework.web.bind.annotation.ExceptionHandler;
import org.springframework.web.bind.annotation.RestControllerAdvice;
@RestControllerAdvice                                           // 使用Restful风格返回
public class GlobalExceptionAdvice {
    @ExceptionHandler(Exception.class)                          // 处理所有异常
    public Object defaultErrorHandler(HttpServletRequest request, Exception e) {
        class ErrorInfo {                                       // 错误提示信息
            private Integer code ;
            private String message ;
            private String url ;
            // setter、getter略
        }
        ErrorInfo info = new ErrorInfo() ;
        info.setCode(HttpStatus.INTERNAL_SERVER_ERROR.value()); // 状态码
        info.setMessage(e.getMessage());                        // 保存错误信息
        info.setUrl(request.getRequestURL().toString());        // 保存错误路径
        return info ;
    }
}
```

本程序使用了@RestControllerAdvice 注解，则此时的异常处理将使用 Restful 风格，程序发生异常之后的运行效果如图 4-7 所示。

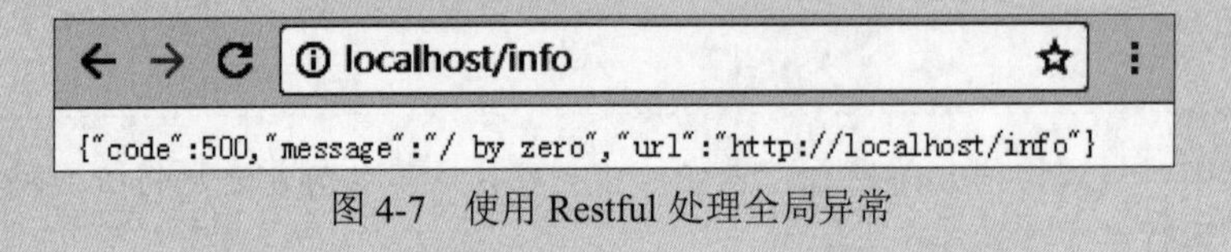

图 4-7　使用 Restful 处理全局异常

4.6　文件上传

文件上传功能是 Web 开发的一项重要技术手段，SpringBoot 本身也支持文件上传操作，并且其实现原理与 SpringMVC 相同，唯一的差异是配置相对减少了许多。

4.6.1　基础上传

SpringBoot 采用 FileUpload 组件实现上传处理，在控制器中可以使用 MultipartFile 类进行接收。

1.【mldnboot-web 项目】建立上传控制器 UploadController，利用 MultipartFile 将上传文件保存在本地磁盘。

```
package cn.mldn.mldnboot.controller;
import java.io.File;
import java.util.HashMap;
import java.util.Map;
import java.util.UUID;
import org.springframework.stereotype.Controller;
import org.springframework.web.bind.annotation.GetMapping;
import org.springframework.web.bind.annotation.PostMapping;
import org.springframework.web.bind.annotation.ResponseBody;
import org.springframework.web.context.request.RequestContextHolder;
import org.springframework.web.context.request.ServletRequestAttributes;
import org.springframework.web.multipart.MultipartFile;
@Controller
public class UploadController {
    @GetMapping("/upload_pre")
    public String uploadPre() {  // 通过model可以实现内容的传递
        return "upload";
    }
    @PostMapping("/upload")
    @ResponseBody
    public Object upload(String name, MultipartFile photo) throws Exception {
        Map<String,Object> map = new HashMap<String,Object>() ;
        if (photo != null) {    // 现在有文件上传
            map.put("name-param", name) ;
            map.put("photo-name", photo.getName()) ;
            map.put("content-type", photo.getContentType()) ;
            map.put("photo-size", photo.getSize()) ;
            String fileName = UUID.randomUUID() + "."
                    + photo.getContentType().substring(
                        photo.getContentType().lastIndexOf("/") + 1);// 创建文件名称
            String filePath = ((ServletRequestAttributes) RequestContextHolder.getRequestAttributes())
                    .getRequest().getServletContext().getRealPath("/") + fileName;
            map.put("photo-path", filePath) ;
            File saveFile = new File(filePath) ;
            photo.transferTo(saveFile);         // 文件保存
            return map ;
        } else {
            return "nothing";
        }
    }
}
```

2.【mldnboot-web 项目】建立 src/main/view/templates/upload.html 页面。

```
<form th:action="@{/upload}" method="post" enctype="multipart/form-data">
    姓名：<input type="text" name="name"/><br/>
    照片：<input type="file" name="photo"/><br/>
    <input type="submit" value="上传"/>
</form>
```

本程序通过表单传递了姓名（文本）和图片（二进制数据）两个数据信息。控制器接收到此请求信息后，如果有上传文件存在，则会直接返回上传信息（开发者也可以根据情况选择将文件保存），程序运行效果如图 4-8 所示。

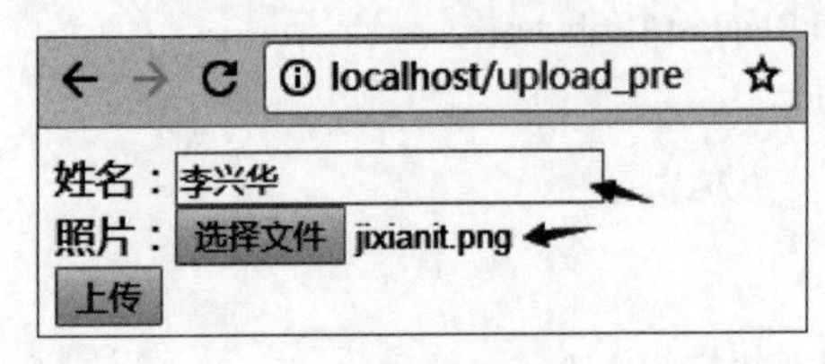

（a）上传表单

localhost/upload

{"photo-size":24000,"content-type":"image/png","photo-path":"C:\\Users\\yootk\\AppData\\Local\\Temp\\tomcat-docbase.7172272715290208496.80\\3c368000-4c16-4a04-b63a-edf1ea63535e.png","photo-name":"photo","name-param":"李兴华"}

（b）接收上传文件

图 4-8　文件上传

4.6.2　上传文件限制

在实际项目开发中，需要对用户上传文件的大小进行限制，这样才可以保证服务器的资源不被浪费。

1.【mldnboot-web 项目】修改 application.yml 配置文件，增加上传限制。

```
spring:
  http:
    multipart:
      enabled: true                    # 启用http上传
      max-file-size: 10MB              # 设置支持的单个上传文件的大小限制
      max-request-size: 20MB           # 设置最大的请求文件大小，设置总体大小请求
      file-size-threshold: 512KB       # 当上传文件达到指定配置量的时候，将文件内容写入磁盘
      location: /                      # 设置上传的临时目录
```

2.【mldnboot-web 项目】对于上传限制，也可以利用 Bean 实现同样的效果。

```
package cn.mldn.mldnboot.config;
import javax.servlet.MultipartConfigElement;
import org.springframework.boot.web.servlet.MultipartConfigFactory;
import org.springframework.context.annotation.Bean;
import org.springframework.context.annotation.Configuration;
@Configuration
public class UploadConfig {
```

```
    @Bean
    public MultipartConfigElement getMultipartConfig() {
        MultipartConfigFactory config = new MultipartConfigFactory() ;
        config.setMaxFileSize("10MB");                    // 设置上传文件的单个大小限制
        config.setMaxRequestSize("100MB");                // 设置总的上传的大小限制
        config.setLocation("/");                          // 设置临时保存目录
        return config.createMultipartConfig() ;           // 创建一个上传配置
    }
}
```

此时如果用户上传的内容超过了配置的限制，就会利用全局异常处理，上传出错后页面执行的效果，如图 4-9 所示。

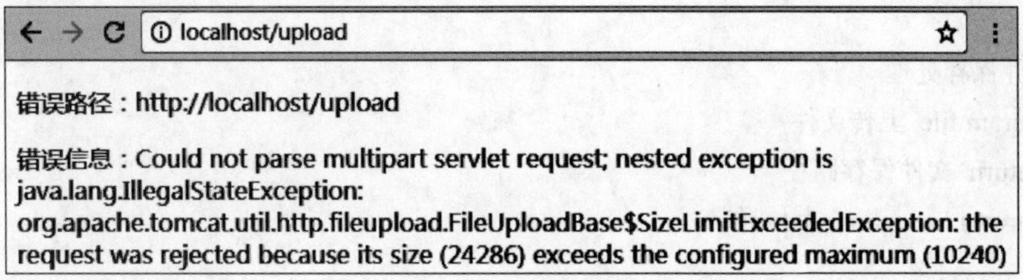

localhost/upload

错误路径：http://localhost/upload

错误信息：Could not parse multipart servlet request; nested exception is java.lang.IllegalStateException: org.apache.tomcat.util.http.fileupload.FileUploadBase$SizeLimitExceededException: the request was rejected because its size (24286) exceeds the configured maximum (10240)

图 4-9　上传限制

4.6.3　上传多个文件

如果要进行多个文件的上传，需要通过 MultipartHttpServletRequest 进行文件接收。

1.【mldnboot-web 项目】修改 upload.html 页面，定义多个文件上传控件。

```
<form th:action="@{/upload}" method="post" enctype="multipart/form-data">
    姓名：<input type="text" name="name"/><br/>
    照片1：<input type="file" name="photo"/><br/>
    照片2：<input type="file" name="photo"/><br/>
    照片3：<input type="file" name="photo"/><br/>
    <input type="submit" value="上传"/>
</form>
```

2.【mldnboot-web 项目】修改 UploadController 控制器，实现多个文件上传。

```
@PostMapping("/upload")
@ResponseBody
public Object upload(String name, HttpServletRequest request) {
    List<String> result = new ArrayList<String>() ;
    if (request instanceof MultipartHttpServletRequest) {
        MultipartHttpServletRequest mrequest = (MultipartHttpServletRequest) request;
        List<MultipartFile> files = mrequest.getFiles("photo");
```

```
            Iterator<MultipartFile> iter = files.iterator();
            while (iter.hasNext()) {
                MultipartFile photo = iter.next() ;          // 取出每一个上传文件
                try {
                    result.add(this.saveFile(photo)) ;  // 保存上传信息
                } catch (Exception e) {
                    e.printStackTrace();
                }
            }
        }
        return result ;
    }
    /**
     * 文件保存处理
     * @param file 上传文件
     * @return 文件保存路径
     * @throws Exception 上传异常
     */
    public String saveFile(MultipartFile file) throws Exception {
        String path = "nothing" ;
        if (file != null) {                                          // 有文件上传
            if (file.getSize() > 0) {
                String fileName = UUID.randomUUID() + "."
                        + file.getContentType().substring(
                            file.getContentType().lastIndexOf("/") + 1);// 创建文件名称
                path = ((ServletRequestAttributes) RequestContextHolder.getRequestAttributes())
                        .getRequest().getServletContext().getRealPath("/") + fileName;
                File saveFile = new File(path) ;
                file.transferTo(saveFile);                   // 文件保存
            }
        }
        return path ;
    }
```

本程序为了方便文件上传，在控制器类中定义了一个 saveFile()方法，以进行文件的保存，同时利用此方法返回了上传文件的保存路径。

提示：上传图片应该保存在图片服务器中。

在本书所讲解的文件上传处理过程中，都是将图片保存到本地的 Web 服务端，但是从实际的开发来讲，这种操作是不可行的。在当今的项目开发中，最流行的设计理念是高可用、高并发、分布式设计，所以在实际项目中需要搭建专门的图片服务器进行上传资源的保存。如图 4-10

所示给读者简单地描述了一个 Web 集群与图片服务器集群的搭建关系。

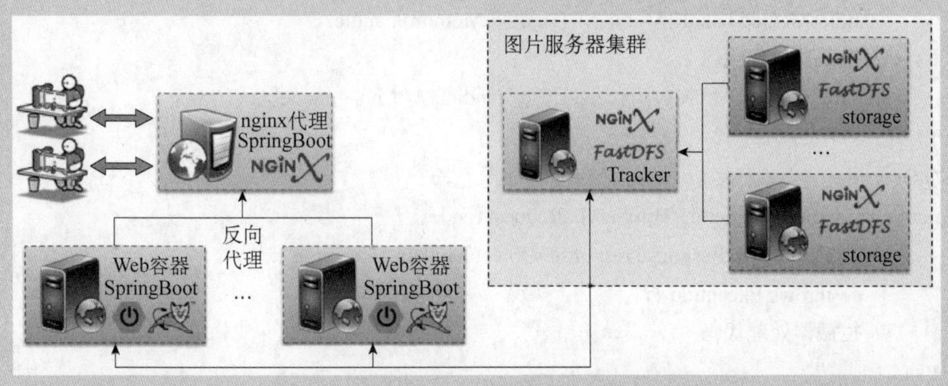

图 4-10　简化的 Web 与图片服务器集群设计

在本书中，由于只涉及 SpringBoot 开发框架的使用，所以不会对此部分的内容进行讲解，有兴趣的读者可以登录 www.mldn.cn 自行学习。

4.7　拦截器

在 Web 请求处理的过程中，拦截器是服务器端进行数据处理的最后一道屏障，可以将所有用户请求的信息在拦截器中进行验证。在 SpringBoot 中可以继续使用 SpringMVC 所提供的拦截器进行处理。

1.【mldnboot-web 项目】定义一个拦截器处理类。

```
package cn.mldn.mldnboot.util.interceptor;
import javax.servlet.http.HttpServletRequest;
import javax.servlet.http.HttpServletResponse;
import org.springframework.web.method.HandlerMethod;
import org.springframework.web.servlet.HandlerInterceptor;
import org.springframework.web.servlet.ModelAndView;
public class MyInterceptor implements HandlerInterceptor {     // 实现拦截器接口
    @Override
    public boolean preHandle(HttpServletRequest request,
            HttpServletResponse response, Object handler) throws Exception {
        HandlerMethod handlerMethod = (HandlerMethod) handler ;
        // 拦截器处理代码
        return true;          // 如果返回false，表示不继续请求；如果返回true，表示继续请求
    }
    @Override
    public void postHandle(HttpServletRequest request,
            HttpServletResponse response, Object handler,
```

```
            ModelAndView modelAndView) throws Exception {
        HandlerMethod handlerMethod = (HandlerMethod) handler ;
        // 拦截器处理代码
        this.log.info("【*** MyInterceptor.postHandle() ***】" + modelAndView);
    }
    @Override
    public void afterCompletion(HttpServletRequest request,
            HttpServletResponse response, Object handler, Exception ex)
            throws Exception {
        // 拦截器处理代码
    }
}
```

在拦截器中最需要用户处理的方法是 preHandle()，此方法会在控制层的方法执行之前进行调用。

2.【mldnboot-web 项目】如果要拦截器生效，则还需要定义一个拦截器的配置类。

```
package cn.mldn.mldnboot.config;
import org.springframework.context.annotation.Configuration;
import org.springframework.web.servlet.config.annotation.InterceptorRegistry;
import org.springframework.web.servlet.config.annotation.WebMvcConfigurerAdapter;
import cn.mldn.mldnboot.util.interceptor.MyInterceptor;
@Configuration
public class MyWebApplicationConfig extends WebMvcConfigurerAdapter {          // 定义MVC配置
    @Override
    public void addInterceptors(InterceptorRegistry registry) {              // 进行拦截器的注册处理操作
        registry.addInterceptor(new MyInterceptor()).addPathPatterns("/**") ;       // 匹配路径
        super.addInterceptors(registry);
    }
}
```

本程序将拦截器配置到了 Web 项目中，配置的访问路径为全部请求路径，这样不管用户如何访问都会先执行拦截器中的处理方法。

4.8 AOP 拦截器

AOP（面向切面编程）是 Spring 提供的重要技术工具，其主要功能是对业务层的方法调用进行拦截处理。SpringBoot 默认情况下并没有配置 AOP 拦截器，开发者需要在项目中手动引入 spring-boot-starter-aop 依赖库后才可以使用。

1.【mldnboot-web 项目】修改 pom.xml 配置文件，配置 spring-boot-starter-aop 依赖库。

```
<dependency>
    <groupId>org.springframework.boot</groupId>
    <artifactId>spring-boot-starter-aop</artifactId>
</dependency>
```

2.【mldnboot-web 项目】定义业务层接口。

```
package cn.mldn.mldnboot.service;
public interface IMessageService {
    public String echo(String msg) ;
}
```

3.【mldnboot-web 项目】定义业务层接口实现子类。

```
package cn.mldn.mldnboot.service.impl;
import org.springframework.stereotype.Service;
import cn.mldn.mldnboot.service.IMessageService;
@Service
public class MessageServiceImpl implements IMessageService {
    @Override
    public String echo(String msg) {
        return "【ECHO】" + msg;
    }
}
```

4.【mldnboot-web 项目】定义 AOP 程序类，对业务方法进行拦截，本例使用环绕通知处理。

```
package cn.mldn.mldnboot.aspect;
import java.util.Arrays;
import org.aspectj.lang.ProceedingJoinPoint;
import org.aspectj.lang.annotation.Around;
import org.aspectj.lang.annotation.Aspect;
import org.slf4j.Logger;
import org.slf4j.LoggerFactory;
import org.springframework.stereotype.Component;
@Aspect
@Component
public class ServiceAspect {                                        // 定义业务层拦截
    private Logger log = LoggerFactory.getLogger(ServiceAspect.class);
    @Around("execution(* cn.mldn..service..*.*(..))")
    public Object arroundInvoke(ProceedingJoinPoint point) throws Throwable {
        this.log.info("【*** Service-Before ***】执行参数：" + Arrays.toString (point.getArgs()));
        Object obj = point.proceed(point.getArgs());                // 进行具体业务调用
        this.log.info("【*** Service-After ***】返回结果：" + obj);
        return obj;
```

```
    }
}
```

5.【mldnboot-web 项目】编写测试类，测试 ServiceAspect 拦截是否生效。

```
package cn.mldn.mldnboot.test;
import org.junit.Test;
import org.junit.runner.RunWith;
import org.springframework.beans.factory.annotation.Autowired;
import org.springframework.boot.test.context.SpringBootTest;
import org.springframework.test.context.junit4.SpringJUnit4ClassRunner;
import org.springframework.test.context.web.WebAppConfiguration;
import cn.mldn.mldnboot.SpringBootStartApplication;
import cn.mldn.mldnboot.service.IMessageService;
@SpringBootTest(classes = SpringBootStartApplication.class)     // 定义要测试的SpringBoot类
@RunWith(SpringJUnit4ClassRunner.class)                          // 使用JUnit进行测试
@WebAppConfiguration                                             // 进行Web应用配置
public class TestMessageService {
    @Autowired
    private IMessageService messageService ;
    @Test
    public void testEcho() {
        System.out.println(this.messageService.echo("www.mldn.cn"));
    }
}
```

程序执行结果	【*** Service-Before ***】执行参数：[www.mldn.cn] 【*** Service-After ***】返回结果：【ECHO】www.mldn.cn 【ECHO】www.mldn.cn

本程序由于将切入点设置在了所有的业务层上，所以在调用 IMessageService 接口方法时会自动执行 AOP 拦截。

4.9 本章小结

1. SpringBoot 项目可以将程序打包为 war 文件，并且部署到 Tomcat 容器上执行。

2. SpringBoot 可以像 web.xml 文件一样设置状态码的错误跳转页，也可以设置异常的错误跳转页。

3. SpringBoot 与 SpringMVC 上传文件的处理形式相同，但是配置更加简化。

4. SpringBoot 可以使用 HandlerInterceptor 拦截器对控制层的请求进行拦截。

5. SpringBoot 可以直接导入 spring-boot-starter-aop 编写 AOP 拦截器，实现业务层拦截。

第 5 章

SpringBoot 服务整合

通过本章学习，可以达到以下目标：

1. 掌握 SpringBoot 与 DataSource 数据源整合。
2. 掌握 SpringBoot 与 MyBatis 开发框架整合。
3. 掌握 SpringBoot 与 SpringDataJPA 开发框架整合。
4. 掌握 SpringBoot 与消息组件（ActiveMQ、RabbitMQ、Kafka）整合。
5. 掌握 SpringBoot 与邮件服务整合。
6. 掌握 SpringBoot 与定时调度服务整合。
7. 掌握 SpringBoot 与 Redis 数据库整合。
8. 掌握 SpringBoot 与 Restful 服务整合。

在实际的项目开发中，Spring 开发框架几乎无处不在，相信有过框架开发经验的读者都有过被成堆的 Spring 配置文件搞疯的经历。幸运的是，SpringBoot 开发框架极大地简化了与第三方框架及第三方服务之间的整合处理。本章将为读者讲解 SpringBoot 与常见开发框架的整合。

提示：本章不涉及具体的服务部署。

众所周知，当前的项目不再由某一种单独的 Web 服务提供支持，而是由一整套的集群服务在为每一个使用者提供支持。本章讲解中将涉及 MyBatis、JPA（SpringDataJPA）、Druid、C3P0、ActiveMQ、RabbitMQ、Kafka、Mail、Redis、线程池等概念，这些内容不再讲解其基本使用与服务部署，对这些概念不熟悉的读者可以登录 www.mldn.cn 自行学习，本章只是讲解服务整合处理。

5.1 SpringBoot 整合数据源

在实际项目开发中任何项目都很难脱离数据库而单独存在，所以为了提高数据库的操作性能，开发者往往会借助于数据库连接池来进行处理，同时在项目中利用 DataSource 进行数据源的连接。常用的有 C3P0 和 Druid 两类数据库连接池，下面一一进行介绍。

5.1.1 SpringBoot 整合 C3P0 数据库连接池

C3P0 是一个开源的 JDBC 连接池，它实现了数据源和 JNDI 绑定，支持 JDBC3 规范和 JDBC2

的标准扩展，同时在 Hibernate、Spring 项目开发中被广泛应用。

1.【mldnboot 项目】修改父 pom.xml 配置文件，追加 C3P0 依赖支持管理。

定义版本属性

```
<c3p0.version>0.9.1.2</c3p0.version>
<mysql-connector-java.version>5.1.21</mysql-connector-java.version>
```

定义配置依赖管理

```
<dependency>
    <groupId>mysql</groupId>
    <artifactId>mysql-connector-java</artifactId>
    <version>${mysql-connector-java.version}</version>
</dependency>
<dependency>
    <groupId>c3p0</groupId>
    <artifactId>c3p0</artifactId>
    <version>${c3p0.version}</version>
</dependency>
```

2.【mldnboot-integration 项目】修改 pom.xml 配置文件，引入 C3P0 的相关依赖支持库。

```
<dependency>
    <groupId>c3p0</groupId>
    <artifactId>c3p0</artifactId>
</dependency>
<dependency>
    <groupId>mysql</groupId>
    <artifactId>mysql-connector-java</artifactId>
</dependency>
```

3.【mldnboot-integration 项目】修改 application.yml 配置文件，追加 C3P0 数据库连接池配置信息。

```
c3p0:                                               # 定义C3P0配置
  jdbcUrl: jdbc:mysql://localhost:3306/mldn         # 数据库连接地址
  user: root                                        # 数据库用户名
  password: mysqladmin                              # 数据库密码
  driverClass: org.gjt.mm.mysql.Driver              # 数据库驱动程序
  minPoolSize: 1                                    # 最小连接数
  maxPoolSize: 1                                    # 最大连接数
  maxIdleTime: 3000                                 # 最大等待时间
  initialPoolSize: 1                                # 初始化连接数
```

4.【mldnboot-integration 项目】建立 C3P0 数据源连接池配置类，此时设置的 Bean 名称为 dataSource。

```
package cn.mldn.mldnboot.config;
import javax.sql.DataSource;
import org.springframework.boot.autoconfigure.jdbc.DataSourceBuilder;
import org.springframework.boot.context.properties.ConfigurationProperties;
import org.springframework.context.annotation.Bean;
import org.springframework.context.annotation.Configuration;
@Configuration
public class C3P0DatasourceConfig {
    @Bean(name = "dataSource")
    @ConfigurationProperties(prefix = "c3p0")                    // 定义资源导入前导标记
    public DataSource dataSource() {
        return DataSourceBuilder.create().type(
                com.mchange.v2.c3p0.ComboPooledDataSource.class).build();
    }
}
```

5.【mldnboot-integration 项目】编写测试类，测试当前 DataSource 配置是否正确。

```
package cn.mldn.mldnboot.test;
import javax.sql.DataSource;
import org.junit.Test;
import org.junit.runner.RunWith;
import org.springframework.beans.factory.annotation.Autowired;
import org.springframework.boot.test.context.SpringBootTest;
import org.springframework.test.context.junit4.SpringJUnit4ClassRunner;
import org.springframework.test.context.web.WebAppConfiguration;
import cn.mldn.mldnboot.SpringBootStartApplication;
@SpringBootTest(classes = SpringBootStartApplication.class)      // 定义要测试的SpringBoot类
@RunWith(SpringJUnit4ClassRunner.class)                          // 使用JUnit进行测试
@WebAppConfiguration                                             // 进行Web应用配置
public class TestDataSource {
    @Autowired
    private DataSource dataSource ;                              // 注入DataSource对象
    @Test
    public void testConnection() throws Exception {
        System.out.println(this.dataSource.getConnection());     // 获取连接
    }
}
```

本程序在确保使用的 MySQL 数据库服务正常开启后，就可以通过 C3P0 工具获取数据库连接。

5.1.2 SpringBoot 整合 Druid 数据库连接池

Druid 是阿里巴巴推出的一款数据库连接池组件（可以理解为 C3P0 的下一代产品），也是一个用于大数据实时查询和分析的高容错、高性能开源分布式系统，可高效处理大规模的数据并实现快速查询和分析。

1.【mldnboot 项目】修改父 pom.xml 文件，引入 Druid 的相关依赖库。

定义版本属性：	`<druid.version>1.1.6</druid.version>`

定义配置依赖管理：	`<dependency>`
	`<groupId>com.alibaba</groupId>`
	`<artifactId>druid</artifactId>`
	`<version>${druid.version}</version>`
	`</dependency>`

2.【mldnboot-integration 项目】修改 pom.xml 配置文件，追加 Druid 依赖配置。

```
<dependency>
    <groupId>com.alibaba</groupId>
    <artifactId>druid</artifactId>
</dependency>
```

3.【mldnboot-integration 项目】修改 application.yml 配置文件，追加 Druid 的连接配置。

```
spring:
  datasource:
    type: com.alibaba.druid.pool.DruidDataSource          # 配置当前要使用的数据源的操作类型
    driver-class-name: org.gjt.mm.mysql.Driver            # 配置MySQL的驱动程序类
    url: jdbc:mysql://localhost:3306/mldn                 # 数据库连接地址
    username: root                                        # 数据库用户名
    password: mysqladmin                                  # 数据库连接密码
    dbcp2:                                                # 进行数据库连接池的配置
      min-idle: 1                                         # 数据库连接池的最小维持连接数
      initial-size: 1                                     # 初始化提供的连接数
      max-total: 1                                        # 最大的连接数
      max-wait-millis: 3000                               # 等待连接获取的最大超时时间
```

此时，项目中就可以采用 Druid 数据库连接池来进行数据库操作了。

提示：进行连接测试前需要导入相应 ORM 框架的依赖支持包。

在本程序中，如果要进行 DataSource 连接测试，则需要导入 ORM 依赖关联包。例如，可以在本程序中导入 MyBatis 的 ORM 开发包。

```
<dependency>
    <groupId>org.mybatis.spring.boot</groupId>
    <artifactId>mybatis-spring-boot-starter</artifactId>
    <version>1.3.1</version>
</dependency>
```

导入以上开发包之后，可以正常测试。如果未导入，则程序测试时会出现 Unsatisfied DependencyException 异常信息。

5.2 SpringBoot 整合 ORM 开发框架

在使用 Spring 整合 ORM 组件的过程中，为了达到简化的目的，往往会进行大量的配置。利用 SpringBoot 可以进一步实现配置的简化，在本章中将为读者讲解两种常用的 ORM 组件的整合：MyBatis 和 JPA。

5.2.1 SpringBoot 整合 MyBatis 开发框架

MyBatis 是一款常用并且配置极为简单的 ORM 开发框架。其与 Spring 结合后，可以利用 Spring 的特征实现 DAO 接口的自动配置。在 SpringBoot 中，又对 MyBatis 框架的整合进行了进一步简化。想实现这种配置，需要在项目中引入 mybatis-spring-boot- starter 依赖支持库。

提示：需要数据库连接池支持。

SpringBoot 与 ORM 开发框架整合时，如无特殊说明，将使用 Druid 作为数据库连接池。同时，数据库名称统一设置为 mldn，数据表也统一使用如下的脚本创建。

```
DROP DATABASE IF EXISTS mldn ;
CREATE DATABASE mldn CHARACTER SET UTF8 ;
USE mldn ;
CREATE TABLE dept (
    deptno          BIGINT          AUTO_INCREMENT ,
    dname           VARCHAR(50) ,
    CONSTRAINT pk_deptno PRIMARY KEY(deptno)
) ;
INSERT INTO dept(dname) VALUES ('开发部') ;
INSERT INTO dept(dname) VALUES ('财务部') ;
INSERT INTO dept(dname) VALUES ('市场部') ;
```

此表表示部门信息，除了自动增长的主键之外，只提供了部门名称（dname）一个字段。

1.【mldnboot-integration项目】建立VO类，并且所有VO类的所在包均为cn.mldn.mldnboot. vo。

```
package cn.mldn.mldnboot.vo;
import java.io.Serializable;
@SuppressWarnings("serial")
public class Dept implements Serializable {
    private Long deptno ;
    private String dname ;
    // setter、getter略
}
```

2.【mldnboot-integration 项目】在 src/main/resources 目录中创建 mybatis/mybatis.cfg.xml 配置文件。

```
<?xml version="1.0" encoding="UTF-8" ?>
<!DOCTYPE configuration
    PUBLIC "-//mybatis.org//DTD Config 3.0//EN"
    "http://mybatis.org/dtd/mybatis-3-config.dtd">
<configuration>         <!-- 进行MyBatis的相应环境的属性定义 -->
    <settings>          <!-- 在本项目之中开启二级缓存 -->
        <setting name="cacheEnabled" value="true"/>
    </settings>
</configuration>
```

3.【mldnboot-integration 项目】修改 application.yml 配置文件，追加 MyBatis 配置。

```
mybatis:
  config-location: classpath:mybatis/mybatis.cfg.xml        # MyBatis配置文件所在路径
  type-aliases-package: cn.mldn.mldnboot.vo                  # 定义所有操作类的别名所在包
```

4.【mldnboot-integration 项目】建立 IDeptDAO 接口，该接口将由 Spring 自动实现。

```
package cn.mldn.mldnboot.dao;
import java.util.List;
import org.apache.ibatis.annotations.Mapper;
import org.apache.ibatis.annotations.Select;
import cn.mldn.mldnboot.vo.Dept;
@Mapper
public interface IDeptDAO {
    @Select("SELECT deptno,dname FROM dept")
    public List<Dept> findAll() ;                      // 查询全部部门信息
}
```

5.【mldnboot-integration 项目】定义 IDeptService 业务层接口。

```
package cn.mldn.mldnboot.service;
import java.util.List;
import cn.mldn.mldnboot.vo.Dept;
public interface IDeptService {
    public List<Dept> list() ;
}
```

6.【mldnboot-integration 项目】定义 DeptServiceImpl 业务层实现子类。

```
package cn.mldn.mldnboot.service.impl;
import java.util.List;
import org.springframework.beans.factory.annotation.Autowired;
import org.springframework.stereotype.Service;
import cn.mldn.mldnboot.dao.IDeptDAO;
import cn.mldn.mldnboot.service.IDeptService;
import cn.mldn.mldnboot.vo.Dept;
@Service
public class DeptServiceImpl implements IDeptService {
    @Autowired
    private IDeptDAO deptDAO ;
    @Override
    public List<Dept> list() {
        return this.deptDAO.findAll() ;
    }
}
```

7.【mldnboot-integration 项目】编写测试类，测试 IDeptService 业务方法。

```
package cn.mldn.mldnboot.test;
import java.util.List;
import org.junit.Test;
import org.junit.runner.RunWith;
import org.springframework.beans.factory.annotation.Autowired;
import org.springframework.boot.test.context.SpringBootTest;
import org.springframework.test.context.junit4.SpringJUnit4ClassRunner;
import org.springframework.test.context.web.WebAppConfiguration;
import cn.mldn.mldnboot.SpringBootStartApplication;
import cn.mldn.mldnboot.service.IDeptService;
import cn.mldn.mldnboot.vo.Dept;
@SpringBootTest(classes = SpringBootStartApplication.class)        // 定义要测试的SpringBoot类
@RunWith(SpringJUnit4ClassRunner.class)                            // 使用JUnit进行测试
@WebAppConfiguration                                               // 进行Web应用配置
```

```
public class TestDeptService {
    @Autowired
    private IDeptService deptService ;                                // 注入业务接口对象
    @Test
    public void testList() {
        List<Dept> allDepts = this.deptService.list() ;
        for (Dept dept : allDepts) {
            System.out.println("部门编号：" + dept.getDeptno() + "、部门名称：" + dept.getDname());
        }
    }
}
```

程序执行结果	部门编号：1、部门名称：开发部 部门编号：2、部门名称：财务部 部门编号：3、部门名称：市场部

读者可以发现，MyBatis 的原始配置并没有做任何改变。整体开发对于 Spring 的整合，只需要修改 application.yml 的几个配置项就可以实现了。

5.2.2 SpringBoot 整合 JPA 开发框架

JPA 是官方推出的 Java 持久层操作标准（现主要使用 Hibernate 实现），使用 SpringData 技术和 JpaRepository 接口技术，也可以达到简化数据层的目的。要在 SpringBoot 中使用 SpringDataJPA，需要 spring-boot-starter-data-jpa 依赖库的支持。

1.【mldnboot-integration 项目】修改 pom.xml 配置文件，引入相关依赖包。

```
<dependency>
    <groupId>org.springframework.boot</groupId>
    <artifactId>spring-boot-starter-data-jpa</artifactId>
</dependency>
<dependency>
    <groupId>org.springframework.boot</groupId>
    <artifactId>spring-boot-starter-cache</artifactId>
</dependency>
<dependency>
    <groupId>org.hibernate</groupId>
    <artifactId>hibernate-ehcache</artifactId>
</dependency>
```

2.【mldnboot-integration 项目】建立持久化类 Dept。

```
package cn.mldn.mldnboot.po;
import java.io.Serializable;
```

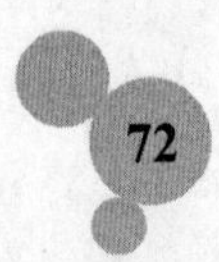

```
import javax.persistence.Cacheable;
import javax.persistence.Entity;
import javax.persistence.GeneratedValue;
import javax.persistence.GenerationType;
import javax.persistence.Id;
@SuppressWarnings("serial")
@Cacheable(true)
@Entity
public class Dept implements Serializable {
	@Id
	@GeneratedValue(strategy = GenerationType.IDENTITY)		// 根据名称引用配置的主键生成器
	private Long deptno;	// 字段的映射（属性名称=字段名称）
	private String dname;
	// setter、getter略
}
```

3.【mldnboot-integration 项目】定义 IDeptDAO 接口，此接口继承 JpaRepository 父接口。

```
package cn.mldn.mldnboot.dao;
import org.springframework.data.jpa.repository.JpaRepository;
import cn.mldn.mldnboot.po.Dept;
public interface IDeptDAO extends JpaRepository<Dept, Long> {	// 包含全部的基础Crud支持
}
```

4.【mldnboot-integration 项目】定义 IDeptService 业务层接口。

```
package cn.mldn.mldnboot.service;
import java.util.List;
import cn.mldn.mldnboot.vo.Dept;
public interface IDeptService {
	public List<Dept> list() ;
}
```

5.【mldnboot-integration 项目】定义 DeptServiceImpl 业务层实现子类。

```
package cn.mldn.mldnboot.service.impl;
import java.util.List;
import org.springframework.beans.factory.annotation.Autowired;
import org.springframework.stereotype.Service;
import cn.mldn.mldnboot.dao.IDeptDAO;
import cn.mldn.mldnboot.service.IDeptService;
import cn.mldn.mldnboot.vo.Dept;
@Service
```

```
public class DeptServiceImpl implements IDeptService {
	@Autowired
	private IDeptDAO deptDAO ;
	@Override
	public List<Dept> list() {
		return this.deptDAO.findAll() ;
	}
}
```

6.【mldnboot-integration 项目】修改程序启动主类，追加 Repository 扫描配置。

```
package cn.mldn.mldnboot;
import org.springframework.boot.SpringApplication;
import org.springframework.boot.autoconfigure.SpringBootApplication;
import org.springframework.data.jpa.repository.config.EnableJpaRepositories;
@SpringBootApplication                                 // 启动SpringBoot程序，而后自带子包扫描
@EnableJpaRepositories(basePackages="cn.mldn.mldnboot.dao")
public class SpringBootStartApplication {              // 必须继承指定的父类
	public static void main(String[] args) throws Exception {
		SpringApplication.run(SpringBootStartApplication.class, args); // 启动SpringBoot程序
	}
}
```

7.【mldnboot-integration 项目】编写测试类，测试 IDeptService 业务方法。

```
package cn.mldn.mldnboot.test;
import java.util.List;
import org.junit.Test;
import org.junit.runner.RunWith;
import org.springframework.beans.factory.annotation.Autowired;
import org.springframework.boot.test.context.SpringBootTest;
import org.springframework.test.context.junit4.SpringJUnit4ClassRunner;
import org.springframework.test.context.web.WebAppConfiguration;
import cn.mldn.mldnboot.SpringBootStartApplication;
import cn.mldn.mldnboot.service.IDeptService;
import cn.mldn.mldnboot.vo.Dept;
@SpringBootTest(classes = SpringBootStartApplication.class)           // 定义要测试的SpringBoot类
@RunWith(SpringJUnit4ClassRunner.class)                              // 使用JUnit进行测试
@WebAppConfiguration                                                 // 进行Web应用配置
public class TestDeptService {
	@Autowired
	private IDeptService deptService ;                                 // 注入业务接口对象
```

```
	@Test
	public void testList() {
		List<Dept> allDepts = this.deptService.list() ;
		for (Dept dept : allDepts) {
			System.out.println("部门编号：" + dept.getDeptno() + "、部门名称：" + dept.getDname());
		}
	}
}
```

程序执行结果	部门编号：1、部门名称：开发部 部门编号：2、部门名称：财务部 部门编号：3、部门名称：市场部

本程序利用 SpringBoot 调用了 SpringDataJPA，并且利用 JpaRepository 实现了 DAO 接口的自动实现。需要注意的是，如果想启用 Repository 配置，则需要在程序启动主类时使用@EnableJpaRepositories 注解配置扫描包，而后才可以正常使用。

5.2.3　事务处理

SpringBoot 中可以使用 PlatformTransactionManager 接口来实现事务的统一控制，而进行控制的时候也可以采用注解或者 AOP 切面配置形式来完成。

1.【mldnboot-integration 项目】在业务层的方法上启用事务控制。

```
public interface IDeptService {
	@Transactional(propagation = Propagation.REQUIRED, readOnly = true)
	public List<Dept> list();
}
```

2.【mldnboot-integration 项目】在程序主类中还需要配置事务管理注解。

```
@SpringBootApplication					// 启动SpringBoot程序，而后自带子包扫描
@EnableTransactionManagement
public class SpringBootStartApplication {	// 必须继承指定的父类
	public static void main(String[] args) throws Exception {
		SpringApplication.run(SpringBootStartApplication.class, args);	// 启动SpringBoot程序
	}
}
```

项目中利用业务层中定义的@Transactional 注解就可以实现事务的控制，但是这样的事务控制过于复杂。在一个大型项目中可能存在成百上千的业务接口，全部使用注解控制必然会造成代码的大量重复。在实际工作中，SpringBoot 与事务结合最好的控制方法就是定义一个事务的配置类。

3.【mldnboot-integration 项目】取消事务注解配置，并定义 TransactionConfig 配置类。

```
package cn.mldn.mldnboot.config;
import java.util.HashMap;
```

```
import java.util.Map;
import org.aspectj.lang.annotation.Aspect;
import org.springframework.aop.Advisor;
import org.springframework.aop.aspectj.AspectJExpressionPointcut;
import org.springframework.aop.support.DefaultPointcutAdvisor;
import org.springframework.beans.factory.annotation.Autowired;
import org.springframework.context.annotation.Bean;
import org.springframework.context.annotation.Configuration;
import org.springframework.transaction.PlatformTransactionManager;
import org.springframework.transaction.TransactionDefinition;
import org.springframework.transaction.interceptor.NameMatchTransactionAttributeSource;
import org.springframework.transaction.interceptor.RuleBasedTransactionAttribute;
import org.springframework.transaction.interceptor.TransactionAttribute;
import org.springframework.transaction.interceptor.TransactionInterceptor;
@Configuration                                                          // 定义配置Bean
@Aspect                                                                 // 采用AOP切面处理
public class TransactionConfig {
    private static final int TRANSACTION_METHOD_TIMEOUT = 5 ;   // 事务超时时间为5秒
    private static final String AOP_POINTCUT_EXPRESSION =
                    "execution (* cn.mldn..service.*.*(..))";           // 定义切面表达式
    @Autowired
    private PlatformTransactionManager transactionManager;              // 注入事务管理对象
    @Bean("txAdvice")                                                   // Bean名称必须为txAdvice
    public TransactionInterceptor transactionConfig() {                 // 定义事务控制切面
        // 定义只读事务控制，该事务不需要启动事务支持
        RuleBasedTransactionAttribute readOnly = new RuleBasedTransactionAttribute();
        readOnly.setReadOnly(true);
        readOnly.setPropagationBehavior(TransactionDefinition.PROPAGATION_NOT_ SUPPORTED);
        // 定义更新事务，同时设置事务操作的超时时间
        RuleBasedTransactionAttribute required = new RuleBasedTransactionAttribute();
        required.setPropagationBehavior(TransactionDefinition.PROPAGATION_REQUIRED);
        required.setTimeout(TRANSACTION_METHOD_TIMEOUT);
        Map<String, TransactionAttribute> transactionMap = new HashMap<>();  // 定义业务切面
        transactionMap.put("add*", required);
        transactionMap.put("edit*", required);
        transactionMap.put("delete*", required);
        transactionMap.put("get*", readOnly);
        transactionMap.put("list*", readOnly);
        NameMatchTransactionAttributeSource source = new NameMatchTransaction AttributeSource();
        source.setNameMap(transactionMap);
        TransactionInterceptor transactionInterceptor = new
```

```
                TransactionInterceptor(transactionManager, source);
        return transactionInterceptor ;
    }
    @Bean
    public Advisor transactionAdviceAdvisor() {
        AspectJExpressionPointcut pointcut = new AspectJExpressionPointcut();
        pointcut.setExpression(AOP_POINTCUT_EXPRESSION);          // 定义切面
        return new DefaultPointcutAdvisor(pointcut, transactionConfig());
    }
}
```

此时程序中的事务控制可以利用 TransactionConfig 类结合 AspectJ 切面与业务层中的方法匹配，而后就不再需要业务方法使用@Transactional 注解重复定义了。

5.3　SpringBoot 整合消息服务组件

在进行分布式系统设计时，经常会使用消息服务组件进行系统整合与异步服务通信，其基本结构为生产者与消费者处理，如图 5-1 所示。常用的消息组件主要包括两类：JMS 标准（ActiveMQ）和 AMQP 标准（RabbitMQ、Kafka），本章将为读者讲解这两类组件与 SpringBoot 的整合。

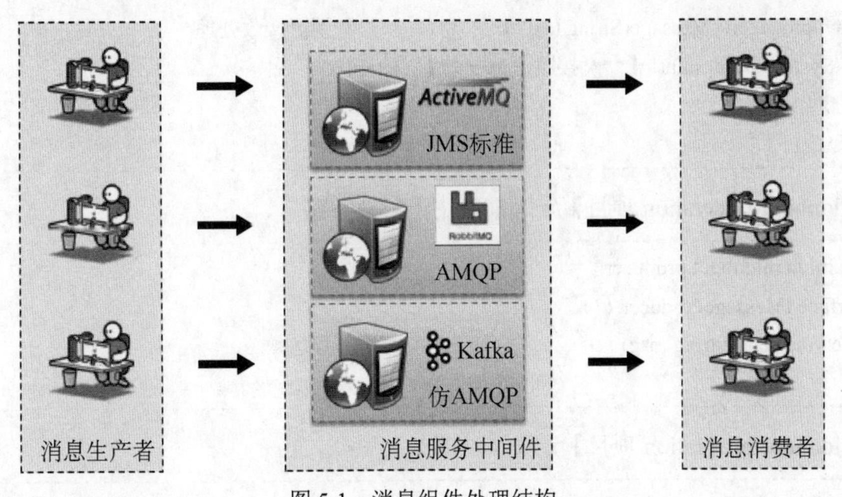

图 5-1　消息组件处理结构

5.3.1　SpringBoot 整合 ActiveMQ 消息组件

ActiveMQ 是 Apache 提供的开源组件，是基于 JMS 标准的实现组件。下面将利用 SpringBoot 整合 ActiveMQ 组件，实现队列消息的发送与接收。

1.【mldnboot-integration 项目】修改 pom.xml 配置文件，追加 spring-boot-starter-activemq 依赖库。

```
<dependency>
    <groupId>org.springframework.boot</groupId>
    <artifactId>spring-boot-starter-activemq</artifactId>
</dependency>
```

2.【mldnboot-integration 项目】修改 application.yml 配置文件，进行 ActiveMQ 的配置。

```
spring:
  jms:
    pub-sub-domain: false                  # 配置消息类型，true表示为topic消息，false表示Queue消息
  activemq:
    user: mldnjava                         # 连接用户名
    password: hello                        # 连接密码
    broker-url: tcp://activemq-server:61616  # 消息组件的连接主机信息
```

3.【mldnboot-integration 项目】定义消息消费监听类。

```
package cn.mldn.mldnboot.consumer;
import org.springframework.jms.annotation.JmsListener;
import org.springframework.stereotype.Service;
@Service
public class MessageConsumer {
    @JmsListener(destination="mldn.msg.queue")       // 定义消息监听队列
    public void receiveMessage(String text) {          // 进行消息接收处理
        System.err.println("【*** 接收消息 ***】" + text);
    }
}
```

4.【mldnboot-integration 项目】定义消息生产者业务接口。

```
package cn.mldn.mldnboot.producer;
public interface IMessageProducer {
    public void send(String msg) ;                    // 消息发送
}
```

5.【mldnboot-integration 项目】定义消息业务实现类。

```
package cn.mldn.mldnboot.producer.impl;
import javax.jms.Queue;
import org.springframework.beans.factory.annotation.Autowired;
import org.springframework.jms.core.JmsMessagingTemplate;
import org.springframework.stereotype.Service;
import cn.mldn.mldnboot.producer.IMessageProducer;
@Service
```

```
public class MessageProducerImpl implements IMessageProducer {
    @Autowired
    private JmsMessagingTemplate jmsMessagingTemplate;              // 消息发送模板
    @Autowired
    private Queue queue;                                            // 注入队列
    @Override
    public void send(String msg) {
        this.jmsMessagingTemplate.convertAndSend(this.queue, msg);  // 消息发送
    }
}
```

6.【mldnboot-integration 项目】定义 JMS 消息发送配置类，该类主要用于配置队列信息。

```
package cn.mldn.mldnboot.config;
import javax.jms.Queue;
import org.apache.activemq.command.ActiveMQQueue;
import org.springframework.context.annotation.Bean;
import org.springframework.context.annotation.Configuration;
import org.springframework.jms.annotation.EnableJms;
@Configuration
@EnableJms
public class ActiveMQConfig {
    @Bean
    public Queue queue() {
        return new ActiveMQQueue("mldn.msg.queue") ;  // 定义队列名称
    }
}
```

本例利用 ActiveMQ 实现了消息的发送与接收处理。每当有消息接收到时，都会自动执行 MessageConsumer 类，进行消息消费。

5.3.2 SpringBoot 整合 RabbitMQ 消息组件

RabbitMQ 是一个在 AMQP 基础上构建的新一代企业级消息系统，该组件由 Pivotal 公司提供，使用 ErLang 语言开发。本小节将为读者讲解如何使用 SpringBoot 实现指定 RoutingKey 的消息处理。

提示：RabbitMQ 与 SpringCloud 整合之中意义重大。

SpringCloud 是在 SpringBoot 基础上构建的微架构技术开发框架，其中的 SpringCloudConfig 自动刷新机制就基于消息组件完成，并且推荐使用 RabbitMQ 消息组件（同属于 Pivotal 公司产品）。在 SpringCloudStream 中也使用 RabbitMQ 作为服务组件。

1.【mldnboot-integration 项目】修改 pom.xml 配置文件，追加 spring-boot-starter-amqp 依赖包。

```
<dependency>
    <groupId>org.springframework.boot</groupId>
    <artifactId>spring-boot-starter-amqp</artifactId>
</dependency>
```

2.【mldnboot-integration 项目】修改 yml.xml 配置文件，进行 RabbitMQ 的相关配置。

```
spring:
  rabbitmq:
    addresses: rabbitmq-server            # RabbitMQ服务主机名称
    username: mldnjava                    # 用户名
    password: hello                       # 密码
    virtual-host: /                       # 虚拟主机
```

3.【mldnboot-integration 项目】为了可以正常使用 RabbitMQ 进行消息处理，还需要做一个消息生产配置类。

```
package cn.mldn.mldnboot.config;
import org.springframework.amqp.core.Binding;
import org.springframework.amqp.core.BindingBuilder;
import org.springframework.amqp.core.DirectExchange;
import org.springframework.amqp.core.Queue;
import org.springframework.context.annotation.Bean;
import org.springframework.context.annotation.Configuration;
@Configuration
public class ProducerConfig {
    public static final String EXCHANGE = "mldn.microboot.exchange";         // 交换空间名称
    public static final String ROUTINGKEY = "mldn.microboot.routingkey";     // 设置路由key
    public static final String QUEUE_NAME = "mldn.microboot.queue";          // 队列名称
    @Bean
    public Binding bindingExchangeQueue(DirectExchange exchange,Queue queue) {
        return BindingBuilder.bind(queue).to(exchange).with(ROUTINGKEY) ;
    }
    @Bean
    public DirectExchange getDirectExchange() {                              // 使用直连的模式
        return new DirectExchange(EXCHANGE, true, true);
    }
    @Bean
    public Queue queue() {                                                   // 队列信息
        return new Queue(QUEUE_NAME);
    }
}
```

4.【mldnboot-integration 项目】建立消息发送接口。

```
package cn.mldn.mldnboot.producer;
public interface IMessageProducer {
    public void send(String msg) ;    // 消息发送
}
```

5.【mldnboot-integration 项目】建立消息业务实现子类。

```
package cn.mldn.mldnboot.producer.impl;
import javax.annotation.Resource;
import org.springframework.amqp.rabbit.core.RabbitTemplate;
import org.springframework.stereotype.Service;
import cn.mldn.mldnboot.config.ProducerConfig;
import cn.mldn.mldnboot.producer.IMessageProducer;
@Service
public class MessageProducerImpl implements IMessageProducer {
    @Resource
    private RabbitTemplate rabbitTemplate;
    @Override
    public void send(String msg) {
        this.rabbitTemplate.convertAndSend(ProducerConfig.EXCHANGE,
                ProducerConfig.ROUTINGKEY, msg);
    }
}
```

6.【mldnboot-integration 项目】建立一个消息消费端的配置程序类。

```
package cn.mldn.mldnboot.config;
import org.springframework.amqp.core.Binding;
import org.springframework.amqp.core.BindingBuilder;
import org.springframework.amqp.core.DirectExchange;
import org.springframework.amqp.core.Queue;
import org.springframework.context.annotation.Bean;
import org.springframework.context.annotation.Configuration;
@Configuration
public class ConsumerConfig {
    public static final String EXCHANGE = "mldn.microboot.exchange";         // 交换空间名称
    public static final String ROUTINGKEY = "mldn.microboot.routingkey";     // 设置路由key
    public static final String QUEUE_NAME = "mldn.microboot.queue";          // 队列名称
    @Bean
    public Queue queue() {                                                    // 队列信息
        return new Queue(QUEUE_NAME);
    }
    @Bean
```

```
    public DirectExchange getDirectExchange() {                              // 使用直连的模式
        return new DirectExchange(EXCHANGE, true, true);
    }
    @Bean
    public Binding bindingExchangeQueue(DirectExchange exchange,Queue queue) {
        return BindingBuilder.bind(queue).to(exchange).with(ROUTINGKEY) ;
    }
}
```

7.【mldnboot-integration 项目】定义监听处理类。

```
package cn.mldn.mldnboot.consumer;
import org.springframework.amqp.rabbit.annotation.RabbitListener;
import org.springframework.stereotype.Service;
@Service
public class MessageConsumer {
    @RabbitListener(queues="mldn.microboot.queue")
    public void receiveMessage(String text) {                              // 进行消息接收处理
        System.err.println("【*** 接收消息 ***】" + text);
    }
}
```

此时程序实现了与 RabbitMQ 消息组件的整合，同时在整个程序中只需要调用 IMessageProducer 接口中的 send()方法就可以正常发送，而后会找到设置同样 ROUTINGKEY 的消费者进行消息消费。

5.3.3 SpringBoot 整合 Kafka 消息组件

Kafka 是新一代的消息系统，也是目前性能最好的消息组件，在数据采集业务中被广泛应用。本程序中配置的 Kafka 将基于 Kerberos 认证实现消息组件处理。

1.【操作系统-Windows】定义一个 Kerberos 客户端文件，路径为 d:\ kafka_client_jaas.conf。

```
KafkaClient {
        org.apache.kafka.common.security.plain.PlainLoginModule required
        username="bob"
        password="bob-pwd";
};
```

2.【mldnboot-integration 项目】修改 pom.xml 配置文件，追加依赖库配置。

```
    <dependency>
        <groupId>org.springframework.kafka</groupId>
        <artifactId>spring-kafka</artifactId>
    </dependency>
```

3.【mldnboot-integration 项目】修改 application.yml 配置文件，进行 Kafka 配置项编写。

```
spring:
  kafka:
    bootstrap-servers:                                  # 定义主机列表
    - kafka-single:9095
    template:
      default-topic: mldn-microboot                     # 定义主题名称
    producer:                                           # 定义生产者配置
      key-serializer: org.apache.kafka.common.serialization.StringSerializer
      value-serializer: org.apache.kafka.common.serialization.StringSerializer
    consumer:                                           # 定义消费者配置
      key-deserializer: org.apache.kafka.common.serialization.StringDeserializer
      value-deserializer: org.apache.kafka.common.serialization.StringDeserializer
      group-id: group-1                                 # 数据分组
    properties:
      sasl.mechanism: PLAIN                             # 安全机制
      security.protocol: SASL_PLAINTEXT                 # 安全协议
```

4.【mldnboot-integration 项目】定义消息业务发送接口。

```
package cn.mldn.mldnboot.producer;
public interface IMessageProducer {
    public void send(String msg) ;                          // 消息发送
}
```

5.【mldnboot-integration 项目】使用 Kafka 消息机制实现消息发送接口。

```
package cn.mldn.mldnboot.producer.impl;
import javax.annotation.Resource;
import org.springframework.kafka.core.KafkaTemplate;
import org.springframework.stereotype.Service;
import cn.mldn.mldnboot.producer.IMessageProducer;
@Service
public class MessageProducerImpl implements IMessageProducer {
    @Resource
    private KafkaTemplate<String, String> kafkaTemplate;        // Kafka消息模板
    @Override
    public void send(String msg) {
        this.kafkaTemplate.sendDefault("mldn-key", msg);        // 发送消息
    }
}
```

6.【mldnboot-integration 项目】建立一个 Kafka 消息的消费程序类。

```
package cn.mldn.mldnboot.consumer;
import org.apache.kafka.clients.consumer.ConsumerRecord;
import org.springframework.kafka.annotation.KafkaListener;
import org.springframework.stereotype.Service;
@Service
public class MessageConsumer {
    @KafkaListener(topics = {"mldn-microboot"})
    public void receiveMessage(ConsumerRecord<String, String> record) { // 进行消息接收处理
        System.err.println("【*** 接收消息 ***】key = " + record.key() + "、value = "
                + record.value());
    }
}
```

7.【mldnboot-integration 项目】由于此时 Kafka 采用 Kerberos 认证，因此需要修改程序启动主类。

```
package cn.mldn.mldnboot;
import org.springframework.boot.SpringApplication;
import org.springframework.boot.autoconfigure.SpringBootApplication;
@SpringBootApplication                          // 启动SpringBoot程序，而后自带子包扫描
public class SpringBootStartApplication {       // 必须继承指定的父类
    static {                                    // 系统环境属性
        System.setProperty("java.security.auth.login.config","d:/kafka_client_jaas.conf");
    }
    public static void main(String[] args) throws Exception {
        SpringApplication.run(SpringBootStartApplication.class, args); // 启动SpringBoot程序
    }
}
```

此时，可以通过测试程序调用 IMessageProducer 接口进行消息发送，由于 Kafka 已经配置了自动创建主题，所以即使现在主题不存在，也不影响程序执行。

5.4 SpringBoot 整合 Redis 数据库

Redis 是当下最流行的用于实现缓存机制的 NoSQL 数据库，其主要通过 key-value 存储，支持高并发访问。在实际工作中，Redis 结合 SpringData 技术后可以方便地实现序列化对象的存储。SpringBoot 很好地支持了 Redis，可以在项目中使用 SpringData 进行 Redis 数据操作。

5.4.1　SpringBoot 整合 RedisTemplate 操作 Redis

RedisTemplate 是 SpringData 提供的 Redis 操作模板，该操作模板主要以 Jedis 驱动程序为实现基础，进行数据操作封装，所以可以直接调用 Redis 中的各种数据处理命令进行数据库操作。

1.【mldnboot-integration 项目】修改项目中的 pom.xml 配置文件，追加 Redis 的依赖引用。

```
<dependency>
    <groupId>org.springframework.boot</groupId>
    <artifactId>spring-boot-starter-data-redis</artifactId>
</dependency>
```

2.【mldnboot-integration 项目】修改 application.yml 配置文件，引入 Redis 相关配置项。

```
spring:
  redis:                                    # Redis相关配置
    host: redis-server                      # 主机名称
    port: 6379                              # 端口号
    password: mldnjava                      # 认证密码
    timeout: 1000                           # 连接超时时间
    database: 0                             # 默认数据库
    pool:                                   # 连接池配置
      max-active: 10                        # 最大连接数
      max-idle: 8                           # 最大维持连接数
      min-idle: 2                           # 最小维持连接数
      max-wait: 100                         # 最大等待连接超时时间
```

3.【mldnboot-integration 项目】在 application.yml 配置文件中定义完 Redis 的相关配置后，就可以通过程序来利用 RedisTemplate 模板进行数据处理了。下面直接编写一个测试类进行测试。

```
package cn.mldn.mldnboot.test;
import org.junit.Test;
import org.junit.runner.RunWith;
import org.springframework.beans.factory.annotation.Autowired;
import org.springframework.boot.test.context.SpringBootTest;
import org.springframework.data.redis.core.RedisTemplate;
import org.springframework.test.context.junit4.SpringJUnit4ClassRunner;
import org.springframework.test.context.web.WebAppConfiguration;
import cn.mldn.mldnboot.SpringBootStartApplication;
@SpringBootTest(classes = SpringBootStartApplication.class)
@RunWith(SpringJUnit4ClassRunner.class)
@WebAppConfiguration
public class TestRedisTemplate {
    @Autowired
    private RedisTemplate<String, String> redisTemplate;          // 引入RedisTemplate
```

```
    @Test
    public void testSet() {
        this.redisTemplate.opsForValue().set("mldn", "java");              // 设置字符串信息
        System.out.println(this.redisTemplate.opsForValue().get("mldn"));  // 根据key获取value
    }
}
```

本程序在测试类中直接注入了 RedisTemplate 模板对象，并且利用模板对象中提供的方法实现了 key-value 数据的保存与获取。

5.4.2 Redis 对象序列化操作

在实际项目开发中，使用 RedisTemplate 操作 Redis 数据库不仅可以方便地进行命令的操作，还可以结合对象序列化操作，实现对象的保存。

1.【mldnboot-integration 项目】定义对象的序列化配置类，以实现 RedisSerializer 接口。

```
package cn.mldn.mldnboot.util.redis;
import org.springframework.core.convert.converter.Converter;
import org.springframework.core.serializer.support.DeserializingConverter;
import org.springframework.core.serializer.support.SerializingConverter;
import org.springframework.data.redis.serializer.RedisSerializer;
import org.springframework.data.redis.serializer.SerializationException;
public class RedisObjectSerializer implements RedisSerializer<Object> {
    // 为了进行对象与字节数组的转换，应准备两个转换器
    private Converter<Object, byte[]> serializingConverter = new SerializingConverter();
    private Converter<byte[], Object> deserializingConverter = new DeserializingConverter();
    private static final byte[] EMPTY_BYTE_ARRAY = new byte[0]; // 做一个空数组，不是null
    @Override
    public byte[] serialize(Object obj) throws SerializationException {
        if (obj == null) { // 如果没有要序列化的对象，则返回一个空数组
            return EMPTY_BYTE_ARRAY;
        }
        return this.serializingConverter.convert(obj); // 将对象变为字节数组
    }
    @Override
    public Object deserialize(byte[] data) throws SerializationException {
        if (data == null || data.length == 0) { // 如果没有对象内容信息
            return null;
        }
        return this.deserializingConverter.convert(data);
    }
}
```

2.【mldnboot-integration 项目】要让建立的对象序列化管理类生效，还需要建立一个 RedisTemplate 的配置类。

```
package cn.mldn.mldnboot.config;
import org.springframework.context.annotation.Bean;
import org.springframework.context.annotation.Configuration;
import org.springframework.data.redis.connection.RedisConnectionFactory;
import org.springframework.data.redis.core.RedisTemplate;
import org.springframework.data.redis.serializer.StringRedisSerializer;
import cn.mldn.mldnboot.util.redis.RedisObjectSerializer;
@Configuration
public class RedisConfig {
    @Bean
    public RedisTemplate<String, Object> getRedisTemplate(
                RedisConnectionFactory factory) {
        RedisTemplate<String, Object> redisTemplate = new RedisTemplate<String, Object>();
        redisTemplate.setConnectionFactory(factory);
        redisTemplate.setKeySerializer(new StringRedisSerializer());         // key的序列化类型
        redisTemplate.setValueSerializer(new RedisObjectSerializer());       // value的序列化类型
        return redisTemplate;
    }
}
```

3.【mldnboot-integration 项目】建立一个待序列化的 VO 类对象。

```
package cn.mldn.mldnboot.vo;
import java.io.Serializable;
@SuppressWarnings("serial")
public class Member implements Serializable {
    private String mid;
    private String name ;
    private Integer age;
    // setter、getter略
}
```

4.【mldnboot-integration 项目】建立测试类，实现对象信息保存。

```
package cn.mldn.mldnboot.test;
import org.junit.Test;
import org.junit.runner.RunWith;
import org.springframework.beans.factory.annotation.Autowired;
import org.springframework.boot.test.context.SpringBootTest;
import org.springframework.data.redis.core.RedisTemplate;
```

```
import org.springframework.test.context.junit4.SpringJUnit4ClassRunner;
import org.springframework.test.context.web.WebAppConfiguration;
import cn.mldn.mldnboot.SpringBootStartApplication;
import cn.mldn.mldnboot.vo.Member;
@SpringBootTest(classes = SpringBootStartApplication.class)
@RunWith(SpringJUnit4ClassRunner.class)
@WebAppConfiguration
public class TestRedisTemplate {
    @Autowired
    private RedisTemplate<String, Object> redisTemplate;              // 引入RedisTemplate
    @Test
    public void testGet() {                                            // 根据key取得数据
        System.out.println(this.redisTemplate.opsForValue().get("mldn"));
    }
    @Test
    public void testSet() {
        Member vo = new Member() ;                                     // 实例化VO对象
        vo.setMid("mldnjava");
        vo.setName("李兴华");
        vo.setAge(19);
        this.redisTemplate.opsForValue().set("mldn", vo);              // 保存数据
    }
}
```

此时的程序可以使用 String 作为 key 类型，Object 作为 value 类型，直接利用 RedisTemplate 可以将对象序列化保存在 Redis 数据库中，也可以利用指定的 key 通过 Redis 获取对应信息。

5.4.3 配置多个 RedisTemplate

SpringBoot 通过配置 application.yml，只能够注入一个 RedisTemplate 对象。从事过实际开发的读者应该清楚，在实际的使用中有可能会在项目中连接多个 Redis 数据源，这时将无法依靠 SpringBoot 的自动配置实现，只能够由用户自己来创建 RedisTemplate 对象。

1.【mldnboot-integration 项目】为了规范配置，需要在 application.yml 中进行两个 Redis 数据库连接的配置。

```
myredis:                                # 自定义Redis连接配置
  redis-one:                            # 第一个Redis连接
    host: redis-server-1                # Redis主机
    port: 6379                          # 连接端口
    password: mldnjava                  # 认证信息
    timeout: 1000                       # 连接超时时间
    database: 0                         # 默认数据库
```

```
    pool:                                        # 连接池配置
      max-active: 10                             # 最大连接数
      max-idle: 8                                # 最大维持连接数
      min-idle: 2                                # 最小维持连接数
      max-wait: 100                              # 最大等待连接超时时间
  redis-two:                                     # 第二个Redis连接
    host: redis-server-2                         # Redis主机
    port: 6379                                   # 连接端口
    password: mldnjava                           # 认证信息
    timeout: 1000                                # 连接超时时间
    database: 1                                  # 默认数据库
    pool:                                        # 连接池配置
      max-active: 10                             # 最大连接数
      max-idle: 8                                # 最大维持连接数
      min-idle: 2                                # 最小维持连接数
      max-wait: 100                              # 最大等待连接超时时间
```

2.【mldnboot-integration 项目】修改 pom.xml 配置文件。

```
<!--  <dependency>          // 取消spring-boot-starter-data-redis依赖库
        <groupId>org.springframework.boot</groupId>
        <artifactId>spring-boot-starter-data-redis</artifactId>
</dependency> -->
```

```
<dependency>
        <groupId>org.springframework.data</groupId>
        <artifactId>spring-data-redis</artifactId>
</dependency>
<dependency>
        <groupId>redis.clients</groupId>
        <artifactId>jedis</artifactId>
</dependency>
```

3.【mldnboot-integration 项目】编写自定义的 Redis 配置类。

```
package cn.mldn.mldnboot.config;
import javax.annotation.Resource;
import org.springframework.beans.factory.annotation.Value;
import org.springframework.context.annotation.Bean;
import org.springframework.context.annotation.Configuration;
import org.springframework.data.redis.connection.RedisConnectionFactory;
import org.springframework.data.redis.connection.jedis.JedisConnectionFactory;
import org.springframework.data.redis.core.RedisTemplate;
```

```
import org.springframework.data.redis.serializer.StringRedisSerializer;
import cn.mldn.mldnboot.util.redis.RedisObjectSerializer;
import redis.clients.jedis.JedisPoolConfig;
@Configuration
public class RedisConfig {                                          // 定义Redis配置类
    @Resource(name="redisConnectionFactory")
    private RedisConnectionFactory redisConnectionFactoryOne ;
    @Resource(name="redisConnectionFactoryTwo")
    private RedisConnectionFactory redisConnectionFactoryTwo ;
    @Bean("redisConnectionFactoryTwo")
    public RedisConnectionFactory getRedisConnectionFactoryTwo(
            @Value("${myredis.redis-two.host}") String hostName,
            @Value("${myredis.redis-two.password}") String password,
            @Value("${myredis.redis-two.port}") int port,
            @Value("${myredis.redis-two.database}") int database,
            @Value("${myredis.redis-two.pool.max-active}") int maxActive,
            @Value("${myredis.redis-two.pool.max-idle}") int maxIdle,
            @Value("${myredis.redis-two.pool.min-idle}") int minIdle,
            @Value("${myredis.redis-two.pool.max-wait}") long maxWait) {
        JedisConnectionFactory jedisFactory = new JedisConnectionFactory();
        jedisFactory.setHostName(hostName);
        jedisFactory.setPort(port);
        jedisFactory.setPassword(password);
        jedisFactory.setDatabase(database);
        JedisPoolConfig poolConfig = new JedisPoolConfig();         // 进行连接池配置
        poolConfig.setMaxTotal(maxActive);
        poolConfig.setMaxIdle(maxIdle);
        poolConfig.setMinIdle(minIdle);
        poolConfig.setMaxWaitMillis(maxWait);
        jedisFactory.setPoolConfig(poolConfig);
        jedisFactory.afterPropertiesSet();                 // 初始化连接池配置
        return jedisFactory;
    }
    @Bean("redisConnectionFactory")                        // 定义RedisConnectionFactory对象名称
    public RedisConnectionFactory getRedisConnectionFactoryOne(
            @Value("${myredis.redis-one.host}") String hostName,
            @Value("${myredis.redis-one.password}") String password,
            @Value("${myredis.redis-one.port}") int port,
            @Value("${myredis.redis-one.database}") int database,
            @Value("${myredis.redis-one.pool.max-active}") int maxActive,
            @Value("${myredis.redis-one.pool.max-idle}") int maxIdle,
```

```
            @Value("${myredis.redis-one.pool.min-idle}") int minIdle,
            @Value("${myredis.redis-one.pool.max-wait}") long maxWait) {
        JedisConnectionFactory jedisFactory = new JedisConnectionFactory();
        jedisFactory.setHostName(hostName);
        jedisFactory.setPort(port);
        jedisFactory.setPassword(password);
        jedisFactory.setDatabase(database);
        JedisPoolConfig poolConfig = new JedisPoolConfig();              // 进行连接池配置
        poolConfig.setMaxTotal(maxActive);
        poolConfig.setMaxIdle(maxIdle);
        poolConfig.setMinIdle(minIdle);
        poolConfig.setMaxWaitMillis(maxWait);
        jedisFactory.setPoolConfig(poolConfig);
        jedisFactory.afterPropertiesSet();                               // 初始化连接池配置
        return jedisFactory;
    }
    @Bean("redisOne")
    public RedisTemplate<String, String> getRedisTemplateOne() {
        RedisTemplate<String, String> redisTemplate = new RedisTemplate<String, String>();
        redisTemplate.setKeySerializer(new StringRedisSerializer());        // key的序列化类型
        redisTemplate.setValueSerializer(new RedisObjectSerializer());      // value的序列化类型
        redisTemplate.setConnectionFactory(this.redisConnectionFactoryOne);
        return redisTemplate;
    }
    @Bean("redisTwo")
    public RedisTemplate<String, String> getRedisTemplateTwo() {
        RedisTemplate<String, String> redisTemplate = new RedisTemplate<String, String>();
        redisTemplate.setKeySerializer(new StringRedisSerializer());        // key的序列化类型
        redisTemplate.setValueSerializer(new RedisObjectSerializer());      // value的序列化类型
        redisTemplate.setConnectionFactory(this.redisConnectionFactoryTwo);
        return redisTemplate;
    }
}
```

4.【mldnboot-integration 项目】编写测试类，使用两个 RedisTemplate 进行数据操作。

```
package cn.mldn.mldnboot.test;
import javax.annotation.Resource;
import org.junit.Test;
import org.junit.runner.RunWith;
import org.springframework.boot.test.context.SpringBootTest;
import org.springframework.data.redis.core.RedisTemplate;
```

```
import org.springframework.test.context.junit4.SpringJUnit4ClassRunner;
import org.springframework.test.context.web.WebAppConfiguration;
import cn.mldn.mldnboot.SpringBootStartApplication;
@SpringBootTest(classes = SpringBootStartApplication.class)
@RunWith(SpringJUnit4ClassRunner.class)
@WebAppConfiguration
public class TestRedisTemplate {
    @Resource(name="redisOne")
    private RedisTemplate<String,String> redisOne ;
    @Resource(name="redisTwo")
    private RedisTemplate<String,String> redisTwo ;
    @Test
    public void testSet() {
        this.redisOne.opsForValue().set("mldnjava", "hello");        // 保存数据
        this.redisTwo.opsForValue().set("jixianit", "hello");        // 保存数据
    }
}
```

本程序利用 RedisConfig 程序类注入了两个 RedisTemplate 对象，因此该程序具备了两个 Redis 数据库的操作能力。

5.5 SpringBoot 整合安全框架

Shiro 是 Apache 推出的新一代认证与授权管理开发框架，可以方便地与第三方的认证机构进行整合。前面的整合过程中已经充分考虑到了分布式会话管理操作，所以本节程序将直接采用自定义缓存类来实现多个 Redis 数据库信息的保存，架构如图 5-2 所示。

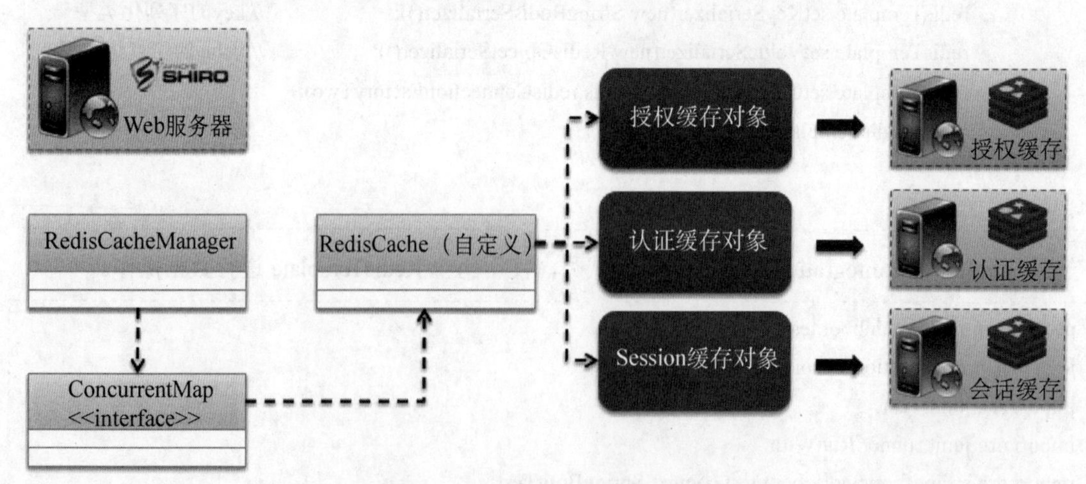

图 5-2　SpringBoot 与 Shiro 整合

5.5.1　SpringBoot 整合 Shiro 开发框架

SpringBoot 与 Shiro 的整合处理，本质上和 Spring 与 Shiro 的整合区别不大，但开发者需要注意以下 3 点：

☑ SpringBoot 可以自动导入一系列的开发包，但是这些开发包里面不包含对 Shiro 的支持，所以还需要配置 shiro 的开发依赖库。

☑ SpringBoot 不提倡使用 spring-shiro.xml 文件进行配置，需要将配置文件转为 Bean 的形式（需要考虑缓存的调度时间问题）。

☑ Shiro 在进行一些 Session 管理以及缓存配置时要用到 shiro-quartz 依赖包，该依赖包使用的是 QuartZ-1.X 版本，而现在能找到的都是 QuartZ-2.x 版本。因此，如果不使用 SpringBoot，那么这样的使用差别不大；如果使用了 SpringBoot 集成，就会产生后台的异常信息。

提示：本例只讲解 Shiro 整合的核心配置。

Shiro 的创建与配置操作有很多，在笔者出版的其他图书中进行了完整讲解。本章只讲解 SpringBoot 与 Shiro 的核心配置，完整代码可以通过本书配套资料获得。不清楚的读者可以学习“名师讲坛”系列的其他图书或登录 www.mldn.cn 自行学习。

考虑到本开发只是在模拟环境中运行，所以并未建立多个 Redis 实例，而是直接使用不同的 Redis 数据库来实现多数据库模拟。

1.【mldnboot 项目】修改 pom.xml 配置文件，追加 Shiro 的相关依赖包。

<table>
<tr><td>定义版本属性</td><td><pre>
<druid.version>1.1.1</druid.version>
<shiro.version>1.3.2</shiro.version>
<thymeleaf-extras-shiro.version>1.2.1</thymeleaf-extras-shiro.version>
</pre></td></tr>
<tr><td>定义配置依赖管理</td><td><pre>
<dependency>
    <groupId>com.github.theborakompanioni</groupId>
    <artifactId>thymeleaf-extras-shiro</artifactId>
    <version>${thymeleaf-extras-shiro.version}</version>
</dependency>
<dependency>
    <groupId>org.apache.shiro</groupId>
    <artifactId>shiro-spring</artifactId>
    <version>${shiro.version}</version>
</dependency>
<dependency>
    <groupId>org.apache.shiro</groupId>
    <artifactId>shiro-core</artifactId>
    <version>${shiro.version}</version>
</dependency>
</pre></td></tr>
</table>

```
<dependency>
	<groupId>org.apache.shiro</groupId>
	<artifactId>shiro-ehcache</artifactId>
	<version>${shiro.version}</version>
</dependency>
<dependency>
	<groupId>org.apache.shiro</groupId>
	<artifactId>shiro-web</artifactId>
	<version>${shiro.version}</version>
</dependency>
<dependency>
	<groupId>org.quartz-scheduler</groupId>
	<artifactId>quartz</artifactId>
	<version>${quartz.version}</version>
</dependency>
```

2.【mldnboot-shiro 项目】修改 pom.xml 配置文件，追加依赖库配置。

```
<dependency>
	<groupId>org.apache.shiro</groupId>
	<artifactId>shiro-spring</artifactId>
</dependency>
<dependency>
	<groupId>org.apache.shiro</groupId>
	<artifactId>shiro-core</artifactId>
</dependency>
<dependency>
	<groupId>org.apache.shiro</groupId>
	<artifactId>shiro-ehcache</artifactId>
</dependency>
<dependency>
	<groupId>org.quartz-scheduler</groupId>
	<artifactId>quartz</artifactId>
</dependency>
```

3.【mldnboot-shiro 项目】建立 ShiroConfig 的配置程序类，将所有 Shiro 的配置项都写在此配置类中。

```
@Configuration
public class ShiroConfig {
	public static final String LOGOUT_URL = "/logout.action" ;		// 退出路径
	public static final String LOGIN_URL = "/loginPage" ;			// 登录路径
```

```
public static final String UNAUTHORIZED_URL = "/unauth" ;          // 未授权错误页
public static final String SUCCESS_URL = "/pages/back/welcome" ;   // 登录成功页
@Resource(name = "redisConnectionFactory")
private RedisConnectionFactory redisConnectionFactoryAuthentication;
@Resource(name = "redisConnectionFactoryAuthorization")
private RedisConnectionFactory redisConnectionFactoryAuthorization;
@Resource(name = "redisConnectionFactoryActiveSessionCache")
private RedisConnectionFactory redisConnectionFactoryActiveSessionCache;
@Bean
public MemberRealm getRealm() {                                     // 定义Realm
    MemberRealm realm = new MemberRealm();
    realm.setCredentialsMatcher(new DefaultCredentialsMatcher());   // 配置缓存
    realm.setAuthenticationCachingEnabled(true);
    realm.setAuthenticationCacheName("authenticationCache");
    realm.setAuthorizationCachingEnabled(true);
    realm.setAuthorizationCacheName("authorizationCache");
    return realm;
}
@Bean(name = "lifecycleBeanPostProcessor")
public LifecycleBeanPostProcessor getLifecycleBeanPostProcessor() {   // Shiro实现控制器处理
    return new LifecycleBeanPostProcessor();
}
@Bean
@DependsOn("lifecycleBeanPostProcessor")
public DefaultAdvisorAutoProxyCreator getDefaultAdvisorAutoProxyCreator() {
    DefaultAdvisorAutoProxyCreator daap = new DefaultAdvisorAutoProxyCreator();
    daap.setProxyTargetClass(true);
    return daap;
}
@Bean
public CacheManager getCacheManager(
        @Qualifier("redisConnectionFactory")
        RedisConnectionFactory redisConnectionFactoryAuthentication ,
        @Qualifier("redisConnectionFactoryAuthorization")
        RedisConnectionFactory redisConnectionFactoryAuthorization ,
        @Qualifier("redisConnectionFactoryActiveSessionCache")
        RedisConnectionFactory redisConnectionFactoryActiveSessionCache) {  // 缓存配置
    RedisCacheManager cacheManager = new RedisCacheManager();     // 缓存集合
    Map<String,RedisConnectionFactory> map = new HashMap<>() ;
    map.put("authenticationCache", redisConnectionFactoryAuthentication) ;
```

```
        map.put("authorizationCache", redisConnectionFactoryAuthorization) ;
        map.put("activeSessionCache", redisConnectionFactoryActiveSessionCache) ;
        cacheManager.setConnectionFactoryMap(map);
        return cacheManager;
    }
    @Bean
    public SessionIdGenerator getSessionIdGenerator() {            // SessionID生成
        return new JavaUuidSessionIdGenerator();
    }
    @Bean
    public SessionDAO getSessionDAO(SessionIdGenerator sessionIdGenerator) {
        EnterpriseCacheSessionDAO sessionDAO = new EnterpriseCacheSessionDAO();
        sessionDAO.setActiveSessionsCacheName("activeSessionCache");
        sessionDAO.setSessionIdGenerator(sessionIdGenerator);
        return sessionDAO;
    }
    @Bean
    public RememberMeManager getRememberManager() {          // 记住我
        CookieRememberMeManager rememberMeManager = new CookieRememberMeManager();
        SimpleCookie cookie = new SimpleCookie("MLDNJAVA-RememberMe");
        cookie.setHttpOnly(true);
        cookie.setMaxAge(3600);
        rememberMeManager.setCookie(cookie);
        return rememberMeManager;
    }
    @Bean
    public DefaultQuartzSessionValidationScheduler getQuartzSessionValidationScheduler(
            DefaultWebSessionManager sessionManager) {
        DefaultQuartzSessionValidationScheduler sessionValidationScheduler =
                new DefaultQuartzSessionValidationScheduler();
        sessionValidationScheduler.setSessionValidationInterval(100000);
        sessionValidationScheduler.setSessionManager(sessionManager);
        return sessionValidationScheduler;
    }
    @Bean
    public AuthorizationAttributeSourceAdvisor getAuthorizationAttributeSourceAdvisor(
            DefaultWebSecurityManager securityManager) {
        AuthorizationAttributeSourceAdvisor aasa = new AuthorizationAttributeSource Advisor();
        aasa.setSecurityManager(securityManager);
        return aasa;
```

```
    }
    @Bean
    public ShiroDialect shiroDialect() {                    // 启动Thymeleaf模板支持
        return new ShiroDialect();
    }
    @Bean
    public DefaultWebSessionManager getSessionManager(SessionDAO sessionDAO) { // Session管理
        DefaultWebSessionManager sessionManager = new DefaultWebSessionManager();
        sessionManager.setDeleteInvalidSessions(true);
        sessionManager.setSessionValidationSchedulerEnabled(true);
        sessionManager.setSessionDAO(sessionDAO);
        SimpleCookie sessionIdCookie = new SimpleCookie("mldn-session-id");
        sessionIdCookie.setHttpOnly(true);
        sessionIdCookie.setMaxAge(-1);
        sessionManager.setSessionIdCookie(sessionIdCookie);
        sessionManager.setSessionIdCookieEnabled(true);
        return sessionManager;
    }
    @Bean
    public DefaultWebSecurityManager getSecurityManager(Realm memberRealm, CacheManager cacheManager,
SessionManager sessionManager, RememberMeManager rememberMeManager) {// 缓存管理
        DefaultWebSecurityManager securityManager = new DefaultWebSecurityManager();
        securityManager.setRealm(memberRealm);
        securityManager.setCacheManager(cacheManager);
        securityManager.setSessionManager(sessionManager);
        securityManager.setRememberMeManager(rememberMeManager);
        return securityManager;
    }
    public FormAuthenticationFilter getLoginFilter() {  // 在ShiroFilterFactoryBean中使用
        FormAuthenticationFilter filter = new FormAuthenticationFilter();
        filter.setUsernameParam("mid");
        filter.setPasswordParam("password");
        filter.setRememberMeParam("rememberMe");
        filter.setLoginUrl(LOGIN_URL);                  // 登录提交页面
        filter.setFailureKeyAttribute("error");
        return filter;
    }
    public LogoutFilter getLogoutFilter() {             // 在ShiroFilterFactoryBean中使用
        LogoutFilter logoutFilter = new LogoutFilter();
        logoutFilter.setRedirectUrl("/");               // 首页路径，登录注销后回到的页面
```

```
        return logoutFilter;
    }
    @Bean
    public ShiroFilterFactoryBean getShiroFilterFactoryBean(DefaultWebSecurityManager
securityManager) {
        ShiroFilterFactoryBean shiroFilterFactoryBean = new ShiroFilterFactoryBean();
        shiroFilterFactoryBean.setSecurityManager(securityManager);      // 设置 SecurityManager
        shiroFilterFactoryBean.setLoginUrl(LOGIN_URL);                   // 设置登录页路径
        shiroFilterFactoryBean.setSuccessUrl(SUCCESS_URL);               // 设置跳转成功页
        shiroFilterFactoryBean.setUnauthorizedUrl(UNAUTHORIZED_URL);     // 授权错误页
        Map<String, Filter> filters = new HashMap<String, Filter>();
        filters.put("authc", this.getLoginFilter());
        filters.put("logout", this.getLogoutFilter());
        shiroFilterFactoryBean.setFilters(filters);
        Map<String, String> filterChainDefinitionMap = new HashMap<String, String>();
        filterChainDefinitionMap.put("/logout.page", "logout");
        filterChainDefinitionMap.put("/loginPage", "authc");             // 定义内置登录处理
        filterChainDefinitionMap.put("/pages/**", "authc");
        filterChainDefinitionMap.put("/*", "anon");
        shiroFilterFactoryBean.setFilterChainDefinitionMap(filterChainDefinitionMap);
        return shiroFilterFactoryBean;
    }
}
```

在本配置程序之中，最为重要的一个配置方法就是 getQuartzSessionValidationScheduler()，这也是 SpringBoot 整合 Shiro 中最为重要的一点。之所以重新配置，主要原因是 SpringBoot 整合 Shiro 时的定时调度组件版本落后，所以才需要由用户自定义一个 SessionValidationScheduler 接口子类。

4.【mldnboot-shiro 项目】在使用 Shiro 的过程中，除了需要对控制层与业务层的拦截过滤之外，对于页面也需要有所支持，而 SpringBoot 本身不提倡使用 JSP 页面，所以就需要引入一个支持 Shiro 处理的 Thymeleaf 命名空间。

```
<html xmlns:th="http://www.thymeleaf.org" xmlns:shiro="http://www.pollix.at/thymeleaf/ shiro">
```

配置完命名空间之后，Shiro 就可以使用<shiro:hasRole/>、<shiro:principal/>这样的标签来进行 Shiro 操作。

5.5.2 SpringBoot 基于 Shiro 整合 OAuth 统一认证

在实际项目开发过程中，随着项目功能不断推出，会出现越来越多的子系统，如图 5-3 所示。很明显，这样就需要使用统一的登录认证处理。在一个良好的系统设计中一般都会存在有一个单点登录，而 OAuth 正是现在最流行的单点登录协议。

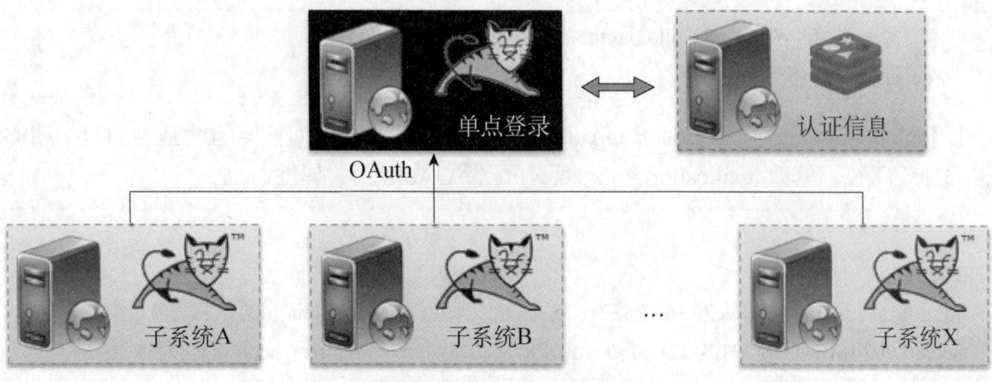

图 5-3　OAuth 单点登录

1.【mldnboot 项目】修改 pom.xml 配置文件，引入 oltu 相关依赖包。

定义版本属性	<oltu.version>1.0.2</oltu.version>
定义配置依赖管理	<dependency> <groupId>org.apache.oltu.oauth2</groupId> <artifactId>org.apache.oltu.oauth2.client</artifactId> <version>${oltu.version}</version> </dependency> <dependency> <groupId>org.apache.oltu.oauth2</groupId> <artifactId>org.apache.oltu.oauth2.authzserver</artifactId> <version>${oltu.version}</version> </dependency> <dependency> <groupId>org.apache.oltu.oauth2</groupId> <artifactId>org.apache.oltu.oauth2.resourceserver</artifactId> <version>${oltu.version}</version> </dependency>

2.【mldnboot-shiro 项目】修改 pom.xml 配置文件，在 SpringBoot 项目中引入相关依赖。

```
<dependency>
    <groupId>org.apache.oltu.oauth2</groupId>
    <artifactId>org.apache.oltu.oauth2.client</artifactId>
</dependency>
<dependency>
    <groupId>org.apache.oltu.oauth2</groupId>
    <artifactId>org.apache.oltu.oauth2.authzserver</artifactId>
</dependency>
<dependency>
    <groupId>org.apache.oltu.oauth2</groupId>
```

```
        <artifactId>org.apache.oltu.oauth2.resourceserver</artifactId>
    </dependency>
```

3.【mldnboot-shiro 项目】对于 OAuth 整合的处理里面，最为重要的就是为项目指明 OAuth 的相关处理路径，修改 application.yml 信息，配置 OAuth 相关属性。

```
oauth:
  client:
    id: d0fde52c-538f-4e06-9c2f-363fe4321c7e               # 保存client_id的信息
    secret: 902be4ff-9a36-331d-9f71-afb604d07787           # 保存client_secret的信息
  token:                                                   # token访问地址
    url: http://www.server.com:80/enterpriseauth-oauth-server/accessToken.action
  memberinfo:                # 获得用户信息的访问地址（此地址要在accessToken获取之后获得）
    url: http://www.server.com:80/enterpriseauth-oauth-server/memberInfo.action
  redirect:                  # 保存返回的地址（此地址要与之前的OAuthFilter对应上）
    uri: http://www.client.com:9090/shiro-oauth
  login:                     # 定义登录访问路径地址
    url: http://www.server.com:80/enterpriseauth-oauth-server/authorize.action
?client_id=d0fde52c-538f-4e06-9c2f-363fe4321c7e&response_type=code
&redirect_uri=http://www.client.com:9090/shiro-oauth
```

4.【mldnboot-shiro 项目】一旦项目中引入 OAuth 处理，则 Realm 一定会发生更改，定义一个新的 OAuthRealm 类（代替之前的 MemberRealm 程序类）。

```
public class OAuthRealm extends AuthorizingRealm {
    @Resource
    private IMemberService memberService;
    private String clientId;                    // 应该由客户服务器申请获得
    private String clientSecret;                // 应该由客户服务器申请获得
    private String redirectUri;                 // 返回地址
    private String accessTokenUrl;              // 进行Token操作的地址定义
    private String memberInfoUrl;               // 获得用户信息的路径
    @Override
    public boolean supports(AuthenticationToken token) {
        return token instanceof OAuthToken; // 只有该类型的Token可以执行此Realm
    }
    @Override
    protected AuthenticationInfo doGetAuthenticationInfo(AuthenticationToken token) throws
AuthenticationException {
        OAuthToken oAuthToken = (OAuthToken) token; // 强制转型为自定义的OAuthToken，里面有Code
        String authCode = (String) oAuthToken.getCredentials(); // 获取OAuth返回的Code数据
        String mid = this.getMemberInfo(authCode);
```

```
        return new SimpleAuthenticationInfo(mid, authCode, "memberRealm");
    }
    @Override
    protected AuthorizationInfo doGetAuthorizationInfo(PrincipalCollection principals) {
        SimpleAuthorizationInfo info = new SimpleAuthorizationInfo(); // 返回授权的信息
        String mid = (String) principals.getPrimaryPrincipal(); // 获得用户名
        Map<String, Set<String>> map = this.memberService.getRoleAndActionByMember(mid);
        info.setRoles(map.get("allRoles")); // 将所有的角色信息保存在授权信息中
        info.setStringPermissions(map.get("allActions")); // 保存所有的权限
        return info;
    }
    private String getMemberInfo(String code) { // 获取用户的信息
        String mid = null;
        try {
            OAuthClient oauthClient = new OAuthClient(new URLConnectionClient());
            OAuthClientRequest accessTokenRequest =
                OAuthClientRequest.tokenLocation(this.accessTokenUrl) // 设置Token的访问地址
                    .setGrantType(GrantType.AUTHORIZATION_CODE).setClientId (this.clientId)
                    .setClientSecret(this.clientSecret).setRedirectURI(this. redirectUri)
                    .setCode(code).buildQueryMessage();
            // 构建一个专门用于进行Token数据回应处理的操作类对象，获得Token的请求是POST
            OAuthJSONAccessTokenResponse oauthResponse = oauthClient.accessToken
(accessTokenRequest,OAuth.HttpMethod.POST);
            String accessToken = oauthResponse.getAccessToken(); // 获得Token
            // 获得AccessToken是为了能够通过此Token获得mid的信息，此时应该构建第二次请求
            // 要想获得请求操作，一定要设置accessToken处理信息
            OAuthClientRequest memberInfoRequest =
                new OAuthBearerClientRequest(this.memberInfoUrl)
                    .setAccessToken(accessToken).buildQueryMessage(); // 创建一个请求操作
            // 要进行指定用户信息请求的回应处理项
            OAuthResourceResponse resouceResponse = oauthClient.resource (memberInfoRequest,
OAuth.HttpMethod.GET, OAuthResourceResponse.class);
            mid = resouceResponse.getBody();        // 获取mid的信息
        } catch (Exception e) {
            e.printStackTrace();
        }
        return mid;
    }
    // setter、getter 略
}
```

5.【mldnboot-shiro 项目】此时基本的 OAuth 整合环境已经配置成功，随后还需要建立一个执行 OAuth 认证的过滤器，在这个过滤器中主要是要获取一个 OAuth-Token 信息（建立一个 OAuthToken 类，该类继承 UsernamePasswordToken 父类，里面保存有 principal、authcode 两个属性信息）。

```
public class OAuthAuthenticatingFilter extends AuthenticatingFilter {
    private String authcodeParam = "code" ;                         // 由OAuth返回的地址提供参数
    private String failureUrl ;                                      // 定义一个失败的跳转页面
    @Override
    protected boolean onAccessDenied(ServletRequest request, ServletResponse response) throws Exception {
        // 随后需要在这个程序之中进行关于OAuth登录处理的相关配置操作
        String error = request.getParameter("error") ;              // 此处要求获得错误的提示信息
        if (!(error == null || "".equals(error))) {                  // 现在出现错误提示信息
            String errorDesc = request.getParameter("error_description") ; // 错误信息
            // 如果此时出现错误信息，则直接跳转到错误页面
            WebUtils.issueRedirect(request, response,
                    this.failureUrl + "?error=" + error + "&error_description" + errorDesc);
            return false ; // 后续的操作不再执行，直接跳转
        }
        Subject subject = super.getSubject(request, response) ;     // 获得Subject
        if (!subject.isAuthenticated()) {                            // 用户现在未进行登录认证
            String code = request.getParameter(this.authcodeParam) ; // 需要接收返回的code数据
            if (code == null || "".equals(code)) {                   // 此时一定是一个错误的处理操作
                super.saveRequestAndRedirectToLogin(request, response); // 跳转到登录页
                return false ;
            }
        }
        return super.executeLogin(request, response);                // 执行登录处理逻辑
    }
    @Override
    protected boolean onLoginSuccess(AuthenticationToken token, Subject subject, ServletRequest request,
ServletResponse response) throws Exception {                         // 登录成功之后跳转到成功页面
        super.issueSuccessRedirect(request, response);               // 跳转到登录成功页面
        return false ;
    }
    @Override
    protected boolean onLoginFailure(AuthenticationToken token, AuthenticationException e, ServletRequest
request, ServletResponse response) {                                 // 登录失败
        Subject subject = super.getSubject(request, response) ;     // 获得当前用户Subject
        if (subject.isAuthenticated() || subject.isRemembered()) {  // 认证判断
            try {                                                    // 登录成功返回到首页上
                super.issueSuccessRedirect(request, response);
```

```
                } catch (Exception e1) {}
            } else {                                                            // 跳转到失败页面
                try {
                    WebUtils.issueRedirect(request, response, this.failureUrl);
                } catch (IOException e1) {}
            }
            return false ;
        }
        public void setAuthcodeParam(String authcodeParam) {
            this.authcodeParam = authcodeParam;
        }
        public void setFailureUrl(String failureUrl) {
            this.failureUrl = failureUrl;
        }
        @Override
        protected AuthenticationToken createToken(ServletRequest request, ServletResponse response) throws
    Exception {        // 传入一个自定义Token信息，保存OAuth返回的数据
            OAuthToken token = new OAuthToken(request.getParameter(this.authcodeParam)) ;
            token.setRememberMe(true);                                          // 设置记住我的功能
            return token ;
        }
    }
```

此时成功地实现了 SpringBoot + Shiro + OAuth 的整合处理，而这样的整合模式也是实际项目开发中的最佳组合。

5.6　SpringBoot 整合邮件服务器

Java 本身提供了 JavaMail 标准以实现邮件的处理，同时用户也可以搭建属于自己的邮件服务器或者直接使用各个邮箱系统实现邮件的发送处理。本节将利用 SpringBoot 整合邮件服务，同时使用 QQ 邮箱系统进行服务整合。

1.【QQ 邮箱】登录 QQ 邮箱，进入邮箱设置页面，如图 5-4 所示。

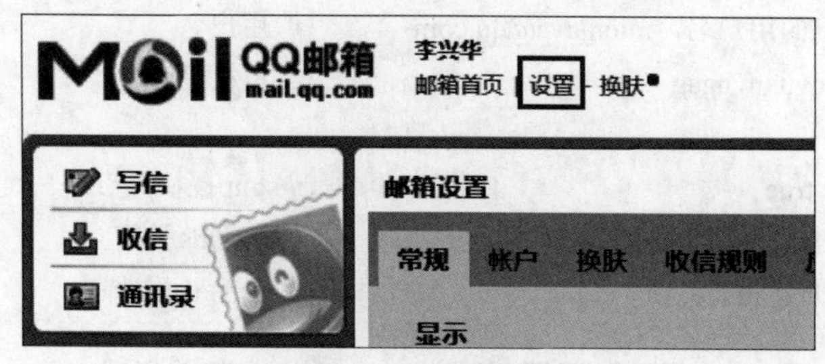

图 5-4　进入 QQ 邮箱设置页面

2.【QQ 邮箱】找到邮件服务配置项，如图 5-5 所示。

POP3/IMAP/SMTP/Exchange/CardDAV/CalDAV服务

开启服务： POP3/SMTP服务（如何使用 Foxmail 等软件收发邮件？） 已关闭 | 开启

IMAP/SMTP服务（什么是 IMAP，它又是如何设置？） 已关闭 | 开启

Exchange服务（什么是Exchange，它又是如何设置？） 已关闭 | 开启

CardDAV/CalDAV服务（什么是CardDAV/CalDAV，它又是如何设置？） 已关闭 | 开启

(POP3/IMAP/SMTP/CardDAV/CalDAV服务均支持SSL连接。如何设置？)

图 5-5　进入 POP 3 服务

3.【QQ 邮箱】开启邮箱的邮件服务后，将得到一个唯一的授权码，如图 5-6 所示。

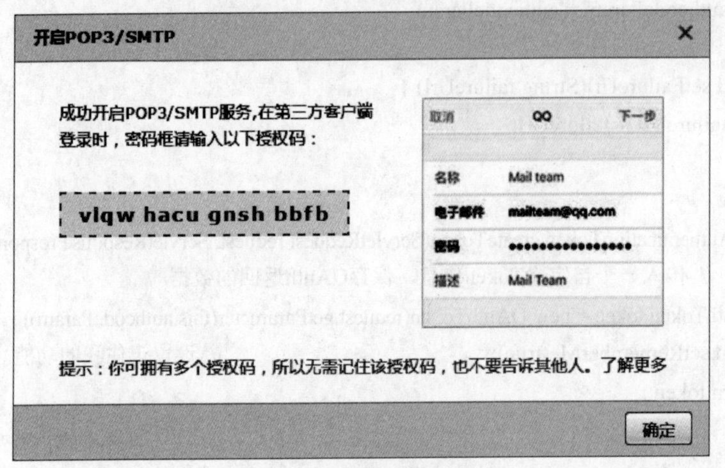

图 5-6　开启邮箱服务

4.【mldnboot-integration 项目】修改 pom.xml 配置文件，引入依赖库。

```
<dependency>
    <groupId>org.springframework.boot</groupId>
    <artifactId>spring-boot-starter-mail</artifactId>
</dependency>
```

5.【mldnboot-integration 项目】修改 application.yml 配置文件，实现邮件配置。

```
spring:
  mail:
    host: smtp.qq.com                                  # 邮箱服务器
    username: 你自己的用户名，mldnjava@qq.com            # 用户名
    password: ansqwtcvuxnvbgag                         # 授权码
    properties:
      mail.smtp.auth: true                             # stmp授权开启
      mail.smtp.starttls.enable: true                  # 启动tls服务
      mail.smtp.starttls.required: true                # 启动 tls 支持
```

6.【mldnboot-integration 项目】编写测试类，进行邮件发送测试。

```
package cn.mldn.mldnboot.test;
import javax.annotation.Resource;
import org.junit.Test;
import org.junit.runner.RunWith;
import org.springframework.boot.test.context.SpringBootTest;
import org.springframework.mail.SimpleMailMessage;
import org.springframework.mail.javamail.JavaMailSender;
import org.springframework.test.context.junit4.SpringJUnit4ClassRunner;
import org.springframework.test.context.web.WebAppConfiguration;
import cn.mldn.mldnboot.SpringBootStartApplication;
@SpringBootTest(classes = SpringBootStartApplication.class)
@RunWith(SpringJUnit4ClassRunner.class)
@WebAppConfiguration
public class TestMail {
    @Resource
    private JavaMailSender javaMailSender ;                          // 注入邮件发送对象
    @Test
    public void testSendMail() {
        SimpleMailMessage message = new SimpleMailMessage() ;         // 要发送的消息内容
        message.setFrom("你自己的邮箱地址，110@qq.com");                // 发送者
        message.setTo("接收者的邮箱地址，220@qq.com");                  // 接收者
        message.setSubject("测试邮件（来自老李的祝福，www.mldn.cn）");  // 邮件主题
        message.setText("好好学习，天天向上，学习就登录www.mldn.cn");    // 邮件内容
        this.javaMailSender.send(message);                            // 发送邮件
    }
}
```

由于 SpringBoot 中已经进行了大量的简化配置，所以此时的程序只需要注入 JavaMailSender 对象，并设置好邮件内容，就可以实现邮件信息的发送。

5.7　定时调度

在企业项目开发中，定时调度是一项重要的技术组成，利用定时调度可以帮助用户实现无人值守程序执行，在 Spring 中提供了简单的 SpringTask 调度执行任务，利用此组件可以实现间隔调度与 CRON 调度处理。

1.【mldnboot-integration 项目】定义一个线程调度类。

```
package cn.mldn.mldnboot.task;
import java.text.SimpleDateFormat;
import java.util.Date;
```

```
import org.springframework.scheduling.annotation.Scheduled;
import org.springframework.stereotype.Component;
@Component
public class MySchedulerTask {
    @Scheduled(fixedRate = 2000)                    // 采用间隔调度，每2秒执行一次
    public void runJobA() {                         // 定义一个要执行的任务
        System.out.println("【*** MyTaskA - 间隔调度 ***】"
                + new SimpleDateFormat("yyyy-MM-dd HH:mm:ss.SSS")
                        .format(new Date()));
    }
    @Scheduled(cron = "* * * * * ?")                // 每秒调用一次
    public void runJobB() {
        System.err.println("【*** MyTaskB - 间隔调度 ***】"
                + new SimpleDateFormat("yyyy-MM-dd HH:mm:ss.SSS")
                        .format(new Date()));
    }
}
```

2.【mldnboot-integration 项目】为了让多个任务并行执行，还需要建立一个定时调度池的配置类。

```
package cn.mldn.mldnboot.config;
import java.util.concurrent.Executors;
import org.springframework.context.annotation.Configuration;
import org.springframework.scheduling.annotation.SchedulingConfigurer;
import org.springframework.scheduling.config.ScheduledTaskRegistrar;
@Configuration                  // 定时调度的配置类一定要实现指定的父接口
public class SchedulerConfig implements SchedulingConfigurer {
    @Override
    public void configureTasks(ScheduledTaskRegistrar taskRegistrar) {      // 开启线程调度池
        taskRegistrar.setScheduler(Executors.newScheduledThreadPool(10));   // 10个线程池
    }
}
```

3.【mldnboot-integration 项目】在程序启动类上追加定时任务配置注解。

```
package cn.mldn.mldnboot;
import org.springframework.boot.SpringApplication;
import org.springframework.boot.autoconfigure.SpringBootApplication;
import org.springframework.scheduling.annotation.EnableScheduling;
@SpringBootApplication                      // 启动SpringBoot程序，而后自带子包扫描
@EnableScheduling                           // 启用调度
```

```
public class SpringBootStartApplication {             // 必须继承指定的父类
    public static void main(String[] args) throws Exception {
        SpringApplication.run(SpringBootStartApplication.class, args);        // 启动SpringBoot程序
    }
}
```

本程序同时启动了两个定时调度，为了使两个线程调度之间不受影响，开辟了一个线程池，可以并行执行多个任务。

5.8　Actuator 监控

Actuator 是 SpringBoot 中一个用来实现系统健康检测的模块，它提供一个 Resetful 的 API 接口，可以将系统运行过程中的磁盘空间、线程数以及程序连接的数据库情况通过 JSON 返回，然后再结合预警、监控模块进行实时系统监控。Actuctor 包括如下访问路径，如表 5-1 所示。

表 5-1　Actuator 监控路径

No.	HTTP 方法	操作路径	描述
01	GET	/autoconfig	提供了一份自动配置报告，记录哪些自动配置条件通过了，哪些没通过
02	GET	/configprops	描述配置属性（包含默认值）如何注入 Bean
03	GET	/beans	描述应用程序上下文里全部的 Bean，以及它们的关系
04	GET	/dump	获取线程活动的快照
05	GET	/env	获取全部环境属性
06	GET	/env/{name}	根据名称获取特定的环境属性值
07	GET	/health	应用程序的健康指标，这些值由 HealthIndicator 的实现类提供
08	GET	/info	获取应用程序的定制信息，这些信息由 info 打头的属性提供
09	GET	/mappings	描述全部 URI 路径，以及它们和控制器（包含 Actuator 端点）的映射关系
10	GET	/metrics	报告各种应用程序度量信息，如内存用量和 HTTP 请求计数
11	GET	/metrics/{name}	报告指定名称的应用程序度量值
12	POST	/shutdown	关闭应用程序，要求 endpoints.shutdown.enabled 设置为 true
13	GET	/trace	提供基本的 HTTP 请求跟踪信息（时间戳、HTTP 头等）

1.【mldnboot-integration 项目】修改 pom.xml 配置文件，追加依赖库。

```
<dependency>
    <groupId>org.springframework.boot</groupId>
    <artifactId>spring-boot-starter-actuator</artifactId>
</dependency>
```

2.【mldnboot-integration 项目】修改 application.yml 配置文件，关闭系统的安全配置。

```
management:
  security:
    enabled: false              # 现在关闭系统的安全配置
```

配置完以上程序代码之后，用户就可以通过表 5-1 给出的路径进行相应信息的查看。

☑ 查看环境信息：http://localhost/env。

☑ 查看配置 Bean：http://localhost/beans。

3.【mldnboot-integration 项目】虽然现在可以实现 Actuator 监控，但却需要关闭安全配置。很显然，这样的配置并不合理，最好的做法是由开发者自行定义相应项目信息。下面将为项目建立一个健康信息。

```
package cn.mldn.mldnboot.actuator;
import org.springframework.boot.actuate.health.Health;
import org.springframework.boot.actuate.health.HealthIndicator;
import org.springframework.stereotype.Component;
@Component
public class MyHealthIndicator implements HealthIndicator {
    @Override
    public Health health() {
        int errorCode = 100 ;           // 这个错误码是通过其他程序获得的
        if (errorCode != 0) {
            return Health.down().withDetail("Error Code", errorCode).build();
        }
        return Health.up().build() ;
    }
}
```

本程序模拟了一个健康状态的处理，开发者可以通过其他程序来生成一个 errorCode 错误状态码，用户可以通过 http://localhost/health 路径进行健康信息访问。

4.【mldnboot-integration 项目】SpringBoot 构建的主要是微服务，由于微服务中需要为开发者或使用者提供大量的信息，为此在 Actuator 中提供了信息访问路径（/info），这些服务信息可以直接通过 application.yml 文件进行配置。

```
info:
  app.name: mldn-microboot                         # 应用名称
  company.name: www.mldn.cn                        # 开发公司
  pom.artifactId: $project.artifactId$             # 项目版本，通过 pom 获得
  pom.version: $project.version$                   # 项目版本，通过 pom 获得
```

5.【mldnboot 项目】信息配置需要 Maven 插件支持，为了让所有子模块都支持这种配置，修改 pom.xml 配置文件。

☑ 添加 maven-resources-plugin 插件。

```
<plugin>
    <groupId>org.apache.maven.plugins</groupId>
    <artifactId>maven-resources-plugin</artifactId>
    <configuration>
        <delimiters>
            <delimiter>$</delimiter>            <!-- 定义描述分割符 -->
        </delimiters>
    </configuration>
</plugin>
```

☑　修改资源操作，启用过滤。

```
<resource>
    <directory>src/main/resources</directory>
    <includes>
        <include>**/*.properties</include>
        <include>**/*.yml</include>
        <include>**/*.xml</include>
        <include>**/*.tld</include>
        <include>**/*.p12</include>
    </includes>
    <filtering>true</filtering>
</resource>
```

此时，程序启动之后，可以输入信息访问路径 http://localhost/info，得到如图 5-7 所示的信息内容。

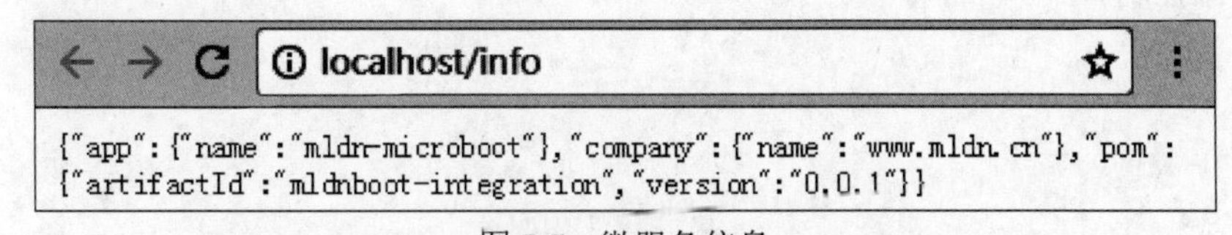

图 5-7　微服务信息

6.【mldnboot 项目】在开发中这种提示信息会成为微服务的重要组成部分，如果重复进行配置文件的定义，那么会比较麻烦。最简单的做法是直接做一个配置程序类，进行信息的配置。

```
package cn.mldn.mldnboot.actuator;
import org.springframework.boot.actuate.info.Info.Builder;
import org.springframework.boot.actuate.info.InfoContributor;
import org.springframework.stereotype.Component;
@Component
public class MicroServiceInfoContributor implements InfoContributor {
    @Override
    public void contribute(Builder builder) {
```

```
        builder.withDetail("company.name", "www.mldn.cn") ;
        builder.withDetail("version", "V1.0") ;
        builder.withDetail("author", "李兴华") ;
    }
}
```

此时可以直接通过配置类获取微服务信息，这里的信息内容如图 5-8 所示。

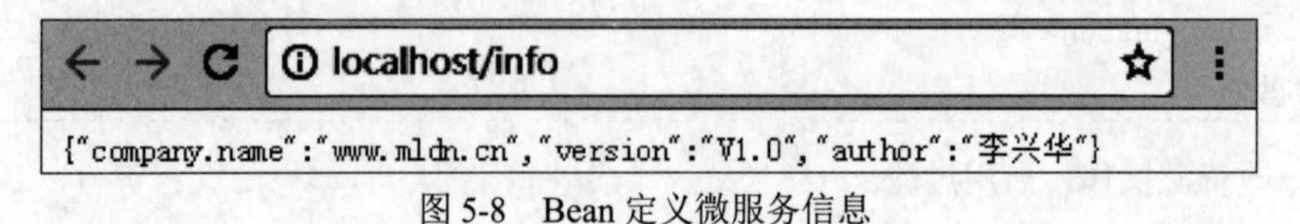

图 5-8 Bean 定义微服务信息

5.9 本章小结

1. SpringBoot 可以方便地与常用 ORM 设计框架整合（MyBatis、JPA），同时也可以实现 DataSource 的自动引入。

2. SpringBoot 整合消息组件时，只需要在 application.yml 配置文件中进行配置即可使用。

3. SpringBoot 整合 Redis 数据库时，可以使用 RedisTemplate 模板进行数据库操作，也可以通过序列化操作，保存对象到数据库之中。

4. SpringBoot 默认只支持单个 Redis 数据库连接的配置，如果需要配置多个 Redis 数据库连接，则需要由开发者自行定义配置程序类来完成。

5. SpringBoot 属于 Web 应用，可以使用 Shiro 实现认证与授权处理，同时也可以结合 OAuth 实现单点登录控制。

第二部分

SpringCloud 篇

- RPC 标准与 SpringCloud
- SpringCloud 与 Restful 访问
- Eureka 注册中心
- Ribbon、Feign、Hystrix 和 Zuul
- SpringCloudConfig 分布式配置管理
- SpringCloudStream
- SpringCloudSleuth

第 6 章

SpringCloud 简介

通过本章学习，可以达到以下目标：

1. 掌握 RPC 技术的主要作用。
2. 了解常见 RPC 开发技术及主要特点。
3. 了解 SpringCloud 开发框架的系统架构。

RPC（Remote Procedure Call，远程过程调用）技术是进行项目业务拆分的重要技术手段，SpringCloud 是新一代的 RPC 技术，其在 SpringBoot 技术上进行构建，基于 Restful 架构，采用标准数据结构（JSON）进行数据交互。本章将为读者讲解 RPC 技术的发展过程及 SpringCloud 的整体架构。

6.1 RPC 分布式开发技术

项目的设计与开发实质上是业务层设计与业务功能的完善，在传统的单主机项目中，业务层的变化一般很少，所以可以直接将业务层定义在 Web 之中，如图 6-1 所示。

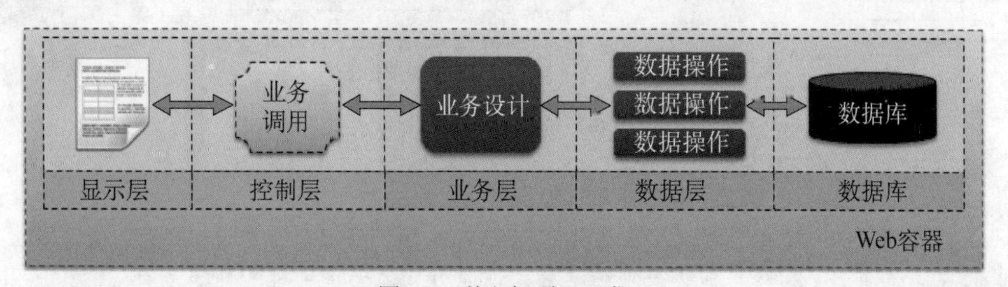

图 6-1　单主机项目开发

在互联网时代，为了应对业务需求的变更以及用户访问量的迅猛增加，同时也为了程序的可维护性，RPC 技术的应用非常广泛。利用 RPC 技术可以方便地帮助企业搭建业务中心，这样所有的 Web 端就可以利用远程接口技术实现业务中心方法的调用，从而单独进行业务中心的维护，如图 6-2 所示。

利用 RPC 技术除了可以实现业务中心的创建之外，最重要的作用是可以为一个业务实现多个 RPC 服务端的创建。当多个 RPC 服务端实现同一个业务时，可以利用代理软件实现负载均衡。这样的设计不仅可以提升性能，也方便了高峰时期的业务端扩展，从而实现高并发、高可用、分布式的项目结构设计，如图 6-3 所示。

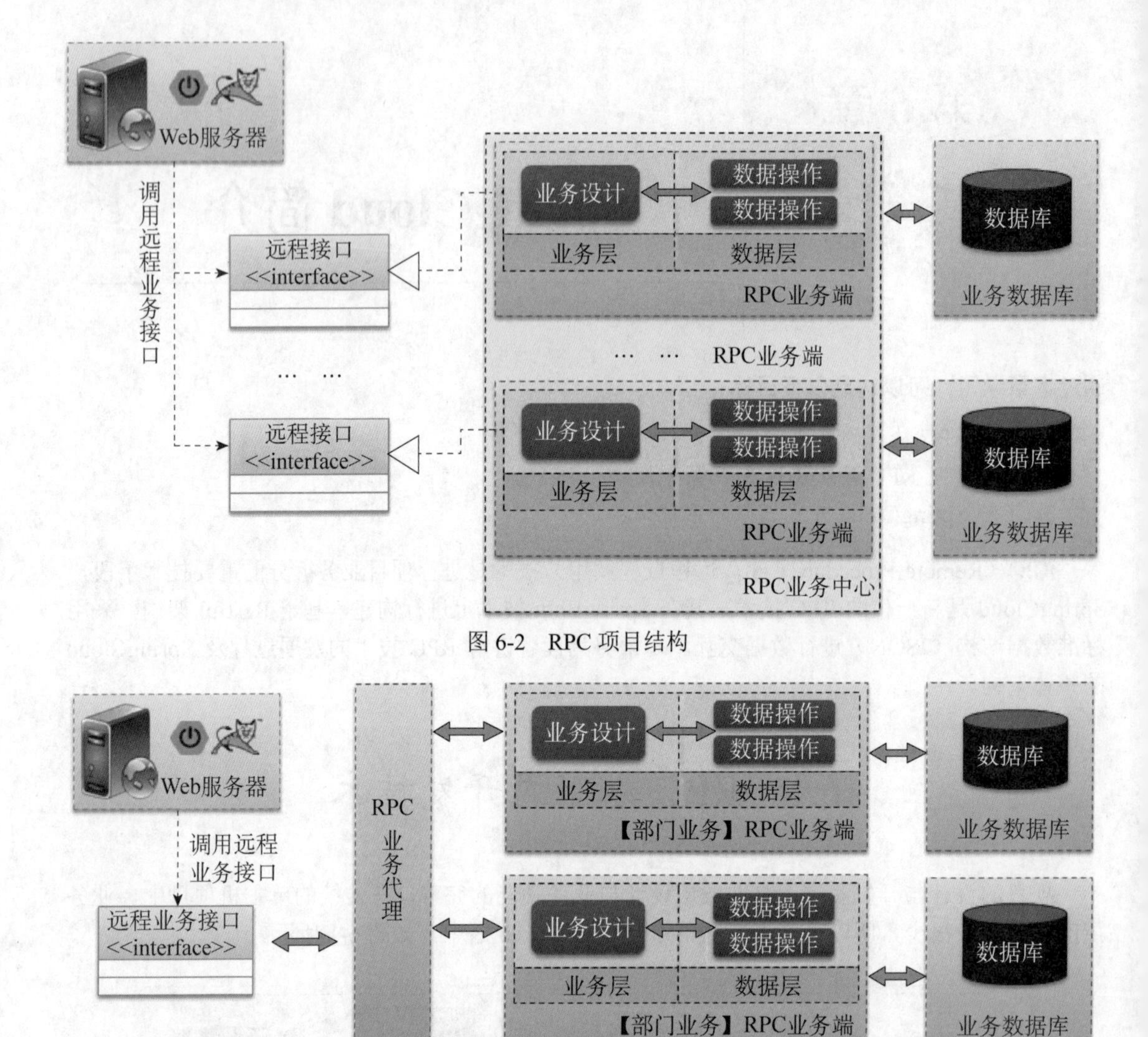

图 6-2　RPC 项目结构

图 6-3　RPC 高可用设计

提示：关于高并发、高可用和分布式设计。

在实际开发中，业务层的调用除了可发生于 Web 端，还可能存在于移动端。为了满足大量的并发访问，必须尽可能提升业务层的处理能力。由于单主机的性能是有限的，所以往往会建立业务层设计集群，使用多台主机共同实现业务层的高性能处理。这种设计中，即使业务集群中的某一台主机出现问题，也不会影响到整体业务的执行，从而实现高可用的处理机制。同时，分布式的业务中心也更便于维护以及业务设计人员进行业务完善。

6.2　RPC 实现技术

RPC 技术采用客户端与服务端的处理模式实现分布式开发调用，其本身是一个标准，并没

有定义任何传输协议，所以用户可以使用各种技术来实现。但随着技术的发展，目前也出现了一些 RPC 的实现技术。

☑ CORBA（Common Object Request Broker Architecture，公共对象请求代理架构）是由 OMG 组织制订的一种标准的面向对象应用程序体系规范。该标准可以使用任何语言实现，同时需要编写 IDL 接口描述文件。

☑ RMI（Remote Method Invocation，远程方法调用）是在 JDK 1.1 版本中提供的分布式开发技术，在实现过程中需要创建骨架与存根后才可以使用。随着 JDK 版本的提升，存根也可以自动生成。RMI 的基础架构如图 6-4 所示。

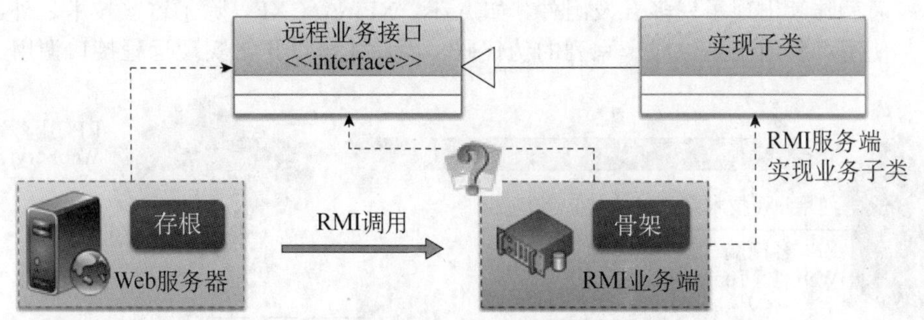

图 6-4　RMI 技术实现架构

☑ EJB（Enterprise JavaBean）是 SUN 公司（已被 Oracle 公司收购）推出的基于 RMI 技术的分布式开发技术，是一个开发基于组件的企业多重应用程序的标准。在 EJB 技术中主要分为 3 种 Bean：会话 Bean（业务层）、实体 Bean（数据层）和消息驱动 Bean（消息队列中间件）。这 3 种 Bean 的对应关系如图 6-5 所示。

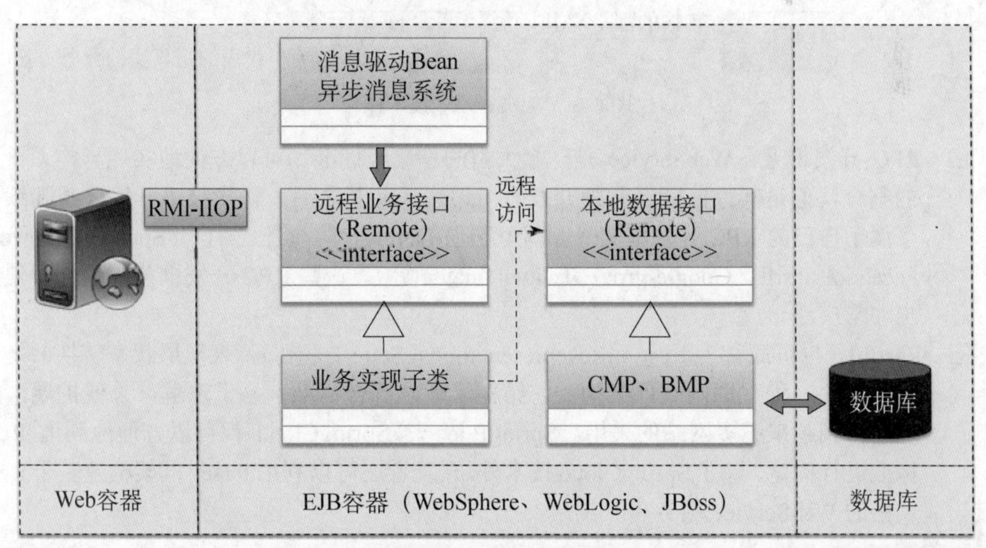

图 6-5　EJB 设计架构

提示：EJB 最后只剩下了理论价值。

EJB 技术是 SUN 公司当年的重头戏，它提出了完善的分布式业务中心设计理念，但由于在实现上存在偏差以及 EJB 容器过于昂贵，EJB 技术并没有真正“火”起来。其后来推出的 EJB 3.x 标准多数情况下还需要与 WebService 相结合，不得不说这是一个失败的实现。但 EJB 并非毫无用处，其设计理念造就了今天开源框架的发展，所以它依然是一个里程碑一样的技术。

☑ WebService（Web 服务）技术使得运行在不同机器上的应用无需借助附加的、专门的第三方软件或硬件，就可相互交换数据或集成。其主要使用 XML 作为接口描述，同时利用 SOAP（Simple Object Access Protocol，简单对象访问协议）进行通信，基本结构如图 6-6 所示。WebService 是一个开发标准，可以直接运行在 Web 容器中，而后陆续出现了许多开发组件，如 Axis、XFire、CXF。除了这些技术之外，还需要使用者利用工具生成一系列的伪代码，而后才可以正常实现远程接口调用。

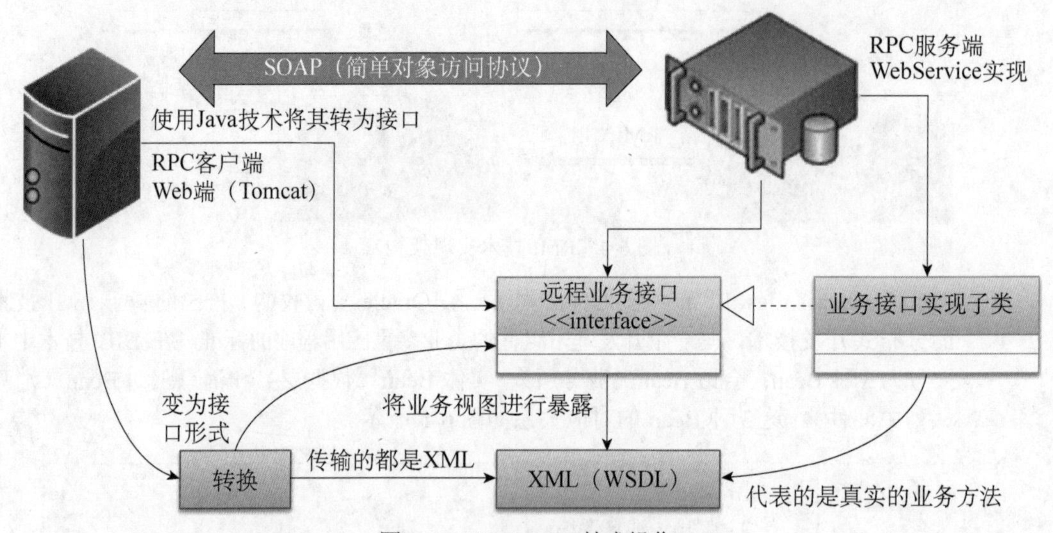

图 6-6 WebService 技术操作

☑ RPC 开发框架：WebService 是一个大型的重量级标准，可以方便地实现异构系统架构的整合，但是其性能却经常被开发者所诟病。在其之后，许多技术开发公司陆续推出了属于自己的 RPC 开发框架（如阿里巴巴-Dubbo、阿里巴巴-High Speed Framework、Facebook-thrift、Google-grpc、Twitter-finagle 等），这些 RPC 开发框架在项目中使用得也非常广泛。

☑ Restful（也可简称为 Rest，指 Representational State Transfer，表现层状态转换）是目前最流行的一种互联网软件架构。它结构清晰，符合标准，易于理解，方便扩展，所以逐渐得到越来越多网站的采用。SpringBoot 与 SpringCloud 都可以方便地利用 Restful 标准进行构建。除了 SpringCloud 技术外，开发者还可以利用 Jersey 框架构建基于 Restful 风格的 WebService 服务。

提示：Rest 提出者 Roy Thomas Fielding。

Rest 是 Roy Thomas Fielding（见图 6-7）在其 2000 年的博士论文中提出的。他是 HTTP（1.0 版和 1.1 版）的主要设计者，Apache 服务器软件的作者之一，Apache 基金会的第一任主席。

图 6-7　Roy Thomas Fielding

6.3　SpringCloud 技术架构

SpringCloud 是一套技术架构（主要使用的是 netflix 技术产品），其整体的架构核心是围绕 Restful 展开的，并且以 SpringBoot 技术为核心基础进行构建的 RPC 分布式技术。在 SpringCloud 技术中，对于 Restful 处理过程中一定要有两个端：服务的提供者（Provider）和服务的消费者（Consumer），基本架构如图 6-8 所示。

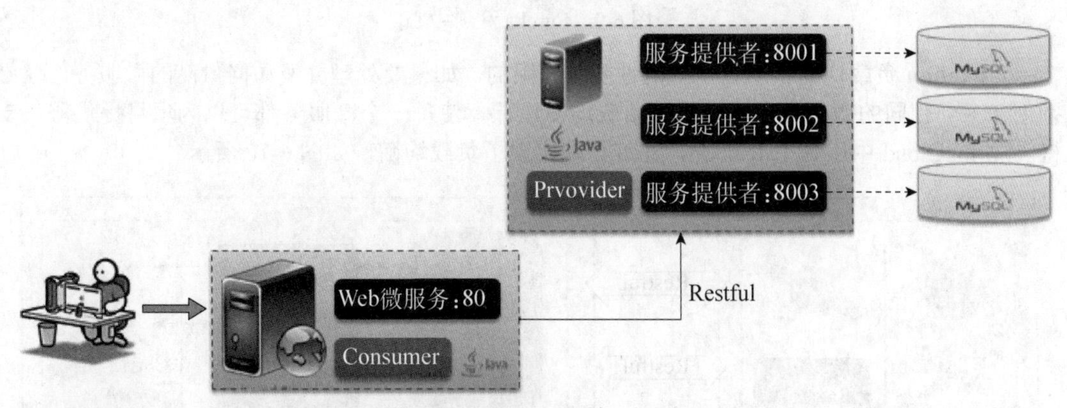

图 6-8　SpringCloud 与 Restful

提示：SpringCloud 以 SpringBoot 为实现基础。

在 SpringBoot 中如果要将控制层的处理方法以 Restful 形式返回，那么往往需要使用@Controller + @ResponseBody 或者@RestController 注解才可以实现。而 SpringCloud 技术是在 SpringBoot 的基础上构建的 RPC 应用，主要以 Restful 设计架构进行异构系统的数据交互，即 SpringCloud 中是不存在控制层跳转到显示页面处理操作的。

SpringCloud 的本质虽然是基于 Restful 的一种应用，但它依然属于 RPC 的一种技术实现。与传统 RPC 不一样的地方在于，SpringCloud 使用了一系列的开源组件进行整合应用，其中包含有如下 5 个基本组件。

1. 微服务注册中心。微服务的核心意义在于将一个总体的业务端拆分到不同的业务主机上，所有微服务的 Restful 访问地址非常多。为了统一管理这些地址信息，可以即时地告诉用户哪些服务不可用，应该准备一个分布式的注册中心，并且该注册中心支持 HA（High Available，高可用性集群）机制。为了高速并且方便地进行所有服务的注册操作，在 SpringCloud 里面提供了一个 Eureka 的注册中心，如图 6-9 所示，所有的微服务都可以在此注册中心中进行注册。

> **提示：HA 机制。**
>
> HA 是保证业务连续性的有效解决方案，一般有两个或两个以上的节点，且分为活动节点及备用节点。可以简单地理解为：如果班长在，则班长负责安排同学；如果班长不在，则副班长顶替班长完成任务。

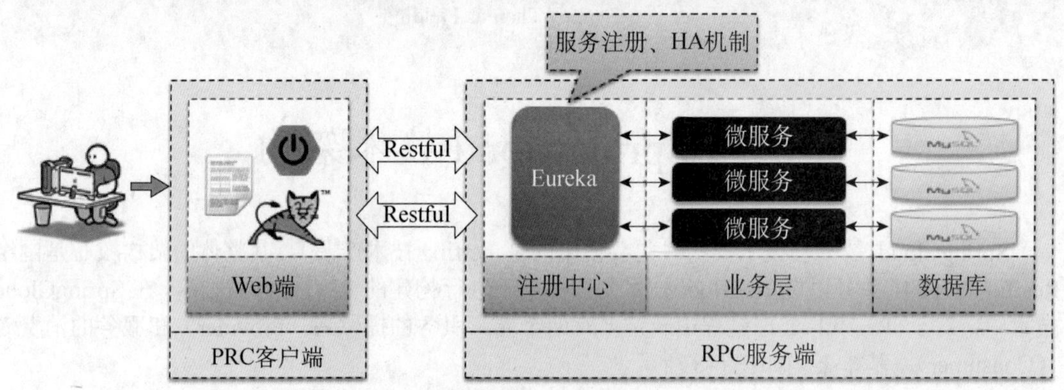

图 6-9　Eureka 注册中心

2. **Ribbon 负载均衡**。单台主机的性能是有限的，如果要处理并发访问量高的微服务，就必须创建多个相同的微服务，同时采用负载均衡设计，使每一个微服务都可以为项目提供服务支持。SpringCloud 中引入了 Ribbon，在客户端实现了负载均衡，如图 6-10 所示。

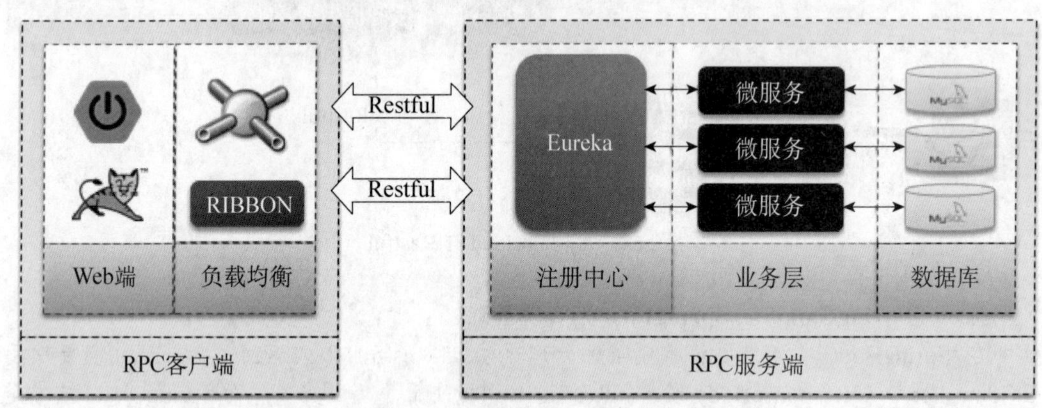

图 6-10　SpringCloud 负载均衡

3. **Feign 接口映射**。RPC 开发技术遵循了客户端与服务端开发模式，且客户端使用远程接口来实现远程业务调用是最为合理的。SpringCloud 技术依赖于 Restful 架构，开发者可以使用 Feign 技术实现远程接口以及远程 Restful 服务的映射处理，如图 6-11 所示。

4. **Zuul 网关代理**。为了保证微服务调用的安全性以及统一性，所有的微服务都可以采用统一的服务网关技术，通过映射名称访问相应的微服务资源。这样更好地体现了 Java 中 key=value 的设计思想。而且所有的微服务通过 Zuul 进行代理后，也可以更加合理地进行名称的隐藏，如图 6-12 所示。

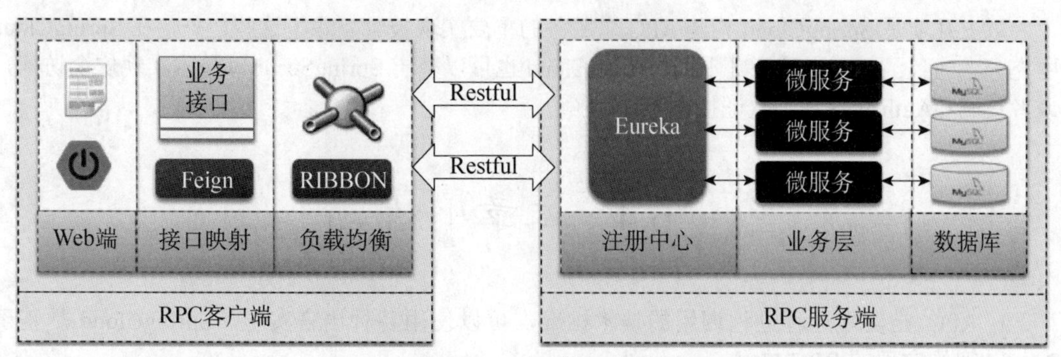

图 6-11　Feign 接口映射

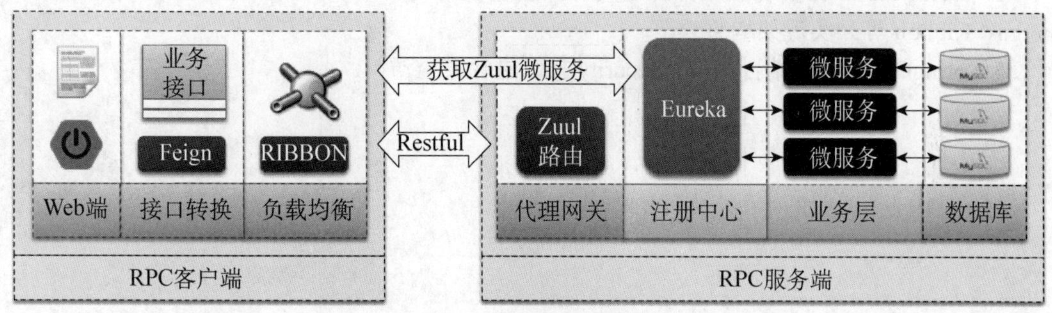

图 6-12　Zuul 代理网关

5. SpringCloudConfig 远程配置。在 SpringCloud 中通常会存在成百上千个微服务（有些大型项目中微服务甚至更多），为保持高效运转，其配置文件需要进行统一管理。在 SpringCloud 技术中实现 SpringCloudConfig 微服务定义，该微服务可以直接注册到 Eureka 中，以实现 HA 机制，而所有提供业务支持的微服务都通过 SpringCloudConfig 从 GIT 或 SVN 服务器上抓取用户配置信息，因此可以将配置资源进行统一管理，而后可以利用 SpringCloudBus 技术实现动态配置更新，如图 6-13 所示。

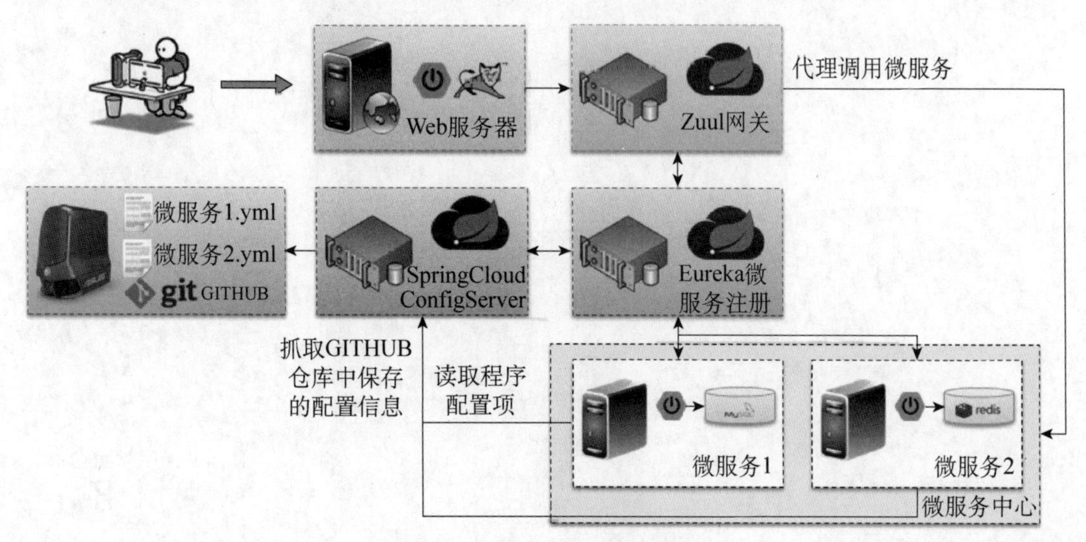

图 6-13　SpringCloudConfig

以上几项是 SpringCloud 构建 RPC 微服务的重要实现技术。除了这些之外，在 SpringCloud 中充分考虑到了 HA 处理机制。同时 SpringCloud 也可以基于 SpringSecurity 技术实现安全访问，或者集成 OAuth 实现统一认证授权管理。

6.4 本章小结

1. RPC 是实现远程过程调用的技术标准，可以使用各种语言实现。SpringCloud 是基于 Restful 架构实现的 RPC 技术。

2. SpringCloud 在实现微服务的定义时，主要使用 Netflix 公司的产品（如 Eureka、Zuul、Feign、Ribbon 等）实现架构整合。

3. SpringCloud 可以结合 SpringSecurity 技术进行安全访问。

第 7 章

SpringCloud 与 Restful

通过本章学习，可以达到以下目标：

1. 掌握 SpringCloud 与 SpringBoot 的关系，利用 SpringCloud 实现 Restful 服务发布。
2. 掌握 RestTemplate 操作类的使用，利用此类实现 Restful 业务调用。
3. 了解 SpringSecurity 与 SpringCloud 结合的意义，使用 SpringSecurity 实现安全认证。
4. 了解 Swagger 工具，可以利用 Swagger 工具实现 Restful 业务接口描述。

SpringCloud 是建立在 SpringBoot 基础上的，所以开发者必须掌握 SpringBoot 开发框架。由于 SpringCloud 是基于 Restful 架构的 RPC 开发实现，所以微架构设计中往往在客户端利用 RestTemplate 来实现远程 Restful 业务调用。为了保证系统安全，也可以使用 SpringSecurity 进行安全访问控制。

7.1 搭建 SpringCloud 项目开发环境

SpringCloud（如图 7-1 所示）技术与 SpringBoot 技术一样，都提供了统一的 pom.xml 配置项，配置好相应的版本之后就可以在各个 Maven 子模块中进行依赖支持库的简单引用。SpringCloud 技术与传统开发不一样的地方在于，其版本号并未采用数字，而是使用了一系列英国的地名作为标注。本例使用的版本为 Edgware，如图 7-2 所示。

图 7-1 SpringCloud 图标

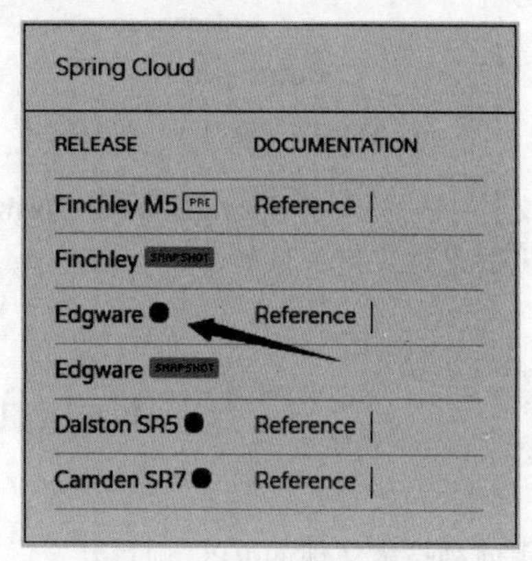

图 7-2 SpringCloud 版本

1.【Eclipse 工具】创建一个新的 Maven 项目 mldncloud，项目类型为 Quick Start，如图 7-3 所示。

2.【mldncloud 项目】修改 pom.xml 配置文件。如果要开发 SpringCloud，则一定要引入 SpringBoot 依赖支持，核心配置内容如下。

Group Id	Artifact Id	Version
org.apache.maven.archetypes	maven-archetype-portlet	1.0.1
org.apache.maven.archetypes	maven-archetype-profiles	1.0-alpha-4
org.apache.maven.archetypes	maven-archetype-quickstart	1.1
org.apache.maven.archetypes	maven-archetype-site	1.1

图 7-3　创建新的 Maven 项目

属性定义

```
<spring-boot-dependencies.version>
    1.5.9.RELEASE</spring-boot-dependencies.version>
<spring-cloud-dependencies.version>
    Edgware.RELEASE</spring-cloud-dependencies.version>
```

依赖库配置

```
<dependencyManagement>
    <dependencies>
        <dependency>        <!-- 定义SpringBoot依赖管理 -->
            <groupId>org.springframework.boot</groupId>
            <artifactId>spring-boot-dependencies</artifactId>
            <version>${spring-boot-dependencies.version}</version>
            <type>pom</type>
            <scope>import</scope>
        </dependency>
        <dependency>        <!-- 进行SpringCloud依赖包的导入处理 -->
            <groupId>org.springframework.cloud</groupId>
            <artifactId>spring-cloud-dependencies</artifactId>
            <version>${spring-cloud-dependencies.version}</version>
            <type>pom</type>
            <scope>import</scope>
        </dependency>
    </dependencies>
</dependencyManagement>
```

本配置文件中只引入了 SpringBoot 与 SpringCloud 的依赖支持包，其他的相关支持包将在讲解过程中根据需求逐步进行依赖配置。

7.2　Restful 基础实现

不管使用何种技术实现的 RPC 项目开发，采用的均为服务端与客户端结构。为了保证服务端定义与客户端访问的标准性，可以单独创建一个远程接口的描述项目。这里建立的项目结构如图 7-4 所示。

提示：Maven 模块项目命名。

在本节讲解过程中，由于 SpringCloud 开发框架需要建立一系列的微服务，为了让读者清楚每一个微服务的作用，将直接在项目名称上做出描述。例如，部门微服务的项目名称为 mldncloud-dept-service-8001，其中，dept 为微服务的作用描述，而 8001 描述的是启动端口。

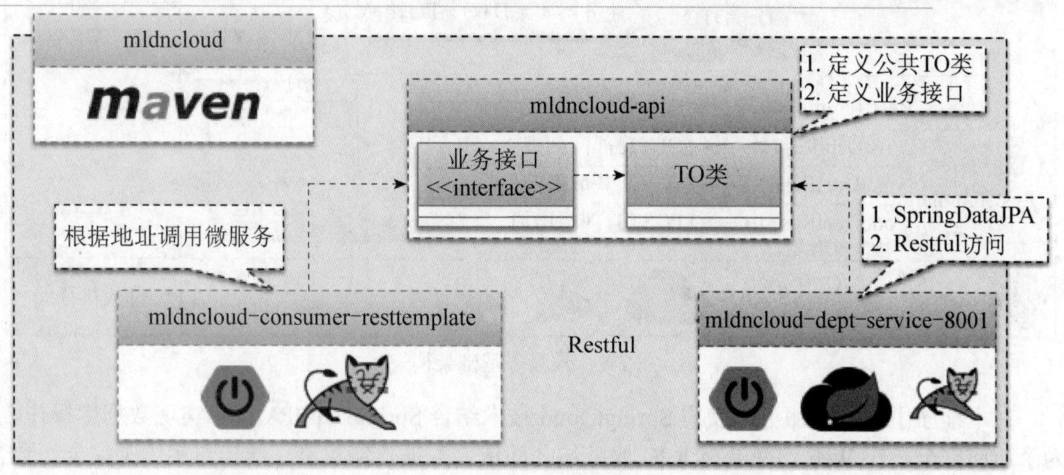

图 7-4　项目描述

提示：关于 TO 类（Transfer Object，数据传输对象）的作用。

在建立 Spring 微服务的时候，数据库的 ORM 开发框架是不可少的。为了方便，这里将使用 JPA 开发框架结合 SpringDataJPA 技术实现数据层的访问。为了对外部隐藏 PO 类，在 mldncloud-api 项目中会建立一个与指定微服务中 PO 类结构相同的 TO 类，进行数据传输。在微服务控制器中只使用 Bean 拷贝技术，实现 PO 与 TO 的内容拷贝处理，这样的实现会更加安全。

由于微服务需要与数据库进行整合处理操作，所以在本程序中将利用 SpringDataJPA 技术实现数据层开发。为了简化，本例将创建如下数据库。

范例：【MySQL 数据库】定义数据库创建脚本。

```
DROP DATABASE IF EXISTS dept8001 ;
CREATE DATABASE dept8001 CHARACTER SET UTF8 ;
USE dept8001 ;
CREATE TABLE dept(
    deptno          bigint  auto_increment ,
    dname           varchar(50) ,
    loc             varchar(50) ,
    constraint pk_deptno primary key(deptno)
) engine=innodb ;
INSERT INTO dept(dname,loc) VALUES ('财务部','北京') ;
INSERT INTO dept(dname,loc) VALUES ('开发部','石家庄') ;
INSERT INTO dept(dname,loc) VALUES ('销售部','上海') ;
```

```
INSERT INTO dept(dname,loc) VALUES ('产品部','广州') ;
INSERT INTO dept(dname,loc) VALUES ('人事部','深圳') ;
```

本数据库名称为 dept8001，而后该数据库将被 SpringCloud 微服务访问，作为服务提供者；而 SpringBoot 将作为服务消费者，实现远程业务调用。本程序的基本处理结构如图 7-5 所示。

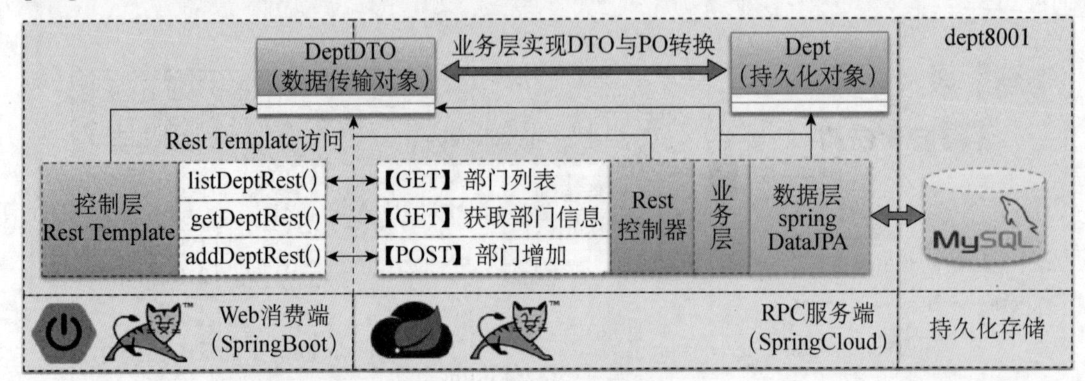

图 7-5 微服务基础架构

在本程序中，RPC 服务端使用 SpringCloud 技术结合 SpringDataJPA 技术实现数据库操作。为了访问安全，在 Web 消费端与 RPC 服务端之间依靠数据传输对象（结构与实体类完全一致）进行数据交互，而后在业务层利用 Spring 开发框架提供的 BeanUtils 类实现对象拷贝处理。

7.2.1 建立公共 API 模块：mldncloud-api

mldncloud-api 项目模块的主要作用是定义所有微服务之中的业务接口与传输类。该类不涉及具体的业务逻辑开发，只是定义了标准。

1.【mldncloud 项目】创建 mldncloud-api 项目模块，如图 7-6 所示。

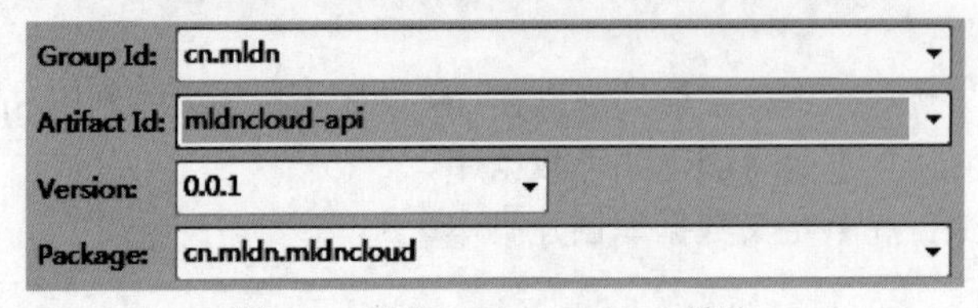

图 7-6 建立 mldncloud-api 项目

2.【mldncloud-api 项目】建立描述部门类的 TO 类。

```
package cn.mldn.mldncloud.dto;
import java.io.Serializable;
@SuppressWarnings("serial")
public class DeptDTO implements Serializable {
    private Long deptno ;
    private String dname ;
    private String loc ;
    // setter、getter略
}
```

3.【mldncloud-api 项目】建立 IDeptService 业务接口。

```
package cn.mldn.mldncloud.service;
import java.util.List;
import cn.mldn.mldncloud.dto.DeptDTO;
public interface IDeptService {
    public DeptDTO add(DeptDTO dto) ;      // 增加新部门
    public DeptDTO get(long deptno) ;      // 根据部门编号获取部门信息
    public List<DeptDTO> list() ;          // 部门信息列表
}
```

公共 API 项目建立完成之后，可以直接在微服务的提供者项目与微服务消费者项目中引入该模块。

7.2.2　建立部门微服务：mldncloud-dept-service-8001

部门微服务项目主要是进行部门业务的实现，同时为了简化实体层开发，本例将直接利用 SpringDataJPA 开发框架实现数据库访问。本例程序要使用的数据库创建脚本如下。

1.【mldncloud 项目】创建新的子模块 mldncloud-dept-service-8001，如图 7-7 所示。

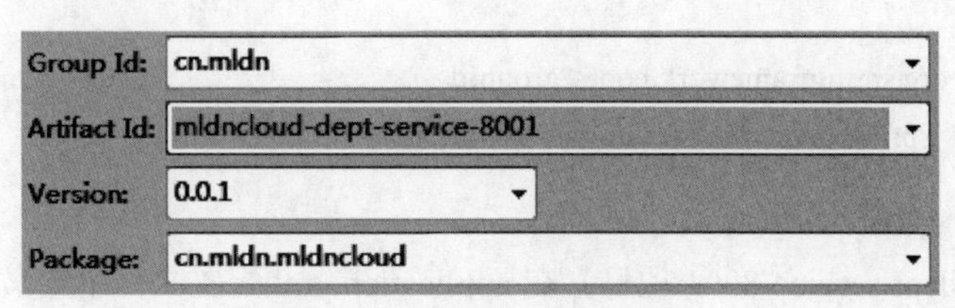

图 7-7　建立部门微服务项目

2.【mldncloud 项目】修改 pom.xml 配置文件，追加数据库与 ORM 框架依赖配置。

定义版本属性	<mldncloud.version>0.0.1</mldncloud.version> <druid.version>1.1.6</druid.version> <mysql-connector-java.version>5.1.21</mysql-connector-java.version>
定义配置依赖管理	<dependency> <groupId>cn.mldn</groupId> <artifactId>mldncloud-api</artifactId> <version>${mldncloud.version}</version> </dependency> <dependency> <groupId>mysql</groupId> <artifactId>mysql-connector-java</artifactId> <version>${mysql-connector-java.version}</version> </dependency> <dependency> <groupId>com.alibaba</groupId>

```
            <artifactId>druid</artifactId>
            <version>${druid.version}</version>
        </dependency>
```

3.【mldncloud-dept-service-8001 项目】修改 pom.xml 配置文件，引入相关依赖包。

```
<dependency>
    <groupId>cn.mldn</groupId>
    <artifactId>mldncloud-api</artifactId>
</dependency>
<dependency>
    <groupId>mysql</groupId>
    <artifactId>mysql-connector-java</artifactId>
</dependency>
<dependency>
    <groupId>com.alibaba</groupId>
    <artifactId>druid</artifactId>
</dependency>
<dependency>
    <groupId>org.springframework.boot</groupId>
    <artifactId>spring-boot-starter-data-jpa</artifactId>
</dependency>
```

4.【mldncloud-dept-service-8001 项目】在 application.yml 配置文件中定义相关配置属性。

```
spring:
  datasource:
    type: com.alibaba.druid.pool.DruidDataSource          # 配置数据源类型
    url: jdbc:mysql://localhost:3306/dept8001             # 数据库的连接地址
    driver-class-name: org.gjt.mm.mysql.Driver            # 驱动程序
    username: root                                        # 用户名
    password: mysqladmin                                  # 密码
    dbcp2:                                                # 数据库连接池配置
      min-idle: 1                                         # 最小维持连接数
      max-idle: 1                                         # 最大维持连接数
      max-total: 1                                        # 最大的可用连接数
      max-wait-millis: 200                                # 最大等待时间
      initial-size: 1                                     # 初始化连接数
  jpa:                                                    # 定义JPA的相关配置项
    show-sql: true                                        # 显示执行的SQL
server:
  port: 8001                                              # 服务端口
```

5.【mldncloud-dept-service-8001 项目】建立持久化对象 Dept。

```
package cn.mldn.mldncloud.po;
import java.io.Serializable;
import javax.persistence.Entity;
import javax.persistence.GeneratedValue;
import javax.persistence.GenerationType;
import javax.persistence.Id;
@SuppressWarnings("serial")
@Entity
public class Dept implements Serializable {
    @Id
    @GeneratedValue(strategy = GenerationType.IDENTITY)
    private Long deptno;
    private String dname ;
    private String loc ;
    // setter、getter略
}
```

6.【mldncloud-dept-service-8001 项目】定义 IDeptDAO 接口，利用 SpringDataJPA 自动进行配置。

```
package cn.mldn.mldncloud.dao;
import org.springframework.data.jpa.repository.JpaRepository;
import cn.mldn.mldncloud.po.Dept;
public interface IDeptDAO extends JpaRepository<Dept,Long> { // 包含全部的基础CRUD支持
}
```

7.【mldncloud-dept-service-8001 项目】定义业务层接口实现子类 DeptServiceImpl。

```
package cn.mldn.mldncloud.service.impl;
import java.util.ArrayList;
import java,util.List;
import org.springframework.beans.BeanUtils;
import org.springframework.beans.factory.annotation.Autowired;
import org.springframework.stereotype.Service;
import cn.mldn.mldncloud.dao.IDeptDAO;
import cn.mldn.mldncloud.dto.DeptDTO;
import cn.mldn.mldncloud.po.Dept;
import cn.mldn.mldncloud.service.IDeptService;
@Service
public class DeptServiceImpl implements IDeptService {
    @Autowired
    private IDeptDAO deptDAO ;                    // 注入DAO接口对象
```

```
    @Override
    public DeptDTO add(DeptDTO dto) {                          // 增加新部门
        Dept po = new Dept() ;                                 // 创建持久化对象
        BeanUtils.copyProperties(dto, po);                     // 将DTO对象内容复制到持久化对象PO之中
        DeptDTO returnTO = new DeptDTO() ;                     // 定义业务方法返回对象
        BeanUtils.copyProperties(this.deptDAO.save(po), returnTO); // 将业务返回结果保存到DTO中
        return returnTO ;
    }
    @Override
    public DeptDTO get(long deptno) {                          // 根据部门编号获取一个部门信息
        Dept po = this.deptDAO.findOne(deptno) ;               // 通过数据层获取持久化对象
        DeptDTO dto = new DeptDTO() ;                          // 定义传输对象
        BeanUtils.copyProperties(po, dto);                     // 将持久化PO对象内容复制到传输对象DTO中
        return dto;
    }
    @Override
    public List<DeptDTO> list() {                              // 部门列表
        List<Dept> allDepts = this.deptDAO.findAll() ;                // 查询全部部门信息
        List<DeptDTO> returnDepts = new ArrayList<DeptDTO>() ;  // 定义返回集合
        allDepts.forEach((dept)->{                             // 将持久化对象集合保存到传输对象集合中
            DeptDTO dto = new DeptDTO() ;
            BeanUtils.copyProperties(dept, dto);               // 复制对象
            returnDepts.add(dto) ;                             // 保存全部DTO对象
        });
        return returnDepts;
    }
}
```

8.【mldncloud-dept-service-8001 项目】定义部门 Rest 控制类。

```
package cn.mldn.mldncloud.rest;
import org.springframework.beans.factory.annotation.Autowired;
import org.springframework.web.bind.annotation.GetMapping;
import org.springframework.web.bind.annotation.PathVariable;
import org.springframework.web.bind.annotation.PostMapping;
import org.springframework.web.bind.annotation.RequestBody;
import org.springframework.web.bind.annotation.RestController;
import cn.mldn.mldncloud.dto.DeptDTO;
import cn.mldn.mldncloud.service.IDeptService;
@RestController
public class DeptRest {
    @Autowired
```

```
    private IDeptService deptService ;                    // 注入部门业务
    @PostMapping("/dept/add")
    public Object add(@RequestBody DeptDTO dept) {
        return this.deptService.add(dept)   ;             // 增加部门信息
    }
    @GetMapping("/dept/get/{deptno}")
    public Object get(@PathVariable("deptno") long deptno) {
        return this.deptService.get(deptno)   ;           // 查询部门信息
    }
    @GetMapping("/dept/list")
    public Object list() {
        return this.deptService.list() ;                  // 部门信息列表
    }
}
```

本程序中为了实现部门数据的增加操作，需要接收消费端传递过来的对象，所以使用了@RequestBody 注解进行声明，同时在根据部门编号获取部门信息时，采用路径参数的方式进行传递，使用了@PathVariable("deptno")注解。

9.【mldncloud-dept-service-8001 项目】定义程序启动类。

```
package cn.mldn.mldncloud;
import org.springframework.boot.SpringApplication;
import org.springframework.boot.autoconfigure.SpringBootApplication;
import org.springframework.data.jpa.repository.config.EnableJpaRepositories;
@SpringBootApplication
@EnableJpaRepositories(basePackages="cn.mldn.mldncloud.dao")
public class StartDeptServiceApplication8001 {
    public static void main(String[] args) {
        SpringApplication.run(StartDeptServiceApplication8001.class, args);
    }
}
```

10.【操作系统】为了方便进行微服务的访问，建议修改 hosts 主机名称，追加主机映射。

```
127.0.0.1   dept-8001.com
```

对于部门微服务，此时主要通过以下 3 个地址进行访问：

☑ **【GET 请求】**获取部门信息地址为 http://dept-8001.com:8001/dept/get/部门编号。

☑ **【GET 请求】**部门列表地址为 http://dept-8001.com:8001/dept/list。

☑ **【POST 请求】**增加部门地址为 http://dept-8001.com:8001/dept/add。

7.2.3　建立 Web 消费端：mldncloud-consumer-resttemplate

在 Web 端如果要进行 Restful 服务端的调用，可以利用 RestTemplate 类实现 GET 或 POST

请求的服务调用。

1.【mldncloud 项目】创建一个新的子模块 mldncloud-consumer-resttemplate，如图 7-8 所示。

2.【mldncloud-consumer-resttemplate 项目】修改 pom.xml 配置文件，除了引入 SpringBoot 的相关依赖包之外，还需要引入公共 mldncloud-api 模块，这样就可以使用 DTO 类进行数据接收。

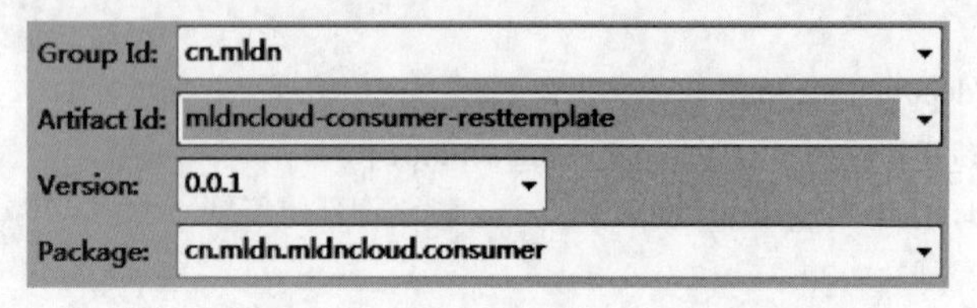

图 7-8 创建新的 Maven 模块

```
<dependency>
    <groupId>cn.mldn</groupId>
    <artifactId>mldncloud-api</artifactId>
</dependency>
```

3.【mldncloud-consumer-resttemplate 项目】建立一个配置类，该类主要用于定义 RestTemplate 对象。

```
package cn.mldn.mldncloud.consumer.config;
import org.springframework.context.annotation.Bean;
import org.springframework.context.annotation.Configuration;
import org.springframework.web.client.RestTemplate;
@Configuration
public class RestfulConfig {
    @Bean
    public RestTemplate getRestTemplate() {
        return new RestTemplate() ;                    // 实例化RestTemplate对象
    }
}
```

4.【mldncloud-consumer-resttemplate 项目】定义控制器，该控制器将利用 RestTemplate 进行访问。

```
package cn.mldn.mldncloud.consumer.controller;
import java.util.List;
import javax.annotation.Resource;
import org.springframework.web.bind.annotation.GetMapping;
import org.springframework.web.bind.annotation.RestController;
import org.springframework.web.client.RestTemplate;
import cn.mldn.mldncloud.dto.DeptDTO;
@RestController                                       // 为方便起见使用Restful风格展示
public class DeptController {
```

```
    public static final String DEPT_ADD_URL = "http://dept-8001.com:8001/dept/add";
    public static final String DEPT_GET_URL = "http://dept-8001.com:8001/dept/get";
    public static final String DEPT_LIST_URL = "http://dept-8001.com:8001/dept/list";
    @Resource
    private RestTemplate restTemplate;                // 注入RestTemplate对象
    @GetMapping("/consumer/dept/list")
    public Object listDeptRest() {
        return this.restTemplate.getForObject(DEPT_LIST_URL, List.class);
    }
    @GetMapping("/consumer/dept/get")
    public Object getDeptRest(long deptno) {
        DeptDTO dept = this.restTemplate.getForObject(DEPT_GET_URL + "/" + deptno,
DeptDTO.class);
        return dept;
    }
    @GetMapping("/consumer/dept/add")
    public Object addDeptRest(DeptDTO dept) {  // 传输DeptDTO对象
        DeptDTO result = this.restTemplate.postForObject(DEPT_ADD_URL, dept, DeptDTO.class);
        return result;
    }
}
```

5.【mldncloud-consumer-resttemplate 项目】建立 src/main/resources/application.yml 配置文件，使程序运行在 80 端口。

```
server:
  port: 80              # 服务端口
```

6.【mldncloud-consumer-resttemplate 项目】建立消费端程序启动类。

```
package cn.mldn.mldncloud.consumer;
import org.springframework.boot.SpringApplication;
import org.springframework.boot.autoconfigure.SpringBootApplication;
@SpringBootApplication
public class StartWebConsumerApplication80 {
    public static void main(String[] args) {
        SpringApplication.run(StartWebConsumerApplication80.class, args);
    }
}
```

7.【操作系统】为了验证方便，修改 hosts 配置文件，追加主机映射。

```
127.0.0.1    consumer.com
```

8.【客户端测试】此时程序准备完毕，开发者可以使用如下的消费端访问路径来调用远程微服务。

【消费端】部门列表	http://consumer.com/consumer/dept/list
【消费端】部门查看	http://consumer.com/consumer/dept/get?deptno=1
【消费端】部门增加	http://consumer.com/consumer/dept/add?dname=MLDN 教学研发部&loc=北京

7.3 Restful 接口描述

利用 SpringCloud 开发技术可以方便地实现 Restful 技术标准。从另外一个方面来说，这些接口如果要给其他消费端程序使用，就需要有良好的接口说明信息，可以明确地将服务接口以及参数的作用告诉使用者，此时就可以利用 Swagger 技术实现。

提示：基于 mldncloud-dept-service-8001 进行复制。

本节讲解的内容属于 SpringCloud 的辅助技术，读者可以用此技术作为接口描述。为了方便读者理解核心配置，本项目中会将 mldncloud-dept-service-8001 项目复制为 mldncloud-dept-service-swagger 项目。

由于 Swagger 技术与 SpringCloud 技术的整体联系不大，所以本例将不再进行 mldncloud 父 pom.xml 文件的配置，只在子项目中进行依赖库引入。

1.【mldncloud-dept-service-swagger 项目】修改 pom.xml 配置文件，引入 Swagger 相关依赖库。

```
<dependency>
        <groupId>io.springfox</groupId>
        <artifactId>springfox-swagger2</artifactId>
        <version>2.7.0</version>
</dependency>
<dependency>
        <groupId>io.springfox</groupId>
        <artifactId>springfox-swagger-ui</artifactId>
        <version>2.7.0</version>
</dependency>
```

2.【mldncloud-dept-service-swagger 项目】建立 Swagger 配置类，定义接口描述基础信息。

```
package cn.mldn.mldncloud.config;
import org.springframework.context.annotation.Bean;
import org.springframework.context.annotation.Configuration;
import springfox.documentation.builders.ApiInfoBuilder;
import springfox.documentation.builders.PathSelectors;
import springfox.documentation.builders.RequestHandlerSelectors;
import springfox.documentation.service.ApiInfo;
```

```
import springfox.documentation.spi.DocumentationType;
import springfox.documentation.spring.web.plugins.Docket;
import springfox.documentation.swagger2.annotations.EnableSwagger2;
@Configuration
@EnableSwagger2
public class Swagger2Config {
    @Bean
    public Docket getDocket() {   // 此类是整个Swagger的配置项，可利用这个类指派扫描包
        return new Docket(DocumentationType.SWAGGER_2)
                .apiInfo(this.getApiInfo()).select()          // 读取描述信息
                .apis(RequestHandlerSelectors
                .basePackage("cn.mldn.mldncloud.rest"))       // 扫描程序包
                .paths(PathSelectors.any()).build();          // 设置文档的显示类型
    }
    private ApiInfo getApiInfo() {                            // 定义相关描述信息
        return new ApiInfoBuilder().title("部门业务微服务")
                .description("更多选择请关注www.mldn.cn")
                .termsOfServiceUrl("http://www.mldn.cn")
                .contact("MLDN-李兴华老师")
                .license("李兴华").version("1.0").build();
    }
}
```

3.【mldncloud-dept-service-swagger 项目】修改 DeptRest 程序类，使用 Swagger 进行接口描述。

```
package cn.mldn.mldncloud.rest;
import org.springframework.beans.factory.annotation.Autowired;
import org.springframework.web.bind.annotation.GetMapping;
import org.springframework.web.bind.annotation.PathVariable;
import org.springframework.web.bind.annotation.PostMapping;
import org.springframework.web.bind.annotation.RequestBody;
import org.springframework.web.bind.annotation.RestController;
import cn.mldn.mldncloud.dto.DeptDTO;
import cn.mldn.mldncloud.service.IDeptService;
import io.swagger.annotations.ApiImplicitParam;
import io.swagger.annotations.ApiImplicitParams;
import io.swagger.annotations.ApiOperation;
@RestController
public class DeptRest {
    @Autowired
    private IDeptService deptService;                         // 注入部门业务
    @ApiOperation(value = "增加部门信息", notes = "增加部门数据，需要传入DTO对象（dname、loc）")
```

```
    @ApiImplicitParams({
        @ApiImplicitParam(name = "dept", value = "部门实体",
                required = true, dataType = "DeptDTO") })
    @PostMapping("/dept/add")
    public Object get(@RequestBody DeptDTO dept) {
        return this.deptService.add(dept);              // 增加部门信息
    }
    @ApiOperation(value = "获取一个部门信息", notes = "根据部门编号获取一个部门的信息")
    @ApiImplicitParams({
            @ApiImplicitParam(name = "deptno", value = "部门编号", required = true,
                    paramType = "path", dataType = "Long") })
    @GetMapping("/dept/get/{deptno}")
    public Object get(@PathVariable("deptno") long deptno) {
        return this.deptService.get(deptno);            // 查询部门信息
    }
    @ApiOperation(value = "部门信息列表", notes = "将进行部门信息的详细列表")
    @GetMapping("/dept/list")
    public Object list() {
        return this.deptService.list();                 // 部门信息列表
    }
}
```

本程序为 Restful 控制器中追加了接口的注解描述，当程序启动后可以通过 swagger-ui.html 地址进行访问，页面运行后的效果如图 7-9 所示。

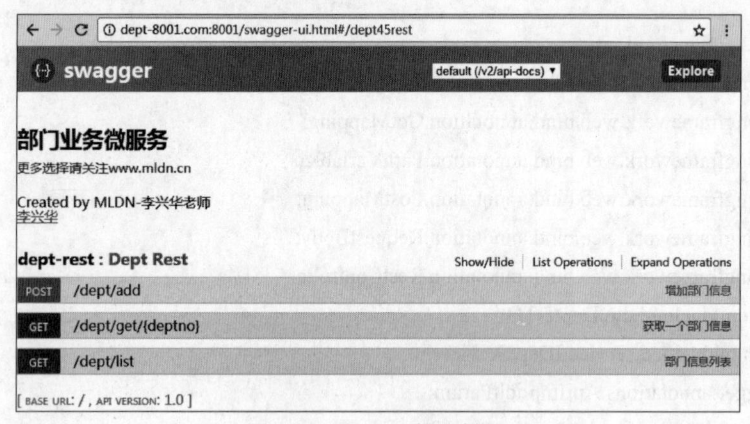

图 7-9　Swagger 接口描述

7.4　SpringSecurity 安全访问

微服务需要通过 Web 服务器进行发布，而所有的业务接口最终都会暴露在公网上。为了保

证业务的安全性，需要引入安全管理机制。在 SpringCloud 中可以利用 SpringSecurity 进行安全访问（见图 7-10），即在访问某一个微服务前首先进行用户认证，认证通过后才可以访问目标微服务；如果认证失败，则返回错误信息。

提示：不使用数据库保存认证信息。

在本节讲解的 SpringSecurity 整合处理中，将采用固定的用户名和密码进行配置，暂不与数据库进行整合处理。在本书第 13 章将利用 OAuth 整合 SpringCloud 进行安全配置。

图 7-10　SpringSecurity 安全访问

7.4.1　微服务安全验证

SpringSecurity 的配置只需要在项目中引入 spring-boot-starter-security 依赖库即可实现。其默认的用户名为 user，由于认证密码会在每一次微服务启动时动态生成，不利于微服务的消费端调用，所以在本例中将配置固定的用户名 mldnjava 与密码 hello。

1.【mldncloud-dept-service-8001 项目】修改 pom.xml 配置文件，追加依赖库配置。

```
<dependency>
    <groupId>org.springframework.boot</groupId>
    <artifactId>spring-boot-starter-security</artifactId>
</dependency>
```

2.【mldncloud-dept-service-8001 项目】修改 application.yml 配置文件，配置用户名和密码。

```
security:                                    # 安全配置
  basic:
    enabled: true                            # 启用SpringSecurity的安全配置项
  user:                                      # 用户认证与授权信息
    name: mldnjava                           # 认证用户名
    password: hello                          # 认证密码
    role:                                    # 授权角色
    - USER
```

本程序配置了微服务的认证信息。用户启动微服务并输入地址（http://dept-8001.com:8001/dept/list）后，可以看见如图 7-11 所示的登录界面，输入配置的用户名和密码，就能访问部门微服务信息，如图 7-12 所示。

在认证微服务访问时，也可以利用地址的方式进行认证信息传递，此时只需要在访问地址前追加用户名和密码即可，访问路径为 http://mldnjava:hello@dept-8001.com:8001/dept/list。

图 7-11　用户认证信息

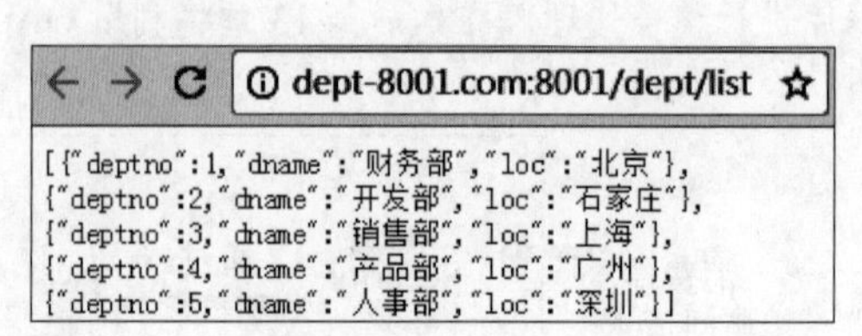

图 7-12　部门列表

7.4.2　消费端安全访问

Restful 服务端配置了安全认证之后，此时的消费端将无法正常访问，需要配置认证请求头信息才可以实现正常的服务调用。

1.【mldncloud-consumer-resttemplate 项目】修改 RestfulConfig 配置类，增加 HTTP 访问头信息配置 Bean。

```
@Bean
public HttpHeaders getHeaders() {                                    // HTTP头信息配置
        HttpHeaders headers = new HttpHeaders();                     // 定义HTTP的头信息
        String auth = "mldnjava:hello";                              // 认证的原始信息
        byte[] encodedAuth = Base64.getEncoder()
                    .encode(auth.getBytes(Charset.forName("US-ASCII")));     // 进行加密处理
        // 在配置授权的头信息内容的时候，加密信息一定要与Basic之间有一个空格
        String authHeader = "Basic " + new String(encodedAuth);
        headers.set("Authorization", authHeader);                    // 保存头信息
        return headers;
}
```

2.【mldncloud-consumer-resttemplate 项目】修改 DeptController 程序类，访问前追加头信息配置。

```
package cn.mldn.mldncloud.consumer.controller;
import java.util.List;
import javax.annotation.Resource;
import org.springframework.http.HttpEntity;
import org.springframework.http.HttpHeaders;
import org.springframework.http.HttpMethod;
import org.springframework.web.bind.annotation.GetMapping;
import org.springframework.web.bind.annotation.RestController;
```

```
import org.springframework.web.client.RestTemplate;
import cn.mldn.mldncloud.dto.DeptDTO;
@RestController                                          // 为方便起见使用Restful风格展示
public class DeptController {
    public static final String DEPT_ADD_URL = "http://dept-8001.com:8001/dept/add";
    public static final String DEPT_GET_URL = "http://dept-8001.com:8001/dept/get";
    public static final String DEPT_LIST_URL = "http://dept-8001.com:8001/dept/list";
    @Resource
    private RestTemplate restTemplate;                   // 注入RestTemplate对象
    @Resource
    private HttpHeaders headers;                         // 注入HTTP头信息对象
    @SuppressWarnings("unchecked")
    @GetMapping("/consumer/dept/list")
    public Object listDeptRest() {
        List<DeptDTO> allDepts = this.restTemplate
                .exchange(DEPT_LIST_URL, HttpMethod.GET,
                        new HttpEntity<Object>(this.headers), List.class)
                .getBody();                              // 访问服务设置头信息
        return allDepts;
    }
    @GetMapping("/consumer/dept/get")
    public Object getDeptRest(long deptno) {
        DeptDTO dept = this.restTemplate
                .exchange(DEPT_GET_URL + "/" + deptno, HttpMethod.GET,
                        new HttpEntity<Object>(this.headers), DeptDTO.class)
                .getBody();                              // 访问服务设置头信息
        return dept;
    }
    @GetMapping("/consumer/dept/add")
    public Object addDeptRest(DeptDTO dept) {  // 传输DeptDTO对象
        DeptDTO result = this.restTemplate.exchange(DEPT_ADD_URL, HttpMethod.POST,
                new HttpEntity<Object>(dept, this.headers), DeptDTO.class)
                .getBody();                              // 访问服务设置头信息
        return result;
    }
}
```

此时消费端配置了认证的头部信息，而后可以正常地实现部门微服务的访问。

7.4.3　StatelessSession

在实际开发中，所有的 Restful 都是基于 HTTP 的一种应用，所有的 Web 应用都会运行在

Web 容器中。Web 容器一般都会提供一个 Session 的管理机制，开发者可以直接通过配置实现以下两种 Session 管理机制。

> **提示：Spring 的 Session 管理策略应以无状态（STATELESS）为主。**
> org.springframework.security.config.http.SessionCreationPolicy 枚举类中定义了 4 种会话管理状态：ALWAYS、NEVER、IF_REQUIRED 和 STATELESS，在实际使用中应该以 STATELESS 为主。在 SpringCloud 中默认的会话状态也是无状态会话（STATELESS）。

【mldncloud-dept-service-8001 项目】修改 application.yml 配置文件，进行 Session 管理配置。

服务端维护 Session 状态	服务端不维护 Session 状态（StatelessSession，无状态）

security:
  sessions: always
	security:
	  sessions: stateless

很明显，让微服务端一直维持着用户请求的会话状态是一件很浪费性能的配置，因为微服务的并发访问量巨大，所以这样的 Session 创建与销毁机制就会带来严重的性能问题。在实际使用中，应该以无状态会话机制为主。

7.4.4 安全配置模块

在进行 Restful 服务开发的时候，为了保证安全，所有程序都需要进行 SpringSecurity 安全认证处理。如果所有的认证处理都需要在 application.yml 配置文件完成，这样的配置明显是重复的。在很多时候进行微服务调用时往往需要统一用户名和密码，所以此时最简单的做法是可以单独创建一个安全配置模块，在需要的时候引入依赖模块配置即可。

1.【mldncloud 项目】建立一个安全配置模块 mldncloud-security，创建的时候需要定义好包名称，如图 7-13 所示。

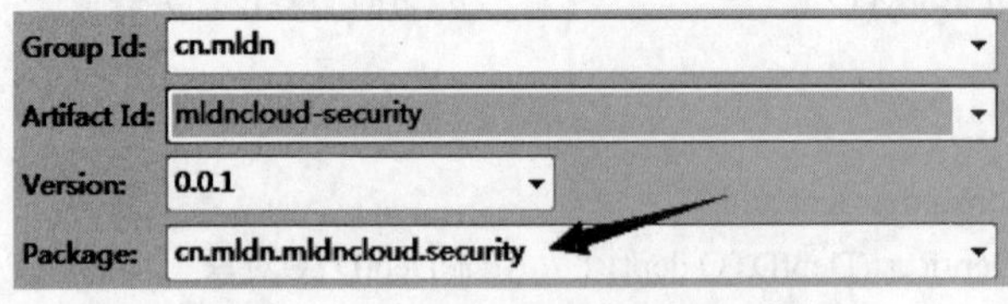

图 7-13 建立 mldncloud-security 模块

2.【mldncloud-security 项目】修改 pom.xml 配置文件，追加依赖库配置。

```
<dependency>
    <groupId>org.springframework.boot</groupId>
    <artifactId>spring-boot-starter-security</artifactId>
</dependency>
```

3.【mldncloud-security 项目】建立统一安全配置类，并且通过程序配置认证与授权信息。

```
package cn.mldn.mldncloud.security;                    // 此包可以为服务扫描子包
import javax.annotation.Resource;
```

```
import org.springframework.context.annotation.Configuration;
import org.springframework.security.config.annotation.authentication.
                builders.AuthenticationManagerBuilder;
import org.springframework.security.config.annotation.web.builders.HttpSecurity;
import org.springframework.security.config.annotation.web.configuration.EnableWebSecurity;
import org.springframework.security.config.annotation.web.
                configuration.WebSecurityConfigurerAdapter;
import org.springframework.security.config.http.SessionCreationPolicy;
@Configuration
@EnableWebSecurity
public class WebSecurityConfig extends WebSecurityConfigurerAdapter {        // 建立安全配置
    @Resource
    public void configGlobal(AuthenticationManagerBuilder auth)
                throws Exception {                                           // 配置用户名与密码
        auth.inMemoryAuthentication().withUser("mldnjava").password("hello")
                .roles("USER").and().withUser("admin").password("hello")
                .roles("USER", "ADMIN");
    }
    @Override
    protected void configure(HttpSecurity http) throws Exception {
        // 表示所有的访问都必须进行认证处理后才可以正常进行
        http.httpBasic().and().authorizeRequests().anyRequest()
                .fullyAuthenticated();
        // 所有的Restful服务一定要设置为无状态，以提升操作性能
        http.sessionManagement()
                .sessionCreationPolicy(SessionCreationPolicy.STATELESS);
    }
}
```

4.【mldncloud 项目】修改 pom.xml 配置文件，追加 mldncloud-security 模块依赖管理。

```
<dependency>
    <groupId>cn.mldn</groupId>
    <artifactId>mldncloud-security</artifactId>
    <version>${mldncloud.version}</version>
</dependency>
```

5.【mldncloud-dept-service-8001 项目】修改 pom.xml 配置文件，引入安全模块。

```
<dependency>
    <groupId>cn.mldn</groupId>
    <artifactId>mldncloud-security</artifactId>
</dependency>
```

由于 mldncloud-security 中包含有 spring-boot-starter-security 依赖库，所以在部门微服务中将不再需要重复引入安全配置，同时由于通过 Bean 进行配置也不再需要通过 application.yml 配置。

7.5 本章小结

1. SpringCloud 是基于 Restful 实现的 RPC 技术，并且需要 SpringBoot 技术支持。
2. 在微服务的创建过程中为了保证服务访问的安全，需要配置 DTO 数据传输类，对请求和返回数据进行封装。
3. RestTemplate 是消费端进行 Restful 服务端访问的程序类，通过地址即可访问。
4. 微服务的信息可以通过 Swagger 框架进行接口描述定义。
5. 为了保证微服务的安全，应该在项目中引入 SpringSecurity，进行用户认证与授权信息配置。
6. 微服务如果要承受高并发访问，则一定要采用无状态（STATELESS）Session 配置。

第 8 章

Eureka 注册服务

通过本章学习，可以达到以下目标：

1. 掌握 Eureka 的主要作用与服务搭建。
2. 掌握微服务注册 Eureka 处理。
3. 掌握 Eureka 服务信息与发现服务配置。
4. 掌握 Eureka-HA 机制实现原理与实现。
5. 掌握 Eureka 服务发布。

在实际的开发环境中会存在大量的微服务，为了更方便地进行微服务的统一管理，则需要使用注册中心。在 SpringCloud 中推荐使用的注册中心为 Eureka，本章将为读者讲解 Eureka 的作用以及服务注册等相关内容。

提示：本章暂不考虑消费端调用。

本章所讲解的程序只针对微服务与 Eureka 注册中心的使用，不涉及消费端的 Restful 接口调用，在第 9 章中将会为读者详细讲解消费端整合处理。

8.1 Eureka 简介

RestTemplate 在进行微服务访问的时候，需要明确地通过微服务的地址进行调用。这样直接利用地址的调用，一旦出现服务端主机地址变更，则消费端就需要进行大量的修改。同时，微服务的主要目的是提高业务处理能力，因此往往会若干个相同业务的微服务一同参与运算。为了解决这样的问题，在微服务的使用中需要采用 Eureka 注册中心对所有微服务进行管理。所有的微服务在启动后需要全部向 Eureka 中进行服务注册，而后客户端直接利用 Eureka 进行服务信息的获得，以实现微服务调用。其基本使用架构如图 8-1 所示。

图 8-1　Eureka 服务注册

8.2 定义 Eureka 服务端

在 SpringCloud 中大量使用了 Netflix 的开源项目，而其中 Eureka 就属于 Netflix 提供的发现服务组件，该应用程序需要由开发者自行定义。

1.【mldncloud 项目】创建新的子模块 mldncloud-eureka-7001，该注册中心将运行在 7001 端口上。

2.【mldncloud-eureka-7001 项目】修改 pom.xml 配置文件，除了引入 SpringBoot 相关依赖库之外，还需要引入 Eureka 相关依赖库。

```
<dependency>
    <groupId>org.springframework.cloud</groupId>
    <artifactId>spring-cloud-starter-eureka-server</artifactId>
</dependency>
```

3.【mldncloud-eureka-7001 项目】修改 application.yml 配置文件，进行 Eureka 服务器配置。

```
server:
  port: 7001                                    # 定义运行端口
eureka:
  instance:                                     # Eureak实例定义
    hostname: eureka-7001.com           # 定义 Eureka 实例所在的主机名称，需要修改 hosts
```

4.【mldncloud-eureka-7001 项目】定义程序启动主类，添加 Eureka 相关注解。

```
package cn.mldn.mldncloud;
import org.springframework.boot.SpringApplication;
import org.springframework.boot.autoconfigure.SpringBootApplication;
import org.springframework.cloud.netflix.eureka.server.EnableEurekaServer;
@SpringBootApplication
@EnableEurekaServer                              // 启用Eureka服务
public class EurekaServerStartApplication7001 {
    public static void main(String[] args) {
        SpringApplication.run(EurekaServerStartApplication7001.class, args);
    }
}
```

5.【操作系统】修改 hosts 配置文件，追加主机配置。

```
127.0.0.1    eureka-7001.com
```

此时配置的主机名称 eureka-7001.com 与 application.yml 中配置的 Eureka 运行主机名称相同。

6.【mldncloud-eureka-7001 项目】启动 Eureka 服务端，随后输入访问地址 http://eureka-7001.com:7001/，可以看见如图 8-2 所示的管理界面。

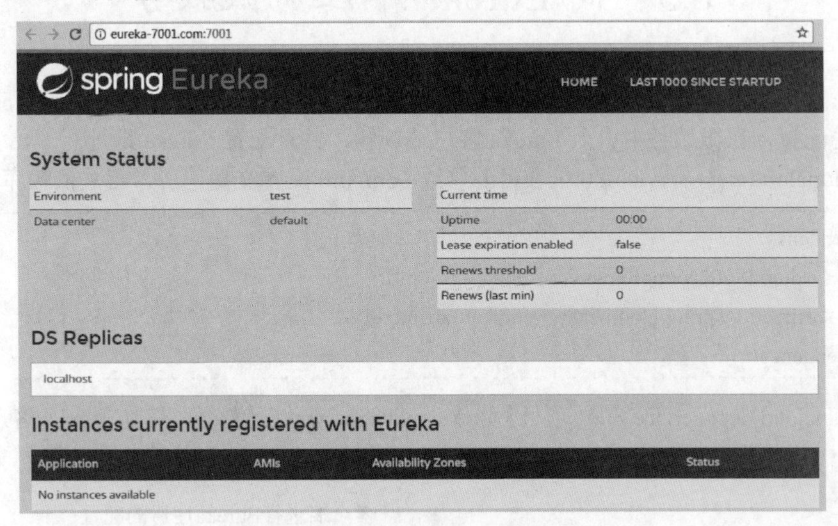

图 8-2　EurekaServer 管理界面

提示：关于程序运行中的 TransportException 异常。

虽然现在已经配置完了 Eureka 注册中心，但在运行中却会发现控制台上会输出如下错误信息：

```
com.netflix.discovery.shared.transport.TransportException: Cannot execute request on any known server
```

之所以会有这些错误信息，主要是因为 Eureka 在默认配置下自己也是一个微服务，并且该微服务应该向 Eureka 中注册，但却无法找到主机所导致的。要想解决这个问题，需要修改 application.yml 配置文件，追加配置项。

【mldncloud-eureka-7001 项目】修改 application.yml 配置文件，追加如下配置。

```
server:
  port: 7001                                    # 定义运行端口
eureka:
  client:                                       # 客户端进行Eureka注册的配置
    service-url:
      defaultZone: http://eureka-7001.com:7001/eureka
    register-with-eureka: false                 # 当前的微服务不注册到Eureka之中
    fetch-registry: false                       # 不通过Eureka获取注册信息
  instance:                                     # Eureak实例定义
    hostname: eureka-7001.com                   # 定义 Eureka 实例所在的主机名称
```

此时的程序配置了微服务要注册的 Eureka 服务地址，但是服务信息注册没有意义，所以配置了 register-with-eureka 与 fetch-registry 选项，不再在 Eureka 注册中心中显示该微服务信息。

8.3 向 Eureka 中注册微服务

Eureka 注册中心搭建成功后，所有的微服务都应该向 Eureka 中进行注册，此时应该进行微服务程序的配置，在微服务中引入 Eureka 客户端依赖，并且配置 Eureka 地址。

1.【mldncloud-dept-service-8001 项目】修改 pom.xml 配置文件，引入相关依赖库。

```
<dependency>
    <groupId>org.springframework.cloud</groupId>
    <artifactId>spring-cloud-starter-eureka</artifactId>
</dependency>
```

2.【mldncloud-dept-service-8001 项目】修改 application.yml 配置文件，追加 Eureka 客户端配置。

```
eureka:
  client:                                        # 客户端进行Eureka注册的配置
    service-url:                                 # 定义Eureka服务地址
      defaultZone: http://eureka-7001.com:7001/eureka
```

3.【mldncloud-dept-service-8001 项目】如果要向 Eureka 中进行微服务注册，还需要为当前微服务定义名称。

```
spring:
  application:
    name: mldncloud-dept-service                 # 定义微服务名称
```

4.【mldncloud-dept-service-8001 项目】修改 StartDeptServiceApplication8001 程序启动类。

```
package cn.mldn.mldncloud;
import org.springframework.boot.SpringApplication;
import org.springframework.boot.autoconfigure.SpringBootApplication;
import org.springframework.cloud.netflix.eureka.EnableEurekaClient;
import org.springframework.data.jpa.repository.config.EnableJpaRepositories;
@SpringBootApplication
@EnableEurekaClient                                // 启用Eureka客户端
@EnableJpaRepositories(basePackages="cn.mldn.mldncloud.dao")
public class StartDeptServiceApplication8001 {
    public static void main(String[] args) {
        SpringApplication.run(StartDeptServiceApplication8001.class, args);
    }
}
```

这里定义了@EnableEurekaClient 注解信息，微服务启动之后，该服务会自动注册到 Eureka

服务器之中。分别启动 Eureka 注册中心微服务与部门微服务之后，就可以通过 Eureka 注册中心观察到所注册的微服务信息，如图 8-3 所示。

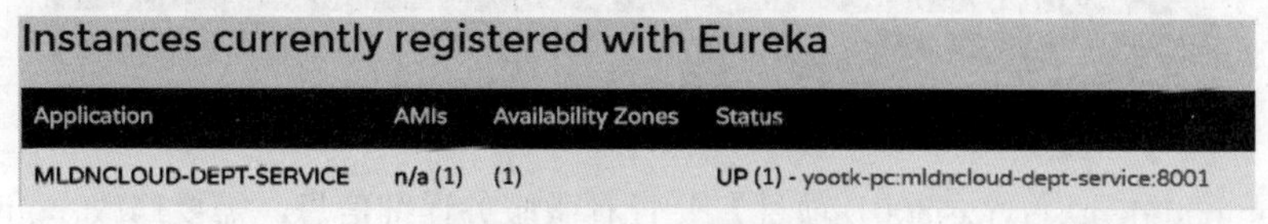

图 8-3　Eureka 中服务信息

8.4　Eureka 服务信息

前面实现了微服务向 Eureka 中的注册处理，但是此时微服务的注册信息并不完整，开发者可以通过微服务的进一步配置，实现更加详细的信息显示。

1.【mldncloud-dept-service-8001 项目】修改 application.yml 配置文件，追加微服务所在主机名称的显示。

```
eureka:
  client:                                          # 客户端进行Eureka注册的配置
    service-url:                                   # 定义Eureka服务地址
      defaultZone: http://eureka-7001.com:7001/eureka
  instance:
    instance-id: dept-8001.com                     # 显示主机名称
```

配置完成后，会在 Eureka 控制中心上显示微服务所在的主机名称，如图 8-4 所示。

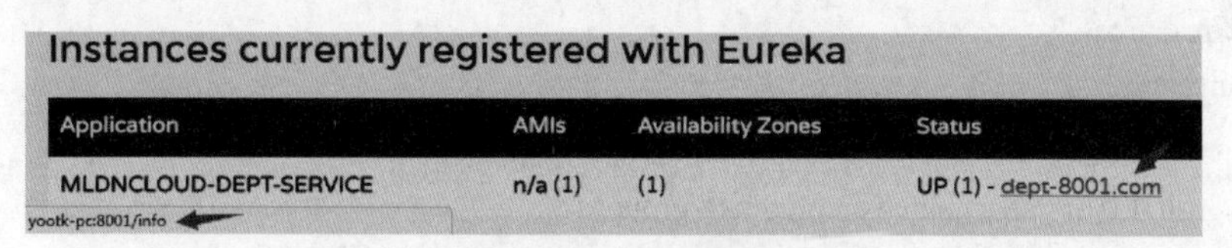

图 8-4　在控制中心显示服务主机信息

2.【mldncloud-dept-service-8001 项目】修改 application.yml，修改服务信息的连接主机为 IP 地址。

```
eureka:
  client:                                          # 客户端进行Eureka注册的配置
    service-url:                                   # 定义Eureka服务地址
      defaultZone: http://eureka-7001.com:7001/eureka
  instance:
    instance-id: dept-8001.com                     # 显示主机名称
    prefer-ip-address: true                        # 访问的路径变为 IP 地址
```

追加 prefer-ip-address 配置项之后，会在显示链接信息处显示 IP 地址，如图 8-5 所示。

用户打开微服务的信息之后，可以使用/info 路径查看信息。由于默认状态下无法显示，此时可以使用 Actuator 显示微服务信息（此配置与 SpringBoot 中的 Actuator 相同）。

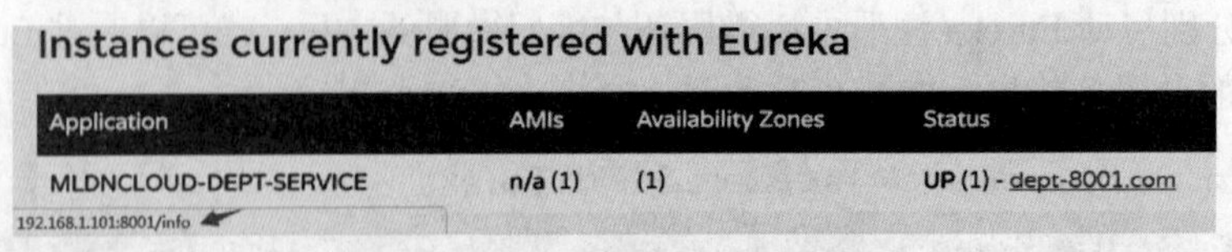

图 8-5　链接信息显示 IP 地址

3.【mldncloud-dept-service-8001 项目】要查看微服务详细信息，需要修改 pom.xml 文件，追加 actuator 依赖。

```
<dependency>
    <groupId>org.springframework.boot</groupId>
    <artifactId>spring-boot-starter-actuator</artifactId>
</dependency>
```

4.【mldncloud 项目】修改 pom.xml 文件，追加信息处理的插件。

```
<resources>
    <resource>
        <directory>src/main/resources</directory>
        <includes>
            <include>**/*.properties</include>
            <include>**/*.yml</include>
            <include>**/*.xml</include>
            <include>**/*.tld</include>
            <include>**/*.p12</include>
        </includes>
        <filtering>true</filtering>
    </resource>
</resources>
```

```
<plugin>
    <groupId>org.apache.maven.plugins</groupId>
    <artifactId>maven-resources-plugin</artifactId>
    <configuration>
        <delimiters>
            <delimiter>$</delimiter>
        </delimiters>
    </configuration>
</plugin>
```

5.【mldncloud-dept-service-8001 项目】修改 application.yml 配置文件，追加 info 的相关信息。

```
info:
  app.name: mldncloud-dept-service
  company.name: www.mldn.cn
  build.artifactId: $project.artifactId$
  build.version: $project.version$
```

此时，当用户通过 Eureka 打开微服务时就可以显示微服务的相应信息，如图 8-6 所示。

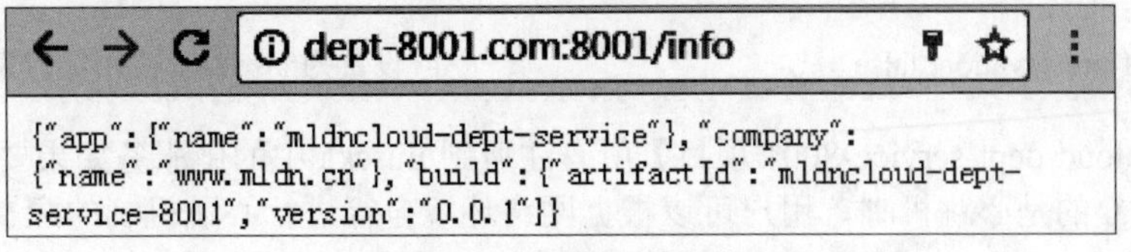

图 8-6 查看微服务信息

8.5 Eureka 发现管理

Eureka 的主要作用是进行微服务注册。在整个微服务的运行过程中，Eureka 也需要对微服务的状态进行监听，对无用的微服务可以进行清除处理，也可以通过发现管理查看 Eureka 信息。

1.【mldncloud-eureka-7001 项目】修改 applicaiton.yml 配置文件，设置微服务清理间隔。

```
eureka:
  server:
    eviction-interval-timer-in-ms: 1000          # 设置清理的间隔时间（毫秒），默认60秒一清理
  client:                                         # 客户端进行Eureka注册的配置
    service-url:
      defaultZone: http://eureka-7001.com:7001/eureka
    register-with-eureka: false                   # 当前的微服务不注册到Eureka之中
    fetch-registry: false                         # 不通过Eureka获取注册信息
  instance:                                       # Eureak实例定义
    hostname: eureka-7001.com                     # 定义 Eureka 实例所在的主机名称
```

本程序为了方便读者观察，将微服务的清除设置为每秒一触发。一旦触发清埋操作后，会在控制台显示如下信息：

```
[a-EvictionTimer] c.n.e.registry.AbstractInstanceRegistry : Running the evict task with compensationTime 0ms
```

过于频繁的清理会造成 Eureka 的性能下降，所以一般情况下建议使用其默认设置。

2.【mldncloud-eureka-7001 项目】在 Eureka 使用过程中经常会看见如下所示的提示文字：

```
EMERGENCY! EUREKA MAY BE INCORRECTLY CLAIMING INSTANCES ARE UP WHEN THEY'RE
NOT. RENEWALS ARE LESSER THAN THRESHOLD AND HENCE THE INSTANCES ARE NOT BEING
EXPIRED JUST TO BE SAFE.
```

该提示的核心意义在于：当某一个微服务不可用时（可能出现了更名或者是宕机等因素），由于所有的微服务提供有保护模式，所以 Eureka 是不会对微服务信息进行清理的。如果希望关闭这种保护模式（一般不推荐），则可以通过修改 application.yml 来实现。在 Eureka 中增加如下配置：

```
eureka:
  server:
    enable-self-preservation: false                    # 设置为false表示关闭保护模式
```

3.【mldncloud-dept-service-8001 项目】所有注册到 Eureka 中的微服务如果要与 Eureka 之间保持联系，依靠的是心跳机制。用户可以根据网络环境自行进行心跳机制的配置，只需要修改微服务中的 application.yml 即可。

```
eureka:
  client:                                              # 客户端进行Eureka注册的配置
    service-url:                                       # 定义Eureka服务地址
      defaultZone: http://eureka-7001.com:7001/eureka
  instance:
    lease-renewal-interval-in-seconds: 2               # 设置心跳的时间间隔（默认是30秒）
    lease-expiration-duration-in-seconds: 5            # 如果现在超过了5秒的间隔（默认是90秒）
    instance-id: dept-8001.com                         # 显示主机名称
```

4.【mldncloud-dept-service-8001 项目】所有注册到 Eureka 中的微服务均可以定义 Eureka 发现信息，只需要在相应的微服务中获取 DiscoveryClient 对象即可。修改 DeptRest 程序类，追加新的方法。

```
    @Autowired
    private DiscoveryClient client ;                   // 进行Eureka的发现服务
    @RequestMapping("/dept/discover")
    public Object discover() {                         // 直接返回发现服务信息
        return this.client ;
    }
```

5.【mldncloud-dept-service-8001 项目】在程序启动类中追加发现服务配置注解。

```
package cn.mldn.mldncloud;
import org.springframework.boot.SpringApplication;
import org.springframework.boot.autoconfigure.SpringBootApplication;
import org.springframework.cloud.client.discovery.EnableDiscoveryClient;
import org.springframework.cloud.netflix.eureka.EnableEurekaClient;
import org.springframework.data.jpa.repository.config.EnableJpaRepositories;
@SpringBootApplication
@EnableEurekaClient                                    // 启用Eureka客户端
@EnableDiscoveryClient
@EnableJpaRepositories(basePackages="cn.mldn.mldncloud.dao")
public class StartDeptServiceApplication8001 {
    public static void main(String[] args) {
        SpringApplication.run(StartDeptServiceApplication8001.class, args);
    }
}
```

程序启动之后通过访问地址 http://mldnjava:hello@dept-8001.com:8001/dept/discover，可以查询到微服务的相关信息，如图 8-7 所示。

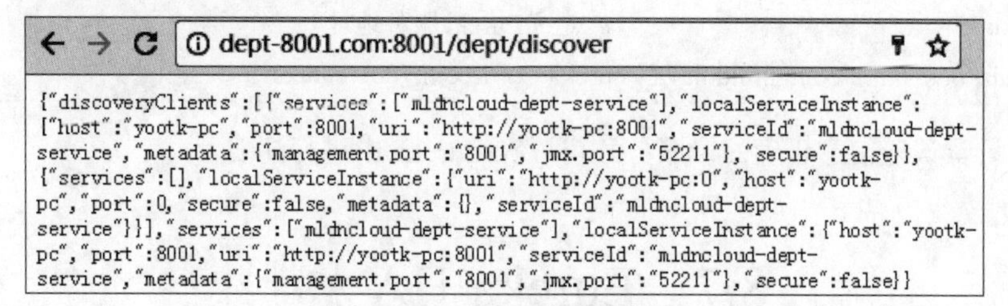

图 8-7　Eureka 发现信息

8.6　Eureka 安全配置

在整个 SpringCloud 微服务架构中，Eureka 是一个重要的注册中心，并且只能够注册自己所需要的微服务。为了保证 Eureka 安全，需要为 Eureka 引入 SpringSecurity 实现安全配置。

1.【mldncloud-eureka-7001 项目】修改 pom.xml 配置文件，引入 SpringSecurity 的依赖包。

```
<dependency>
    <groupId>org.springframework.boot</groupId>
    <artifactId>spring-boot-starter-security</artifactId>
</dependency>
```

2.【mldncloud-eureka-7001 项目】修改 application.yml 配置文件，追加安全配置。

```
security:
  basic:
    enabled: true                          # 启用安全认证处理
  user:
    name: edmin                            # 用户名
    password: mldnjava                     # 密码
eureka:
  client:                                  # 客户端进行Eureka注册的配置
    service-url:
      defaultZone: http://edmin:mldnjava@eureka-7001.com:7001/eureka
    register-with-eureka: false            # 当前的微服务不注册到Eureka之中
    fetch-registry: false                  # 不通过Eureka获取注册信息
  instance:                                # Eureak实例定义
    hostname: eureka-7001.com              # 定义 Eureka 实例所在的主机名称
```

在本配置中追加了新的用户名 edmin/mldnajva，而 Eureka 微服务本身也需要设置一个 Eureka 服务器的访问地址，所以要修改 defaultZone 的访问路径，追加认证信息。

3.【mldncloud-dept-service-8001 项目】修改 application.yml 配置文件，进行授权的注册连接。

```
eureka:
  client:                                          # 客户端进行Eureka注册的配置
    service-url:                                   # 定义Eureka服务地址
      defaultZone: http://edmin:mldnjava@eureka-7001.com:7001/eureka
```

此时，只有认证信息正确的微服务才可以在 Eureka 中进行注册。

8.7 Eureka-HA 机制

Eureka 是整个微服务架构中的核心组件，如果 Eureka 服务器出现问题，则所有的微服务都无法注册，这样整个项目就会彻底瘫痪。为了避免出现这样的问题，可采用 Eureka 集群的模式来处理，即使用多台主机共同实现 Eureka 注册服务，这样即使有一台主机出现问题，另外的主机也可以正常提供服务支持。如图 8-8 所示给出了 2 台 Eureka 主机的集群搭建，如图 8-9 所示给出了 3 台 Eureka 主机的集群搭建。

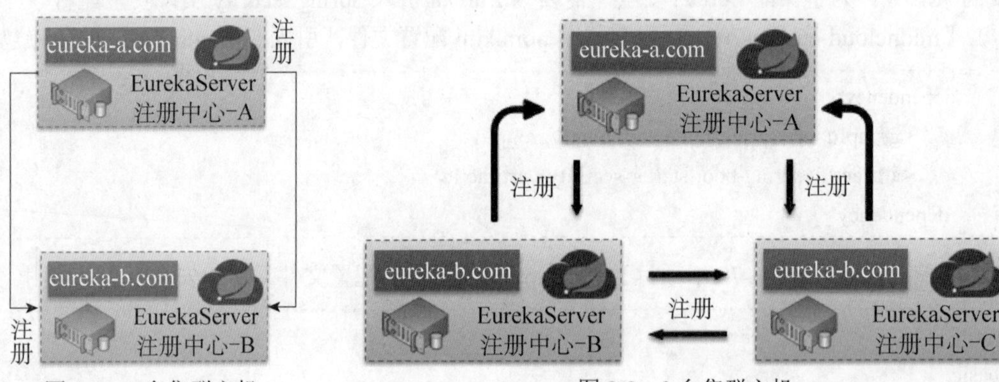

图 8-8　2 台集群主机　　　　图 8-9　3 台集群主机

通过图 8-8 与图 8-9 可以发现，在实现 Eureka 集群时最重要的实现形式就是某一台 Eureka 主机（客户端）需要向其他的 Eureka 主机进行注册。本例将为读者演示 3 台 Eureka 主机的 HA 集群配置。

1.【操作系统】修改 hosts 主机文件，追加 3 个主机名称，与要创建的 3 个 Eureka 微服务对应。

No.	hosts 主机配置	Eureka 微服务项目	服务端口
1	127.0.0.1　eureka-a.com	mldncloud-eureka-server-a	7101
2	127.0.0.1　eureka-b.com	mldncloud-eureka-server-b	7102
3	127.0.0.1　eureka-c.com	mldncloud-eureka-server-c	7103

2.【mldncloud-eureka-server-a 项目】修改 application.yml 配置文件，追加集群环境配置。

```
server:
  port: 7101                                       # 定义运行端口
security:
  basic:
```

```
    enabled: true                              # 启用安全认证处理
  user:
    name: edmin                                # 用户名
    password: mldnjava                         # 密码
eureka:
  client:                                      # 客户端进行Eureka注册的配置
    service-url:
      defaultZone: http://edmin:mldnjava@eureka-b.com:7102/eureka,
                   http://edmin:mldnjava@eureka-c.com:7103/eureka
    register-with-eureka: false                # 当前的微服务不注册到Eureka之中
    fetch-registry: false                      # 不通过Eureka获取注册信息
  instance:                                    # Eureak实例定义
    hostname: eureka-a.com                     # 定义 Eureka 实例所在的主机名称
```

3.【mldncloud-eureka-server-b 项目】修改 application.yml 配置文件，追加集群环境配置。

```
server:
  port: 7102                                   # 定义运行端口
security:
  basic:
    enabled: true                              # 启用安全认证处理
  user:
    name: edmin                                # 用户名
    password: mldnjava                         # 密码
eureka:
  client:                                      # 客户端进行Eureka注册的配置
    service-url:
      defaultZone:
      defaultZone: http://edmin:mldnjava@eureka-a.com:7101/eureka,
                   http://edmin:mldnjava@eureka-c.com:7103/eureka
    register-with-eureka: false                # 当前的微服务不注册到Eureka之中
    fetch-registry: false                      # 不通过Eureka获取注册信息
  instance:                                    # Eureak实例定义
    hostname: eureka-b.com                     # 定义 Eureka 实例所在的主机名称
```

4.【mldncloud-eureka-server-c 项目】修改 application.yml 配置文件，追加集群环境配置。

```
server:
  port: 7103                                   # 定义运行端口
security:
  basic:
    enabled: true                              # 启用安全认证处理
```

```
  user:
    name: edmin                                   # 用户名
    password: mldnjava                            # 密码
eureka:
  client:                                         # 客户端进行Eureka注册的配置
    service-url:
      defaultZone: http://edmin:mldnjava@eureka-a.com:7101/eureka,
                   http://edmin:mldnjava@eureka-b.com:7102/eureka
    register-with-eureka: false                   # 当前的微服务不注册到Eureka之中
    fetch-registry: false                         # 不通过Eureka获取注册信息
  instance:                                       # Eureak实例定义
    hostname: eureka-c.com                        # 定义 Eureka 实例所在的主机名称
```

5.【mldncloud-dept-service-8001 项目】修改 application.yml 配置文件，向 3 台主机同时注册微服务。

```
eureka:
  client:                                         # 客户端进行Eureka注册的配置
    service-url:                                  # 定义Eureka服务地址
      defaultZone:
            http://edmin:mldnjava@eureka-a.com:7101/eureka,
            http://edmin:mldnjava@eureka-b.com:7102/eureka,
            http://edmin:mldnjava@eureka-c.com:7103/eureka
```

6.【mldncloud-eureka-server-*项目】启动所有的 Eureka 服务，登录其中任意一台 Eureka 控制台，就可以看见 Eureka-HA 集群主机，同时注册的微服务会在 3 台主机上同时存在，如图 8-10 所示。

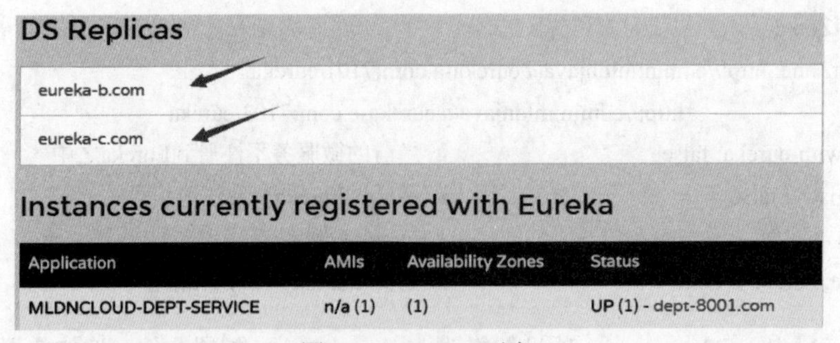

图 8-10　Eureka 副本

提示：后续讲解依然以单节点 Eureka 为主。

Eureka-HA 机制需要多台主机共同支持才可以实现微服务注册，这一机制只在实际的项目环境中存在。因此本书后续讲解时将继续以单节点 Eureka 为主，使用 mldncloud-eureka-7001 项目。

8.8　Eureka 服务发布

Eureka 作为独立的微服务存在，也需要进行项目打包与部署。在实际项目环境中，由于 Eureka 需要 HA 机制的支撑，所以本节将利用 profile 实现多个环境的配置。

1.【mldncloud-eureka-profile 项目】修改 application.yml 配置文件，设置多个 profile 配置。

```
spring:
  profiles:
    active:
    - product-7101                                  # 默认profile
---
spring:
  profiles: product-7101                            # 定义profile名称
server:
  port: 7101                                        # 定义运行端口
security:
  basic:
    enabled: true                                   # 启用安全认证处理
  user:
    name: edmin                                     # 用户名
    password: mldnjava                              # 密码
eureka:
  client:                                           # 客户端进行Eureka注册的配置
    service-url:
      defaultZone:
          http://edmin:mldnjava@eureka-b.com:7102/eureka,
          http://edmin:mldnjava@eureka-c.com:7103/eureka
    register-with-eureka: false                     # 当前的微服务不注册到Eureka之中
    fetch-registry: false                           # 不通过Eureka获取注册信息
  instance:                                         # Eureak实例定义
    hostname: eureka-a.com                          # 定义Eureka实例所在的主机名称
---
spring:
  profiles: product-7102                            # 定义profile名称
server:
  port: 7102                                        # 定义运行端口
security:
  basic:
    enabled: true                                   # 启用安全认证处理
```

```
  user:
    name: edmin                              # 用户名
    password: mldnjava                       # 密码
eureka:
  client:                                    # 客户端进行Eureka注册的配置
    service-url:
      defaultZone:
          http://edmin:mldnjava@eureka-a.com:7101/eureka,
          http://edmin:mldnjava@eureka-c.com:7103/eureka
    register-with-eureka: false              # 当前的微服务不注册到Eureka之中
    fetch-registry: false                    # 不通过Eureka获取注册信息
  instance:                                  # Eureak实例定义
    hostname: eureka-b.com                   # 定义Eureka实例所在的主机名称
---
spring:
  profiles: product-7103                     # 定义profile名称
server:
  port: 7103                                 # 定义运行端口
security:
  basic:
    enabled: true                            # 启用安全认证处理
  user:
    name: edmin                              # 用户名
    password: mldnjava                       # 密码
eureka:
  client:                                    # 客户端进行Eureka注册的配置
    service-url:
      defaultZone:
          http://edmin:mldnjava@eureka-a.com:7101/eureka,
          http://edmin:mldnjava@eureka-b.com:7102/eureka
    register-with-eureka: false              # 当前的微服务不注册到Eureka之中
    fetch-registry: false                    # 不通过Eureka获取注册信息
  instance:                                  # Eureak实例定义
    hostname: eureka-c.com                   # 定义 Eureka 实例所在的主机名称
```

2.【mldncloud-eureka-profile 项目】修改 pom.xml 配置文件，追加打包插件。

```
<build>
    <finalName>eureka-server</finalName>
    <plugins>
        <plugin>                    <!-- 该插件的主要功能是进行项目的打包发布处理 -->
```

```
				<groupId>org.springframework.boot</groupId>
				<artifactId>spring-boot-maven-plugin</artifactId>
				<configuration>				<!-- 设置程序执行的主类 -->
					<mainClass>cn.mldn.mldncloud.EurekaServerStartApplication</mainClass>
				</configuration>
				<executions>
					<execution>
						<goals>
							<goal>repackage</goal>
						</goals>
					</execution>
				</executions>
			</plugin>
		</plugins>
	</build>
```

3.【mldncloud-eureka-profile 项目】通过 Maven 打包 clean package，如图 8-11 所示。随后就可以在项目的目录中发现生成的 eureka-server.jar 文件。

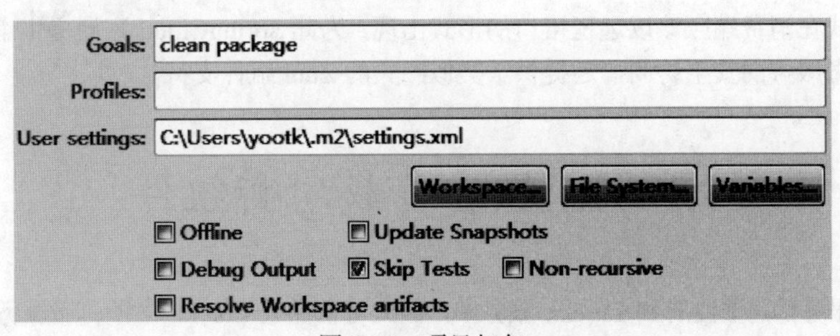

图 8-11　项目打包

4.【操作系统】使用默认配置启动 Eureka 服务 java -jar eureka-server.jar。

5.【操作系统】使用其他 profile 启动 Eureka。

☑　运行 product-7102：java -jar eureka-server.jar --spring.profiles.active=product-7102。

☑　运行 product-7103：java -jar eureka-server.jar --spring.profiles.active=product-7103。

8.9　本章小结

1. Eureka 提供微服务的注册服务，所有的微服务都需要在 Eureka 中注册并进行服务发布。
2. Eureka 提供发现管理，利用 DiscoveryClient 类可以实现发现信息。
3. Eureka 提供默认服务注册，开发者也可以根据实际情况配置心跳时间、清除时间等。
4. Eureka 使用 SpringSecurity 实现安全机制，以保证注册中心的安全。
5. Eureka 支持 HA 机制，以避免单节点导致的服务瘫痪问题。

第 9 章 SpringCloud 服务组件

通过本章学习，可以达到以下目标：

1. 掌握客户端负载均衡组件 Ribbon 的使用。
2. 掌握 Restful 接口转换 Feign 组件的使用。
3. 理解 Hystrix 熔断机制的作用与配置。
4. 掌握 Zuul 组件的使用与代理转换处理。
5. 掌握 Zuul 与上传微服务的使用。

SpringCloud 微架构开发中存在着众多的微服务，这些微服务之间也会存在互相的调用关联，为了防止某一个微服务不可用时关联微服务出现问题，需要引入 Hystrix 熔断处理机制。同时，微服务的调用形式在消费端应该以远程接口的形式出现，为此 SpringCloud 家族提供了 Feign 转换技术。为了保证微服务的安全访问，还提供了类似网关的 Zuul 组件支持。

9.1 Ribbon 负载均衡组件

所有的微服务都需要注册到 Eureka 服务中，因此可以通过 Eureka 对所有微服务进行管理。消费端应该通过 Eureka 来进行微服务接口调用，这种调用可以利用 Ribbon 技术来实现。

9.1.1 Ribbon 基本使用

Ribbon 是一个与 Eureka 结合的组件，其主要作用是进行 Eureka 中的服务调用。要使用 Ribbon，需要在项目中配置 spring-cloud-starter-ribbon 依赖库。同时对于所有注册到 Eureka 中的微服务也要求有微服务的名称，如图 9-1 所示，在消费端将通过微服务的名称进行微服务调用。

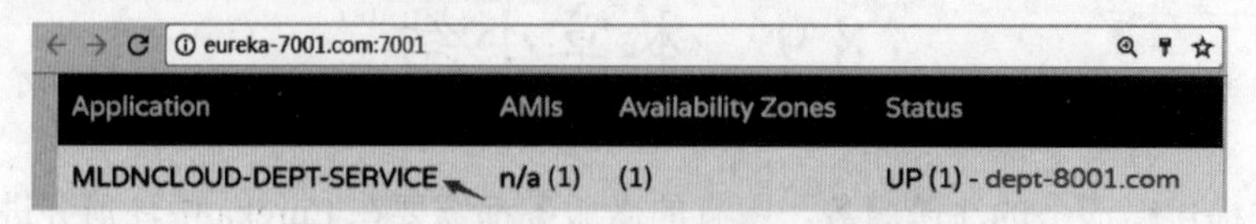

图 9-1 Eureka 中注册的微服务名称

1.【mldncloud-consumer-ribbon 项目】修改 pom.xml 配置文件，引入 Ribbon 依赖库。

```
<dependency>
    <groupId>org.springframework.cloud</groupId>
```

```
        <artifactId>spring-cloud-starter-ribbon</artifactId>
    </dependency>
    <dependency>
        <groupId>org.springframework.cloud</groupId>
        <artifactId>spring-cloud-starter-eureka</artifactId>
    </dependency>
```

2.【mldncloud-consumer-ribbon 项目】修改 RestfulConfig 配置类，追加 Ribbon 注解。

```
    @Bean
    @LoadBalanced                                    // Ribbon提供的负载均衡注解
    public RestTemplate getRestTemplate() {
        return new RestTemplate() ;                  // 实例化RestTemplate对象
    }
```

3.【mldncloud-consumer-ribbon 项目】修改 application.yml 配置文件，追加 Eureka 访问地址。

```
eureka:
  client:                              # 客户端进行Eureka注册的配置
    service-url:
      defaultZone: http://edmin:mldnjava@eureka-7001.com:7001/eureka
    register-with-eureka: false        # 当前的微服务不注册到 Eureka 之中
```

4.【mldncloud-consumer-ribbon 项目】在 DeptController 控制器类中通过微服务名称调用微服务，此时不再需要知道微服务的具体主机信息。

```
@RestController                                          // 为方便起见使用Restful风格展示
public class DeptController {
    public static final String DEPT_ADD_URL = "http://MLDNCLOUD-DEPT-SERVICE/dept/add";
    public static final String DEPT_GET_URL = "http://MLDNCLOUD-DEPT-SERVICE/dept/get";
    public static final String DEPT_LIST_URL = "http://MLDNCLOUD-DEPT-SERVICE/dept/list";
    @Autowired
    private RestTemplate restTemplate;                   // 注入RestTemplate对象
    @Autowired
    private HttpHeaders headers;                         // 注入HTTP头信息对象
    @SuppressWarnings("unchecked")
    @GetMapping("/consumer/dept/list")
    public Object listDeptRest() {
        List<DeptDTO> allDepts = this.restTemplate
                .exchange(DEPT_LIST_URL, HttpMethod.GET,
                        new HttpEntity<Object>(this.headers), List.class)
                .getBody();                              // 访问服务设置头信息
        return allDepts;
```

```
    }
    @GetMapping("/consumer/dept/get")
    public Object getDeptRest(long deptno) {
        DeptDTO dept = this.restTemplate
                    .exchange(DEPT_GET_URL + "/" + deptno, HttpMethod.GET,
                            new HttpEntity<Object>(this.headers), DeptDTO.class)
                    .getBody();                             // 访问服务设置头信息
        return dept;
    }
    @GetMapping("/consumer/dept/add")
    public Object addDeptRest(DeptDTO dept) { // 传输DeptDTO对象
        DeptDTO result = this.restTemplate.exchange(DEPT_ADD_URL, HttpMethod.POST,
                    new HttpEntity<Object>(dept, this.headers), DeptDTO.class)
                    .getBody();                             // 访问服务设置头信息
        return result;
    }
}
```

5. 【mldncloud-consumer-ribbon 项目】修改程序启动类，追加 Eureka 客户端注解。

```
@SpringBootApplication
@EnableEurekaClient
public class StartWebConsumerApplication80 {
    public static void main(String[] args) {
        SpringApplication.run(StartWebConsumerApplication80.class, args);
    }
}
```

此时，消费端就可以实现 Eureka 中注册微服务的调用，并且在消费端通过名称实现微服务调用。以消费端调用部门微服务的部门列表（dept/list）操作为例，如图 9-2 所示。

图 9-2　消费端调用部门列表微服务

9.1.2　Ribbon 负载均衡

微服务搭建的业务中心可以通过多台业务功能相同的微服务构建微服务集群，所有的微服务为了可以动态维护，都需要将其注册到 Eureka 之中，这样消费端就可以利用 Ribbon 与 Eureka 的服务主机列表实现微服务轮询调用，以实现负载均衡，如图 9-3 所示。需要注意的是，Ribbon 提供的是一种客户端的负载均衡配置。

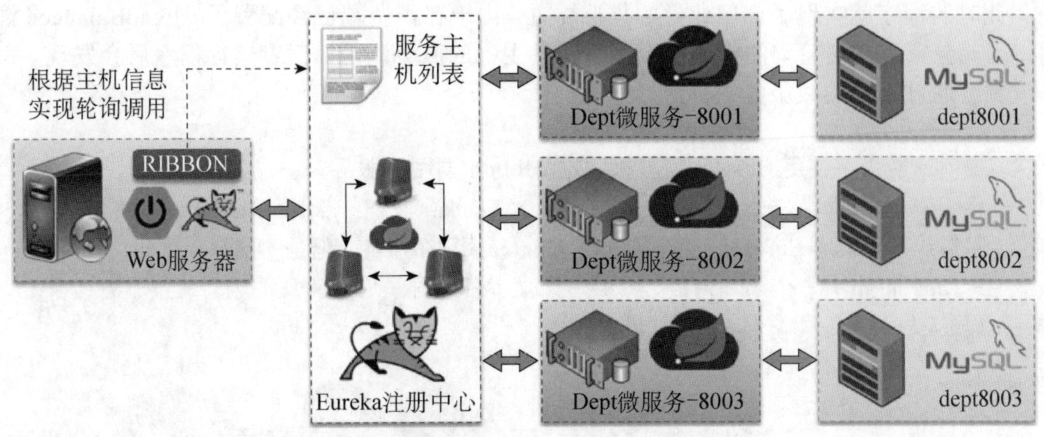

图 9-3　Ribbon 负载均衡

提示：关于项目搭建。

为了使读者观察方便，本节的程序将进行模拟环境下的集群配置，即将部门微服务项目复制两份，这样就可以有 3 个不同的微服务实现相同的业务功能，然后 3 个微服务应该拥有各自不同的数据库（数据库脚本在各个项目中已经提供），但是数据库的信息应该相同。本节使用的项目基本信息如表 9-1 所示。

表 9-1　部门业务负载均衡

No.	项目名称	运行端口	数据库名称	hosts 主机名称
1	mldncloud-dept-service-8001	8001	dept8001	dept-8001.com
2	mldncloud-dept-service-8002	8002	dept8002	dept-8002.com
3	mldncloud-dept-service-8003	8003	dept8003	dept-8003.com

项目创建后，修改各自的 application.yml 配置文件，修改运行端口号，并且在 hosts 配置文件中追加新的主机名称。另外需要提醒读者的是，本程序暂不去考虑数据库之间的数据同步与负载均衡问题。

1.【mldncloud-dept-service-800*项目】要实现负载均衡，首先要保证注册到 Eureka 中的所有微服务的名称相同。修改 application.yml 配置文件，实现服务名称定义。

```
spring:
  application:
    name: mldncloud-dept-service                    # 定义微服务名称，3 个微服务名称必须相同
```

2.【mldncloud-dept-service-800*】启动所有的微服务，并且同时向 Eureka 中进行注册。通过如图 9-4 所示的结果可以发现，针对这些相同名称的微服务，会有 3 台主机提供服务支持。

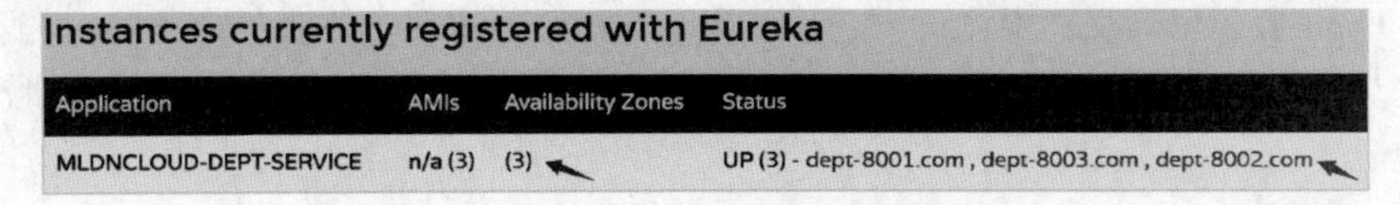

图 9-4　Eureka 中同一微服务注册多台主机

此时的程序就实现了部门业务的集群配置。由于在消费端已经配置了@LoadBalanced 注解，因此会采用自动轮询的模式实现不同业务主机的服务调用。读者运行程序后会发现，每一次都会通过不同的微服务主机执行业务。

提问：能否不使用 Eureka 而直接通过 Ribbon 调用微服务？

所有的微服务都在 Eureka 中注册，如果不通过 Eureka，能否直接使用 Ribbon 进行微服务调用？

回答：可以禁用 Eureka，而直接利用 Ribbon 调用微服务，但是不推荐。

在 Ribbon 中有一个服务器信息列表，开发者可以利用它配置所要访问的微服务列表，以实现微服务的调用，如图 9-5 所示。

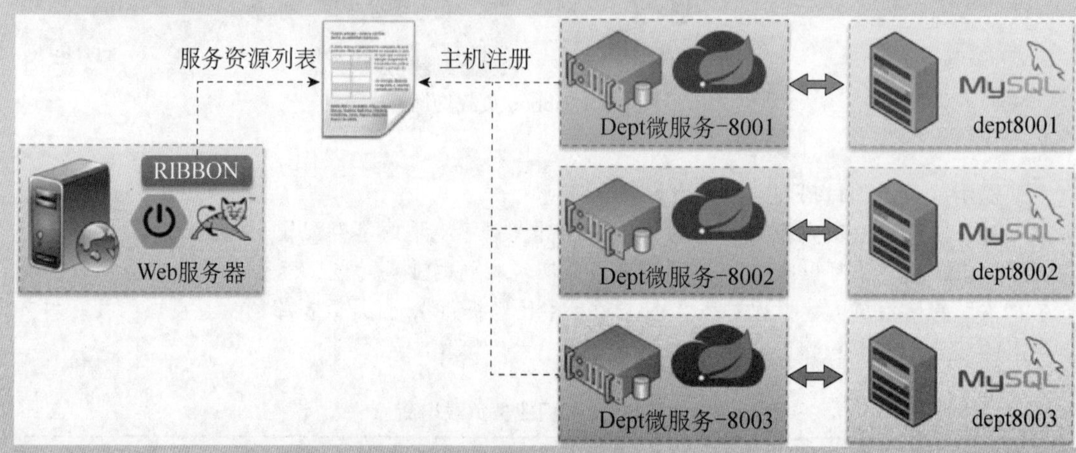

图 9-5　Ribbon 直接访问微服务

然后需要在消费端的 application.yml 配置文件中进行如下配置。

范例：直接使用 Ribbon 访问微服务。

```
ribbon:
  eureka:
    enabled: false                        # 禁用Eureka配置
mldncloud-dept-service:                   # 定义微服务名称
  ribbon:                                 # 定义主机列表
    listOfServers:
          http://dept-8001.com:8001,
          http://dept-8002.com:8002,
          http://dept-8003.com:8003
```

此时直接在消费端配置了所有的微服务信息列表，而调用的形式也必须通过 Ribbon 特定的 LoadBalancerClient 类才可以完成访问。这里面最麻烦的问题在于：使用过程中如果有某台微服务主机出现宕机现象，Ribbon 会自动将其剔除，但是在其恢复之后，开发者需要手动将其添加到 Ribbon 服务器列表之中才可以继续使用。从这一点来讲，并不如 Eureka 智能。在进行微服务的开发中，强烈建议使用 Eureka 来负责所有微服务的注册，这样可以实现服务端列表的动态更新。

9.1.3　Ribbon 负载均衡策略

默认情况下，Ribbon 中采用服务列表的顺序模式实现负载均衡处理，开发者也可以根据自身的情况实现自定义的负载均衡配置。Ribbon 中，有如下 3 种核心配置策略（接口和类都在 com.netflix.loadbalancer 包中）。

- ☑ **.IRule**：Ribbon 的负载均衡策略（所有的负载均衡策略均继承自 IRule 接口，常用子类如表 9-2 所示），默认采用 ZoneAvoidanceRule 实现，该策略能够在多区域环境下选出最佳区域的实例进行访问。
- ☑ **IPing**：Ribbon 的实例检查策略，默认采用 NoOpPing 子类实现。该检查策略是一个特殊的实现，实际上它并不会检查实例是否可用，而是始终返回 true，默认认为所有服务实例都是可用的。如果用户有需要，也可以更换为 PingUrl 子类。
- ☑ **ILoadBalancer**：负载均衡器，默认采用 ZoneAwareLoadBalancer 实现，具备区域感知的能力。

表 9-2　IRule 接口常用子类

No.	子类	描述
1	com.netflix.loadbalancer.RoundRobinRule	轮询策略，利用余数判断下一次要调用的主机信息
2	com.netflix.loadbalancer.RandomRule	随机策略，会随机抽取一台主机执行业务调用
3	com.netflix.loadbalancer.BestAvailableRule	对所有服务实例进行迭代，而后调用连接数最少的主机
4	com.netflix.loadbalancer.RetryRule	基于重试的轮询策略
5	com.netflix.loadbalancer.WeightedResponseTimeRule	实例初始化的时候会开启一个定时任务，通过定时任务来获取服务响应时间定期维护每个服务的权重
6	com.netflix.loadbalancer.AvailabilityFilteringRule	根据可用性进行筛选
7	com.netflix.loadbalancer.ZoneAvoidanceRule	根据区域和可用性进行筛选、默认策略

由于 Ribbon 是工作在消费端的程序，所以本例中进行负载均衡策略配置时，只需要在消费端进行处理。注意，不要将配置类放在 SpringBoot 程序启动时可以扫描到的子包中。

1.【mldncloud-consumer-ribbon 项目】追加一个 LoadBalance 的配置类，此类要放在 Spring 启动时无法扫描到的包中。

```
package cn.mldn.commons.config;
import org.springframework.context.annotation.Bean;
import com.netflix.loadbalancer.IPing;
import com.netflix.loadbalancer.IRule;
public class RibbonLoadBalanceConfig {
    @Bean
    public IRule ribbonRule() {                                    // 其中IRule就是所有规则的标准
```

```
        return new com.netflix.loadbalancer.RandomRule();        // 随机的访问策略
    }
    @Bean
    public IPing ribbonPing() {                                  // 定义Ping策略
        return new com.netflix.loadbalancer.PingUrl() ;
    }
}
```

2.【mldncloud-consumer-ribbon 项目】在程序启动类中使用@RibbonClient 注解引入配置。

```
package cn.mldn.mldncloud.consumer;
import org.springframework.boot.SpringApplication;
import org.springframework.boot.autoconfigure.SpringBootApplication;
import org.springframework.cloud.netflix.eureka.EnableEurekaClient;
import org.springframework.cloud.netflix.ribbon.RibbonClient;
import cn.mldn.commons.config.RibbonLoadBalanceConfig;
@SpringBootApplication
@EnableEurekaClient
@RibbonClient(name="ribbonClient",configuration=RibbonLoadBalanceConfig.class)
public class StartWebConsumerApplication80 {
    public static void main(String[] args) {
        SpringApplication.run(StartWebConsumerApplication80.class, args);
    }
}
```

这里采用自定义的负载均衡策略与主机检测策略实现了微服务调用。

3.【mldncloud-consumer-ribbon 项目】在通过 Ribbon 调用微服务过程中，还可以利用 LoadBalancerClient 获取要调用的微服务端信息。修改 DeptController 控制器程序类，追加资源注入。

```
    @Autowired
    private LoadBalancerClient loadBalancerClient ;     // 客户端信息
    @GetMapping("/consumer/client")
    public Object client() {
        // 获取指定名称的微服务实例对象
        ServiceInstance serviceInstance =
            this.loadBalancerClient.choose("MICROCLOUD-DEPT-SERVICE") ;
        Map<String,Object> info = new HashMap<String,Object>() ;
        info.put("host", serviceInstance.getHost()) ;
        info.put("port", serviceInstance.getPort()) ;
        info.put("serviceId", serviceInstance.getServiceId()) ;
        return info ;
    }
```

Restful信息（consumer.com/consumer/client）	{ "port":8002, "host":"yootk-pc", "serviceId":"MLDNCLOUD-DEPT-SERVICE"}

本程序使用了 LoadBalancerClient 类进行客户端信息的注入，并且利用此对象根据微服务的名称 MICROCLOUD-DEPT-SERVICE 来获取客户端要调用的微服务的基础信息。

9.2　Feign 远程接口映射

SpringCloud 是以 Restful 为基础实现的开发框架，在整体调用过程中，即使引入了 Eureka，也需要消费端使用完整的路径才可以正常访问远程接口，同时还需要开发者手动利用 RestTemplate 进行调用与返回结果的转换。为了解决这种复杂的调用逻辑，在 SpringCloud 中提供 Feign 技术（依赖于 Ribbon 技术支持），利用此技术可以将远程的 Restful 服务映射为远程接口，消费端可通过远程接口实现远程方法调用，如图 9-6 所示。

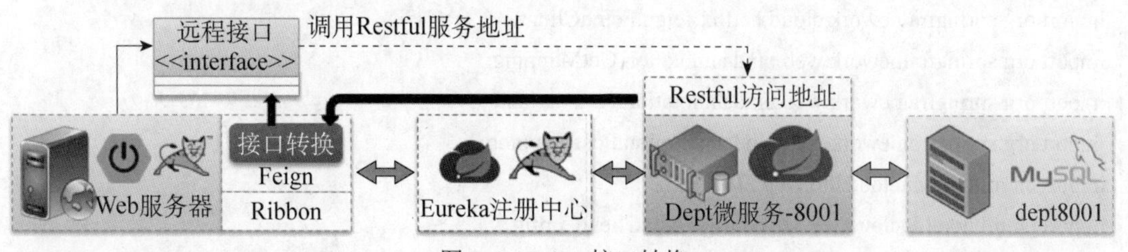

图 9-6　Feign 接口转换

9.2.1　Feign 接口转换

Feign 接口转换技术是针对 Restful 访问地址的封装，即同一组 Restful 访问地址应该变为一个远程接口中的业务方法，提供给消费端使用。

> **提示：在 mldncloud-api 项目中进行远程接口描述。**
>
> 在 SpringCloud 项目中为了明确地描述接口定义，专门提供了 mldncloud-api 项目模块，不仅 Restful 服务提供者可以使用此模块，消费端也可以使用此模块，所以本次的讲解直接在 api 模块中进行功能扩充。而在实际开发中，开发者可以根据业务功能再定义其他的项目模块。

1.【mldncloud-api 项目】修改 pom.xml 配置文件，引入 Feign 依赖库（会自动引入 Ribbon 依赖）。

```
<dependency>
    <groupId>org.springframework.cloud</groupId>
    <artifactId>spring-cloud-starter-feign</artifactId>
</dependency>
```

2.【mldncloud-api 项目】由于此时服务端需要通过认证访问，因此需要建立一个 Feign 的配置类，配置认证请求。

```
package cn.mldn.mldncloud.service.config;
import org.springframework.context.annotation.Bean;
import org.springframework.context.annotation.Configuration;
import feign.auth.BasicAuthRequestInterceptor;
@Configuration
public class FeignClientConfig {
    @Bean
    public BasicAuthRequestInterceptor getBasicAuthRequestInterceptor() {
        return new BasicAuthRequestInterceptor("mldnjava", "hello");
    }
}
```

3.【mldncloud-api 项目】修改 IDeptService 接口定义，追加 Feign 处理相关注解。

```
package cn.mldn.mldncloud.service;
import java.util.List;
import org.springframework.cloud.netflix.feign.FeignClient;
import org.springframework.web.bind.annotation.GetMapping;
import org.springframework.web.bind.annotation.PathVariable;
import org.springframework.web.bind.annotation.PostMapping;
import cn.mldn.mldncloud.dto.DeptDTO;
import cn.mldn.mldncloud.service.config.FeignClientConfig;
@FeignClient(value="MLDNCLOUD-DEPT-SERVICE",configuration=FeignClientConfig.class)
public interface IDeptService {
    @PostMapping("/dept/add")
    public DeptDTO add(DeptDTO dto) ;                                   // 增加新部门
    @GetMapping("/dept/get/{deptno}")
    public DeptDTO get(@PathVariable("deptno") long deptno) ;          // 根据部门编号获取部门信息
    @GetMapping("/dept/list")
    public List<DeptDTO> list() ;                                       // 部门信息列表
}
```

4.【mldncloud-consumer-feign 项目】此时消费端不再需要通过 RestTemplate 来进行 Restful 服务访问。直接在控制器中注入 IDeptService，即可实现微服务调用。

```
package cn.mldn.mldncloud.consumer.controller;
import org.springframework.beans.factory.annotation.Autowired;
import org.springframework.web.bind.annotation.GetMapping;
import org.springframework.web.bind.annotation.RestController;
import cn.mldn.mldncloud.dto.DeptDTO;
import cn.mldn.mldncloud.service.IDeptService;
@RestController                                   // 为方便起见使用Restful风格展示
```

```
public class DeptController {
    @Autowired
    private IDeptService deptService ;                          // 注入远程业务接口对象
    @GetMapping("/consumer/dept/list")
    public Object listDeptRest() {
        return this.deptService.list() ;                        // 调用Restful业务方法
    }
    @GetMapping("/consumer/dept/get")
    public Object getDeptRest(long deptno) {
        return this.deptService.get(deptno);                    // 调用Restful业务方法
    }
    @GetMapping("/consumer/dept/add")
    public Object addDeptRest(DeptDTO dept) {                   // 传输DeptDTO对象
        return this.deptService.add(dept);                      // 调用Restful业务方法
    }
}
```

5.【mldncloud-consumer-feign 项目】修改启动类，配置 Feign 转换接口扫描包。

```
package cn.mldn.mldncloud.consumer;
import org.springframework.boot.SpringApplication;
import org.springframework.boot.autoconfigure.SpringBootApplication;
import org.springframework.cloud.netflix.eureka.EnableEurekaClient;
import org.springframework.cloud.netflix.feign.EnableFeignClients;
@SpringBootApplication
@EnableEurekaClient
@EnableFeignClients(basePackages={"cn.mldn.mldncloud.service"})        // 定义Feign接口扫描包
public class StartWebConsumerApplication80 {
    public static void main(String[] args) {
        SpringApplication.run(StartWebConsumerApplication80.class, args);
    }
}
```

在消费端进行远程业务调用时，所有的访问地址都与 IDeptService 接口中的方法对应，这样消费端调用远程操作中就感觉像在本地调用一样。

注意：远程 POST 请求时需要关闭 CSRF。

在本程序中对于部门增加业务使用了 POST 请求模式处理，这样就会出现跨域访问问题（consumer.com 访问 dept-8001.com）。为了实现调用，需要修改 SpringSecurity 配置类，关闭 CSRF（Cross-Site Request Forgery，跨站请求伪造）校验检测。

范例：【mldncloud-security 项目】修改 SpringSecurity 配置类。

```
@Override
protected void configure(HttpSecurity http) throws Exception {
    // 表示所有的访问都必须进行认证处理后才可以正常进行
    http.httpBasic().and().authorizeRequests().anyRequest()
            .fullyAuthenticated().and().csrf().disable();
    // 所有的Restful服务一定要设置为无状态，以提升操作性能
    http.sessionManagement()
            .sessionCreationPolicy(SessionCreationPolicy.STATELESS);
}
```

此时，程序就可以实现跨域的 POST 请求访问，业务接口中的方法就可以正常实现 Restful 接口映射。

9.2.2 Feign 相关配置

Feign 的核心作用是将 Restful 服务的信息转换为接口，在整体的处理过程中依然需要进行 JSON（或者 XML、文本传输）数据的传递。为了避免长时间占用网络带宽，提升数据传输效率，往往需要对数据进行压缩。

1. 【mldncloud-consumer-feign 项目】修改 application.yml，进行数据压缩配置。

```
feign:
  compression:
    request:
      mime-types:                          # 可以被压缩的类型
      - text/xml
      - application/xml
      - application/json
      min-request-size: 2048               # 超过 2048 的字节进行压缩
```

2. 【mldncloud-consumer-feign 项目】在 SpringBoot 项目启动过程中，对于 Feign 接口与远程 Restful 地址的映射也可以通过日志信息进行详细显示，修改 application.yml 进行日志级别变更。

```
logging:
  level:
    cn.mldn.mldncloud.service: DEBUG      # 定义显示转换信息的开发包与日志级别
```

3. 【mldncloud-api 项目】修改 FeignClientConfig 配置类，追加日志配置。

```
@Bean
public feign.Logger.Level getFeignLoggerLevel() {
    return feign.Logger.Level.FULL ;
}
```

配置完成后重新启动消费端项目（mldncloud-consumer-feign），在第一次进行接口调用时，可以通过控制台看到如下的重要提示信息：

```
[IDeptService#get] <--- HTTP/1.1 200 (1155ms)
[IDeptService#get] cache-control: no-cache, no-store, max-age=0, must-revalidate
[IDeptService#get] content-type: application/json;charset=UTF-8
[IDeptService#get] expires: 0
[IDeptService#get] pragma: no-cache
[IDeptService#get] transfer-encoding: chunked
[IDeptService#get] x-application-context: mldncloud-dept-service:8001
[IDeptService#get] x-content-type-options: nosniff
[IDeptService#get] x-frame-options: DENY
[IDeptService#get] x-xss-protection: 1; mode=block
[IDeptService#get] {"deptno":1,"dname":"财务部-8001","loc":"北京"}
[IDeptService#get] <--- END HTTP (52-byte body)
Flipping property: MLDNCLOUD-DEPT-SERVICE.ribbon.ActiveConnectionsLimit to use NEXT property:
niws.loadbalancer.availabilityFilteringRule.activeConnectionsLimit = 2147483647
```

通过上述提示信息可以发现，Feign 在进行接口转换时集成了 Ribbon 负载均衡机制，微服务消费端和提供端之间的信息采用 JSON 结构进行传递，并且可以自动将相应的返回数据变为目标类型。

9.3　Hystrix 熔断机制

在实际项目中，由于业务功能的不断扩充，会出现大量的微服务互相调用的情况。如图 9-7 所示，微服务 1 要想完成功能，需要调用微服务 2、微服务 3、微服务 4，一旦这个时候微服务 4 出现问题（其他微服务没有问题），则微服务 1、2、3 就有可能出现错误。这样的问题在微服务开发中称为雪崩效应。

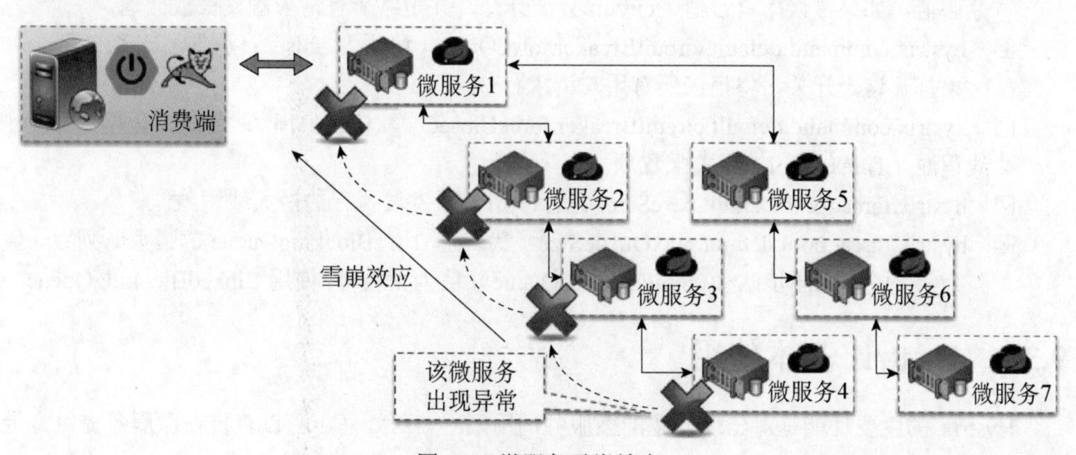

图 9-7　微服务雪崩效应

为了防止这种雪崩效应的出现，在 SpringCloud 中引入了 Hystrix 熔断机制。在大部分开发状态下，开发者可以直接使用 Hystrix 的默认配置。如果有需要，开发者也可以使用如下几类常用配置项。

1. 微服务执行相关配置项。

☑ hystrix.command.default.execution.isolation.strategy（默认为 thread）：隔离策略，可选用 thread 或 semaphore。

☑ hystrix.command.default.execution.isolation.thread.timeoutInMilliseconds（默认为 1000ms）：命令执行超时时间。

☑ hystrix.command.default.execution.timeout.enabled（默认为 true）：执行是否启用超时配置。

☑ hystrix.command.default.execution.isolation.thread.interruptOnTimeout（默认为 true）：发生超时时是否中断。

☑ hystrix.command.default.execution.isolation.semaphore.maxConcurrentRequests（默认为 10）：最大并发请求数，该参数在使用 ExecutionIsolationStrategy.SEMAPHORE 策略时才有效。如果达到最大并发请求数，请求会被拒绝。理论上选择 semaphore size 和选择 thread size 一致，但选用 semaphore 时每次执行的单元要比较小且执行速度较快（ms 量级），否则应该选用 thread。semaphore 一般占整个容器（Tomcat 或 Jetty）线程池的一小部分。

2. 失败回退（fallback）相关配置项。

☑ hystrix.command.default.fallback.isolation.semaphore.maxConcurrentRequests（默认为 10）：如果并发数达到该设置值，请求会被拒绝，抛出异常，并且失败回退，不会被调用。

☑ hystrix.command.default.fallback.enabled（默认为 true）：当执行失败或者请求被拒绝时，是否会调用 fallback 方法。

3. 熔断处理相关的配置项。

☑ hystrix.command.default.circuitBreaker.enabled（默认为 true）：跟踪 circuit 的健康性，如果出现问题，则请求熔断。

☑ hystrix.command.default.circuitBreaker.sleepWindowInMilliseconds（默认为 5000）：触发熔断时间。

☑ hystrix.command.default.circuitBreaker.errorThresholdPercentage（默认为 50）：错误比率阀值，如果错误率≥该值，circuit 会被打开，并短路所有请求触发失败回退。

☑ hystrix.command.default.circuitBreaker.forceOpen（默认为 false）：强制打开熔断器。如果打开这个开关，将拒绝所有用户请求。

☑ hystrix.command.default.circuitBreaker.forceClosed（默认为 false）：强制关闭熔断器。

4. 线程池（ThreadPool）相关配置项。

☑ hystrix.threadpool.default.coreSize（默认为 10）：并发执行的最大线程数。

☑ hystrix.threadpool.default.maxQueueSize（默认为-1）：BlockingQueue 的最大队列数。值为-1 时，使用同步队列（SynchronousQueue）；值为正数时，使用 LinkedBlcokingQueue。

9.3.1 Hystrix 基本使用

Hystrix 的主要功能是对出现问题的微服务调用采用熔断处理，可以直接在微服务提供方上进行配置。

1.【mldncloud-dept-service-8001 项目】修改 pom.xml 配置文件，追加 Hystrix 依赖配置。

```
<dependency>
        <groupId>org.springframework.cloud</groupId>
        <artifactId>spring-cloud-starter-hystrix</artifactId>
</dependency>
```

2.【mldncloud-dept-service-8001 项目】修改程序启动主类，增加熔断注解配置。

```
package cn.mldn.mldncloud;
import org.springframework.boot.SpringApplication;
import org.springframework.boot.autoconfigure.SpringBootApplication;
import org.springframework.cloud.client.circuitbreaker.EnableCircuitBreaker;
import org.springframework.cloud.client.discovery.EnableDiscoveryClient;
import org.springframework.cloud.netflix.eureka.EnableEurekaClient;
import org.springframework.data.jpa.repository.config.EnableJpaRepositories;
@SpringBootApplication
@EnableEurekaClient                                  // 启用Eureka客户端
@EnableDiscoveryClient
@EnableCircuitBreaker                                // 启用熔断机制
@EnableJpaRepositories(basePackages="cn.mldn.mldncloud.dao")
public class StartDeptServiceApplication8001 {
        public static void main(String[] args) {
                SpringApplication.run(StartDeptServiceApplication8001.class, args);
        }
}
```

此时的程序配置了熔断机制，这样即使有更多层级的微服务调用，也不会因为某一个微服务出现问题而导致所有的微服务均不可用。

9.3.2　失败回退

失败回退（fallback）也被称为服务降级，指的是当某个服务不可用时默认执行的处理操作。Hystrix 中的失败回退是在客户端实现的一种处理机制。

1.【mldncloud-api 项目】如果要定义失败回退处理，建议通过 FallbackFactory 接口来进行实现。

```
package cn.mldn.mldncloud.service.fallback;          // 该程序包需要配置扫描
import java.util.ArrayList;
import java.util.List;
import org.springframework.stereotype.Component;
import cn.mldn.mldncloud.dto.DeptDTO;
import cn.mldn.mldncloud.service.IDeptService;
import feign.hystrix.FallbackFactory;
```

```
@Component
public class DeptServiceFallbackFactory implements FallbackFactory<IDeptService> {
    @Override
    public IDeptService create(Throwable exp) {
        return new IDeptService() {
            @Override
            public DeptDTO add(DeptDTO dto) {
                DeptDTO returnDto = new DeptDTO();
                returnDto.setDeptno(-1L);
                returnDto.setDname("部门名称 - Fallback");
                returnDto.setLoc("部门位置 - Fallback");
                return returnDto;
            }
            @Override
            public DeptDTO get(long deptno) {
                DeptDTO returnDto = new DeptDTO();
                returnDto.setDeptno(deptno);
                returnDto.setDname("部门名称 - Fallback");
                returnDto.setLoc("部门位置 - Fallback");
                return returnDto;
            }
            @Override
            public List<DeptDTO> list() {
                return new ArrayList<DeptDTO>();
            }
        };
    }
}
```

2.【mldncloud-api 项目】修改 IDeptService 接口定义，追加 fallbackFactory 处理。

```
package cn.mldn.mldncloud.service;
import java.util.List;
import org.springframework.cloud.netflix.feign.FeignClient;
import org.springframework.web.bind.annotation.GetMapping;
import org.springframework.web.bind.annotation.PathVariable;
import org.springframework.web.bind.annotation.PostMapping;
import cn.mldn.mldncloud.dto.DeptDTO;
import cn.mldn.mldncloud.service.config.FeignClientConfig;
import cn.mldn.mldncloud.service.fallback.DeptServiceFallbackFactory;
@FeignClient(value = "MLDNCLOUD-DEPT-SERVICE", configuration = FeignClientConfig.class, fallback
Factory=DeptServiceFallbackFactory.class)
```

```
public interface IDeptService {
    @PostMapping("/dept/add")
    public DeptDTO add(DeptDTO dto) ;                              // 增加新部门
    @GetMapping("/dept/get/{deptno}")
    public DeptDTO get(@PathVariable("deptno") long deptno) ;      // 根据部门编号获取部门信息
    @GetMapping("/dept/list")
    public List<DeptDTO> list() ;                                  // 部门信息列表
}
```

在进行 Feign 接口转换中，使用 fallback 设置当微服务不可用时的返回处理执行类。这样调用失败后，会返回 DeptServiceFallback 子类中所实现的方法内容。

提示：也可以单独定义为一个接口定义 Fallback 处理类。

在进行 Fallback 类定义时，用户还可以直接创建 IDeptService 的失败回退子类（DeptServiceFallback），而后在通过@FeignClient 注解中的 fallback 属性（fallback=DeptServiceFallback.class）进行配置。

这种子类配置有可能造成业务接口对象的注入混淆，所以不建议使用。

3.【mldncloud-consumer-feign 项目】修改程序启动主类，追加扫描包配置，需要将配置的 Fallback 处理类进行配置。

```
package cn.mldn.mldncloud.consumer;                    // 无法扫描到Fallback类
import org.springframework.boot.SpringApplication;
import org.springframework.boot.autoconfigure.SpringBootApplication;
import org.springframework.cloud.netflix.eureka.EnableEurekaClient;
import org.springframework.cloud.netflix.feign.EnableFeignClients;
import org.springframework.context.annotation.ComponentScan;
@SpringBootApplication
@EnableEurekaClient
@ComponentScan("cn.mldn.mldncloud.service,cn.mldn.mldncloud.consumer")
@EnableFeignClients(basePackages={"cn.mldn.mldncloud.service"})     // 定义Feign接口扫描包
public class StartWebConsumerApplication80 {
    public static void main(String[] args) {
        SpringApplication.run(StartWebConsumerApplication80.class, args);
    }
}
```

这样当微服务关闭之后，由于服务提供方不再可用，所以此时会自动调用 DeptServiceFallback 类中的相应方法进行处理，返回的都是固定的“失败”信息，如图 9-8 所示。

图 9-8　调用 fallback 处理

9.3.3 HystrixDashboard

Hystrix 提供了监控功能，这个功能就是 Hystrix Dashboard，可以利用它来进行某一个微服务的监控操作。

1.【mldncloud-hystrix-dashboard、mldncloud-dept-service-8001 项目】修改 pom.xml 配置文件，追加依赖库。

```
<dependency>
    <groupId>org.springframework.cloud</groupId>
    <artifactId>spring-cloud-starter-hystrix-dashboard</artifactId>
</dependency>
```

2.【mldncloud-dept-service-8001 项目】微服务如果需要被监控，则要引入 actuator 依赖库。

```
<dependency>
    <groupId>org.springframework.boot</groupId>
    <artifactId>spring-boot-starter-actuator</artifactId>
</dependency>
```

3.【mldncloud-dept-service-8001 项目】在需要进行监控的控制器方法上追加@HystrixCommand 注解。

```
@RestController
public class DeptRest {
    @Autowired
    private IDeptService deptService ;                        // 注入部门业务
    @PostMapping("/dept/add")
    @HystrixCommand                                           // Hystrix监控注解
    public Object add(@RequestBody DeptDTO dept) {
        return this.deptService.add(dept)    ;                // 增加部门信息
    }
    @GetMapping("/dept/get/{deptno}")
    @HystrixCommand                                           // Hystrix监控注解
    public Object get(@PathVariable("deptno") long deptno) {
        return this.deptService.get(deptno)    ;              // 查询部门信息
    }
    @GetMapping("/dept/list")
    @HystrixCommand                                           // Hystrix监控注解
    public Object list() {
        return this.deptService.list() ;                      // 部门信息列表
    }
}
```

本程序在控制器中的 add()、get()、list()3 个方法上使用了@HystrixCommand 注解，这样只有这 3 个方法的状态可以被监控到。

提示：@HystrixCommand 也可以配置失败回退处理。

对于失败回退，也可以直接在控制层进行定义，此时只需要在控制层的相应方法上使用@HystrixCommand 注解中的 fallbackMethod 属性定义。

范例：在控制层上定义失败回退。

```
@GetMapping("/dept/list")
@HystrixCommand(fallbackMethod="listFallback")        // Hystrix监控注解
public Object list() {
    return this.deptService.list() ;                  // 部门信息列表
}
public Object listFallback() {                        // 失败回退
    return new ArrayList<DeptDTO>() ;
}
```

此时，当 list()方法执行有问题时，会自动调用 listFallback()方法进行失败处理。

4. 【mldncloud-dept-service-8001 项目】修改微服务启动类，追加 Hystrix 支持。

```
@SpringBootApplication
@EnableEurekaClient                          // 启用Eureka客户端
@EnableDiscoveryClient
@EnableCircuitBreaker                        // 启用熔断机制
@EnableHystrix                               // 启用Hystrix支持
@EnableJpaRepositories(basePackages="cn.mldn.mldncloud.dao")
public class StartDeptServiceApplication8001 {
    public static void main(String[] args) {
        SpringApplication.run(StartDeptServiceApplication8001.class, args);
    }
}
```

5. 【mldncloud-hystrix-dashboard 项目】修改 application.yml 配置文件，修改运行端口。

```
server:
  port: 9001      # 运行端口
```

6. 【mldncloud-hystrix-dashboard 项目】定义程序启动主类。

```
package cn.mldn.mldncloud.consumer;
import org.springframework.boot.SpringApplication;
import org.springframework.boot.autoconfigure.SpringBootApplication;
import org.springframework.cloud.netflix.hystrix.dashboard.EnableHystrixDashboard;
```

```
@SpringBootApplication
@EnableHystrixDashboard                         // 启动HystrixDashboard
public class StartHystrixDashboardApplication {
    public static void main(String[] args) {
        SpringApplication.run(StartHystrixDashboardApplication.class, args);
    }
}
```

7. 【操作系统】修改 hosts 配置文件，追加主机信息。

```
127.0.0.1   dashboard.com
```

程序配置完成后分别启用所需要的 Eureka 微服务（mldncloud-eureka-7001）、部门微服务（mldncloud-dept-service-8001）、HystrixDashboard 微服务（mldncloud-hystrix-dashboard）、消费端微服务（mldncloud-consumer-feign）。服务启动后通过 Dashboard 访问地址 http://dashboard.com:9001/hystrix，并且输入监控地址 http://mldnjava:hello@dept-8001.com:8001/hystrix.stream，界面如图 9-9 所示。当通过消费端访问微服务之后，会针对微服务的状态进行跟踪，如图 9-10 所示。

图 9-9　HystrixDashboard 启动界面

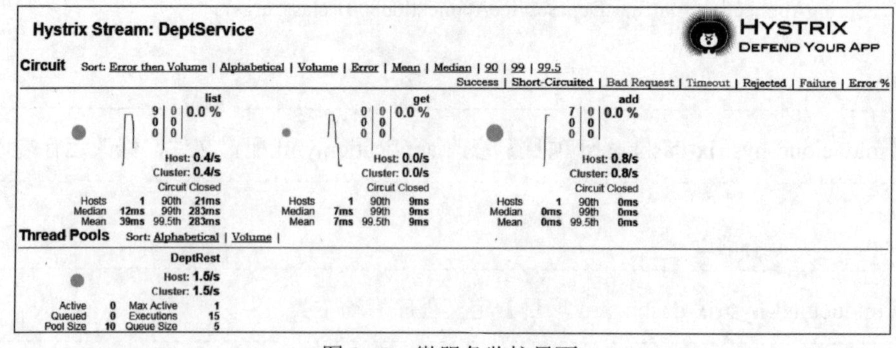

图 9-10　微服务监控界面

9.3.4　Turbine 聚合监控

HystrixDashboard 只能够针对某一个微服务进行监控，如果项目中有许多个微服务，且需要

对所有微服务统一监控的时候，就可以使用 Turbine 来实现聚合监控。

1.【mldncloud-hystrix-turbine 项目】修改 pom.xml 配置文件，引入 Turbine 依赖库。

```
<dependency>
    <groupId>org.springframework.cloud</groupId>
    <artifactId>spring-cloud-starter-turbine</artifactId>
</dependency>
```

2.【mldncloud-hystrix-turbine 项目】修改 application.yml，进行 Turbine 聚合配置。

```
server:
  port: 9101                                # 运行端口
eureka:
  client:                                   # 客户端进行Eureka注册的配置
    service-url:
      defaultZone: http://edmin:mldnjava@eureka-7001.com:7001/eureka
    register-with-eureka: false             # 当前的微服务不注册到Eureka之中
    fetch-registry: true                    # 通过Eureka获取注册信息
turbine:
  app-config: MLDNCLOUD-DEPT-SERVICE        # 监控的微服务信息，多个微服务使用“,”分割
  cluster-name-expression: new String("default")  # 监控的表达式，此表达式表示要获取监控信息名称
```

3.【mldncloud-security 项目】如果要对所有的微服务进行监控，则在定义微服务时需要配置认证信息。由于这种认证信息只能够在访问微服务的路径中进行配置，所以需要修改安全配置类，取消对监控路径的安全限制。

```
@Override
public void configure(WebSecurity web) throws Exception {                  // 覆写Web安全配置
    web.ignoring().antMatchers("/hystrix.stream","/turbine.stream") ;      // 定义忽略验证路径
}
```

application.yml等同配置	security:
	  ignored:
	  - /hystrix.stream
	  - /turbine.stream

在本配置中，忽略了 Web 安全访问（WebSecurity）下/hystrix.stream、/turbine.stream 两个路径的认证，所以对这两个路径不再进行安全认证处理。

4.【mldncloud-hystrix-turbine 项目】定义程序启动类，使用 Turbine 注解。

```
package cn.mldn.mldncloud.consumer;
import org.springframework.boot.SpringApplication;
import org.springframework.boot.autoconfigure.SpringBootApplication;
import org.springframework.cloud.netflix.hystrix.dashboard.EnableHystrixDashboard;
```

```
import org.springframework.cloud.netflix.turbine.EnableTurbine;
@SpringBootApplication
@EnableTurbine                                          // 启用Turbine
public class StartTurbineApplication {
    public static void main(String[] args) {
        SpringApplication.run(StartTurbineApplication.class, args);
    }
}
```

本程序在启动主类上使用@EnableTurbine 注解，这样就可以启动 Turebine 聚合监控了。

5.【操作系统】修改 hosts 主机配置，增加新的主机名称。

```
127.0.0.1    turbine.com
```

启动所有相关的微服务，随后通过 HystrixDashboard 启动监控程序，如图 9-11 所示，输入 Turbine 的监控路径（http://turbine.com:9101/turbine.stream）并且利用消费端访问相应的微服务信息，就可以得到如图 9-12 所示的监控结果。

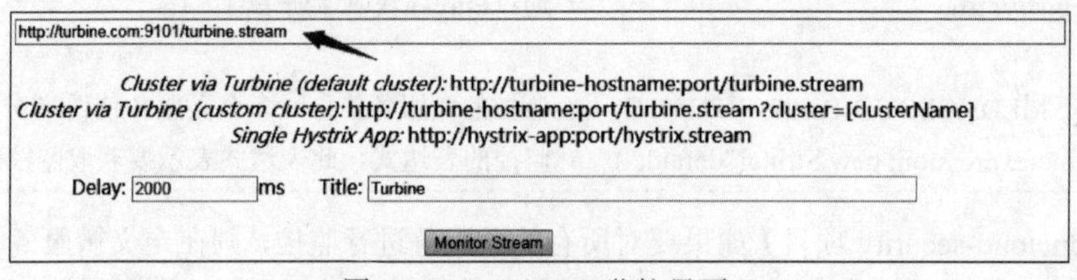

图 9-11　Dashboard 监控界面

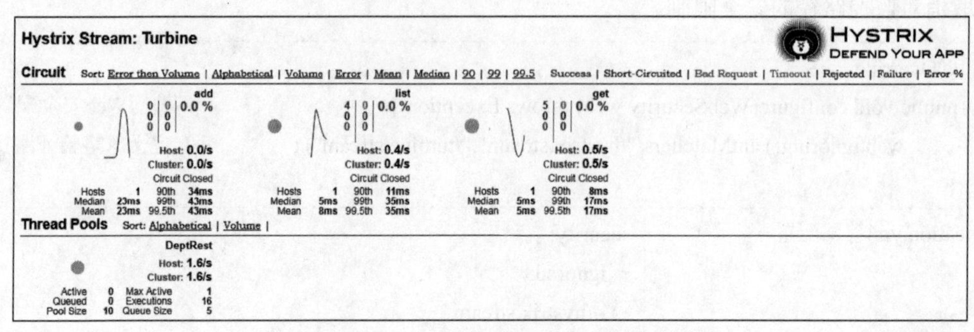

图 9-12　利用 Turebine 对微服务进行聚合监控

9.4　Zuul 路由网关

微服务创建是一个庞大的系统工程，在一个整体项目中往往会存在着若干类的微服务。例如，要开发一个企业管理程序，有可能会用到 3 类微服务集群：内部员工操作微服务集群、外部客户操作微服务集群和网站管理操作微服务集群，如图 9-13 所示。而此时就可以采用网关的概念来对相关微服务集群进行隔离，这就是 Zuul 的主要作用。

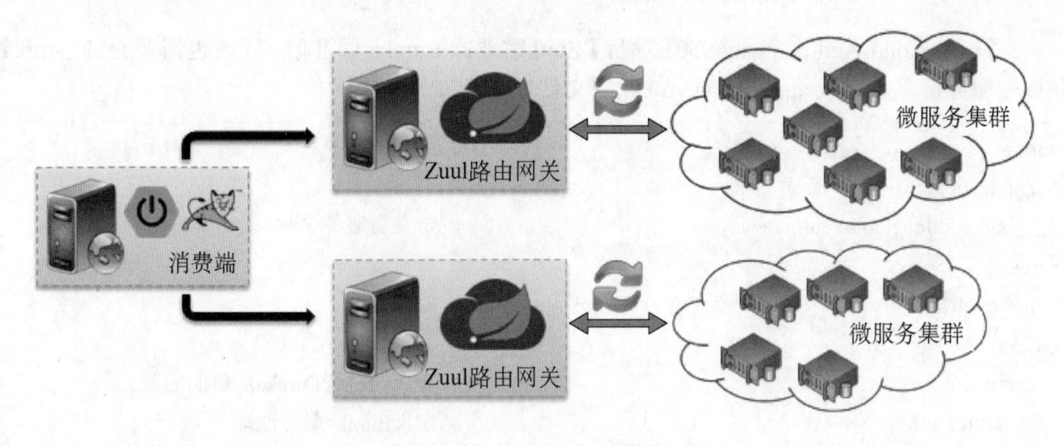

图 9-13　Zuul 路由网关

9.4.1　Zuul 整合微服务

Zuul 实现的是路由网关微服务，为了方便 Zuul 的统一管理，所有的 Zuul 微服务需要向 Eureka 注册，然后 Zuul 才可以利用 Eureka 获取所有的微服务信息，而后客户端再通过 Zuul 调用微服务，整体流程如图 9-14 所示。

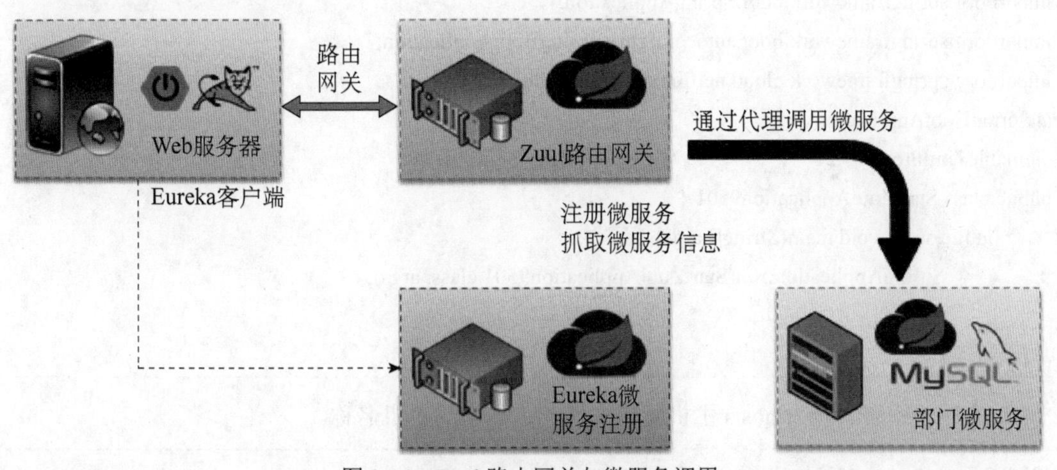

图 9-14　Zuul 路由网关与微服务调用

提示：本程序暂时取消部门微服务的认证处理。

通过 Zuul 代理访问微服务时，无法指定微服务认证信息的传递，必须通过 Zuul 的过滤访问来进行。为了帮助读者更好地理解问题，本例中部门微服务暂时取消掉 mldncloud-security 配置模块，即暂不对部门微服务进行认证保护。

1.【mldncloud-zuul-gateway-9501 项目】修改 pom.xml 配置文件，追加 Zuul 相关依赖包。

```
<dependency>
    <groupId>org.springframework.cloud</groupId>
    <artifactId>spring-cloud-starter-zuul</artifactId>
</dependency>
```

2.【mldncloud-zuul-gateway-9501 项目】Zuul 需要向 Eureka 中注册，同时也需要通过 Eureka 获取微服务信息。修改 application.yml 配置文件如下：

```
spring:
  application:
    name: mldncloud-zuul-gateway                    # 定义微服务名称
server:
  port: 9501                                        # 服务端口
eureka:
  client:                                           # 客户端进行Eureka注册的配置
    service-url:                                    # 定义Eureka服务地址
      defaultZone: http://edmin:mldnjava@eureka-7001.com:7001/eureka
  instance:
    instance-id: gateway-9501.com                   # 显示主机名称
```

3.【mldncloud-zuul-gateway-9501 项目】修改程序启动主类，追加 Zuul 注解配置。

```
package cn.mldn.mldncloud;
import org.springframework.boot.SpringApplication;
import org.springframework.boot.autoconfigure.SpringBootApplication;
import org.springframework.cloud.netflix.zuul.EnableZuulProxy;
@SpringBootApplication
@EnableZuulProxy
public class StartZuulApplication9501 {
    public static void main(String[] args) {
        SpringApplication.run(StartZuulApplication9501.class, args);
    }
}
```

4.【操作系统】修改 hosts 主机配置文件，追加新的主机名称。

```
127.0.0.1   gateway-9501.com
```

5.【mldncloud-*】启动相应微服务。通过 Eureka 控制中心可以发现，Zuul 微服务信息依然会保存到 Eureka 注册中心，如图 9-15 所示。

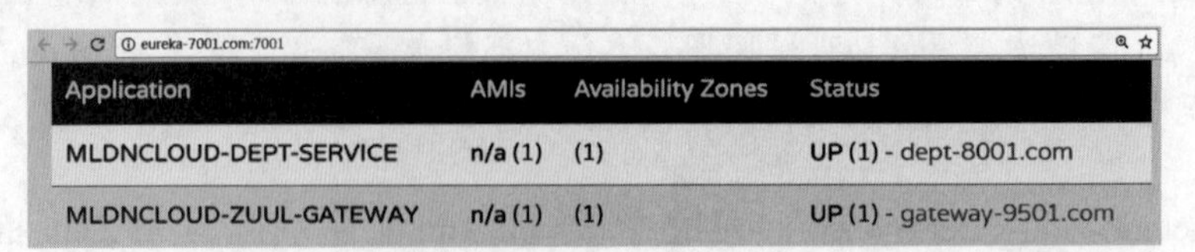

图 9-15　Eureka 注册中心

6.【mldncloud-zuul-gateway-9501 项目】微服务可以通过 Zuul 代理访问。由于此时没有进行任何配置，所以可以直接通过 Eureka 注册的微服务代理 http://gateway-9501.com:9501/mldncloud-dept-service/dept/list 进行访问，提示信息如下：

```
Mapped URL path [/mldncloud-dept-service/**] onto handler of type [class org.springframework.cloud.netflix.
zuul.web.ZuulController]
```

可以发现，当进行 Zuul 代理访问时，默认情况下采用的就是 Eureka 的注册名称。

9.4.2　Zuul 访问过滤

Zuul 本质上就属于一个网关代理操作。在实际使用中，所有的微服务都要有自己的认证信息，因此，如果用户当前所代理的微服务具有认证信息，就必须在其访问前追加认证的头部操作。这样的功能需要通过 Zuul 的过滤操作完成。

提示：关于 Zuul 网关代理认证设计。

在本例讲解过程中将首先恢复部门微服务中的认证管理，同时将通过 Zuul 进行微服务认证的配置。程序进行认证信息定义与处理的过程中采用的基本流程为：Zuul 通过过滤器配置微服务的认证信息，而后 Zuul 再通过 SpringSecurity 定义 Zuul 的认证信息，如图 9-16 所示。

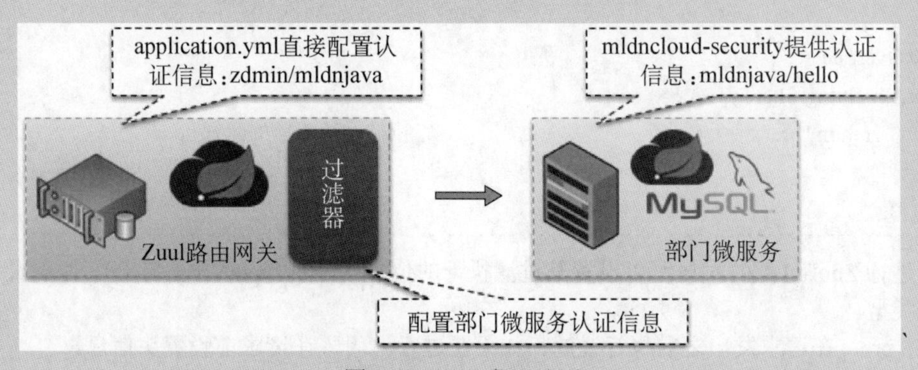

图 9-16　Zuul 代理认证

这样的设计结构，密码管理会非常混乱。如果是小型开发，则可以使用。如果是大型开发，则整体项目一旦出现问题，维护成本是相当高的，所以这种认证机制将在后续通过 OAuth 代替。

1.【mldncloud-zuul-gateway-9501 项目】建立认证请求过滤器，该过滤器必须继承 ZuulFilter 父类。

```
package cn.mldn.mldncloud.filter;
import java.nio.charset.Charset;
import java.util.Base64;
import com.netflix.zuul.ZuulFilter;
import com.netflix.zuul.context.RequestContext;
public class AuthenticationRequestZuulFilter extends ZuulFilter {  // 定义认证过滤器
    @Override
    public Object run() {
        RequestContext currentContext = RequestContext.getCurrentContext() ; // 获取请求上下文
```

```
        String auth = "mldnjava:hello";                 // 认证的原始信息
        byte[] encodedAuth = Base64.getEncoder()
                .encode(auth.getBytes(Charset.forName("US-ASCII"))); // 进行一个加密的处理
        // 在进行授权的头信息内容配置的时候，加密的信息一定要与Basic之间有一个空格
        String authHeader = "Basic " + new String(encodedAuth);
        currentContext.addZuulRequestHeader("Authorization", authHeader);
        return currentContext;
    }
    @Override
    public boolean shouldFilter() {
        return true;
    }
    @Override
    public int filterOrder() {
        return 0;                                       // 设置优先级，数字越大，优先级越低
    }
    @Override
    public String filterType() {
        return "pre";                                   // 执行过滤器前的配置
    }
}
```

在进行 Zuul 过滤的时候可以设置其过滤执行的位置（filterType()方法定义），那么此时有如下 4 种类型。

☑ pre：在请求发出之前执行过滤。如果要进行访问，在请求前设置头信息。
☑ route：在进行路由请求的时候被调用。
☑ post：在路由之后发送请求信息的时候被调用。
☑ error：出现错误之后进行调用。

2.【mldncloud-zuul-gateway-9501 项目】定义 Zuul 路由配置 Bean。

```
package cn.mldn.mldncloud.config;
import org.springframework.context.annotation.Bean;
import org.springframework.context.annotation.Configuration;
import cn.mldn.mldncloud.filter.AuthenticationRequestZuulFilter;
@Configuration
public class ZuulConfig {
    @Bean
    public AuthenticationRequestZuulFilter getAuthorizedRequestFilter() {
        return new AuthenticationRequestZuulFilter() ;          // 定义过滤器配置Bean
    }
}
```

3.【mldncloud-zuul-gateway-9501 项目】可以对微服务的访问密码进行过滤配置，此时通过

Zuul 代理的微服务将不再需要进行认证密码的配置。这样的配置是不安全的，所以下面要对 Zuul 代理进行安全配置。修改 pom.xml 配置文件，引入 spring-boot-starter-security 依赖库。

```
<dependency>
    <groupId>org.springframework.boot</groupId>
    <artifactId>spring-boot-starter-security</artifactId>
</dependency>
```

4.【mldncloud-zuul-gateway-9501 项目】修改 application.yml 配置文件，追加认证信息。

```
security:
  basic:
    enabled: true
  user:
    name: zdmin
    password: mldnjava
```

微服务访问地址为 http://zdmin:mldnjava@gateway-9501.com:9501/mldncloud-dept-service/dept/list。此时通过 Zuul 代理访问时就可以直接使用以上认证信息，而后 Zuul 会通过配置的 ZuulFilter 自动匹配微服务的认证信息。

提示：可以通过配置文件实现过滤器的启用或关闭。

本程序通过配置类实现了 ZuulFilter 配置，这个时候如果要停用 ZuulFilter，也可以修改 application.yml，进行指定过滤器的禁用。

范例：【mldncloud-zuul-gateway-9501 项目】禁用过滤器。

```
zuul:
  AuthenticationRequestZuulFilter:            # 配置要使用的过滤器
    pre:                                      # pre阶段
      disable: true                           # 停用 Filter
```

此时将禁止 AuthenticationRequestZuulFilter 的使用。

9.4.3　Zuul 路由配置

在默认情况下，Zuul 是直接通过 Eureka 中注册的微服务名称进行代理访问的（/mldncloud-dept-service/*）。这样直接通过微服务的名称进行访问是不安全的，这个时候可以通过 Zuul 为具体的访问微服务定义别名。

1.【mldncloud-zuul-gateway-9501 项目】修改 application.yml 配置文件，追加微服务名称代理映射。

```
zuul:
  routes:
    mldncloud-dept-service: /dept-proxy/**             # 指定微服务代理名称
```

本程序为 mldncloud-dept-service 微服务定义了一个映射路径/dept-proxy/**，这样就可以通过映射路径进行代理访问 http://zdmin:mldnjava@gateway-9501.com:9501/dept-proxy/dept/list。同时可以在控制台发现如下信息：

```
Mapped URL path [/dept-proxy/**] onto handler of type [class org.springframework.cloud.netflix.zuul.web.ZuulController]
Mapped URL path [/mldncloud-dept-service/**] onto handler of type [class org.springframework.cloud.netflix.zuul.web.ZuulController]
```

2.【mldncloud-zuul-gateway-9501 项目】通过提示信息可以发现，虽然此时配置了代理路径，但依然可以通过原始的微服务名称进行代理访问。为了停用这种通过微服务名称访问的操作，修改 application.yml 配置文件。

禁用全部微服务名称访问	禁用某个微服务名称访问

zuul:
  ignored-services: # 禁用微服务列表
    "*"
  routes:
    mldncloud-dept-service: /dept-proxy/**
	zuul:
	  ignored-services: # 禁用微服务列表
	    mldncloud-dept-service
	  routes:
	    mldncloud-dept-service: /dept-proxy/**

此程序启动之后，只能够通过配置的代理名称来进行微服务访问。

提示：对于路由代理地址访问的其他配置形式。

对于代理地址的配置，还有如下两种做法。

范例：【mldncloud-zuul-gateway-9501 项目】定义微服务路径与服务名称。

```
zuul:
  ignored-services: # 禁用微服务列表
    "*"
  routes:
    mydept.path: /dept-proxy/**
    mydept.serviceId: mldncloud-dept-service
```

访问地址为 http://zdmin:mldnjava@gateway-9501.com:9501/dept-proxy/dept/list。

范例：【mldncloud-zuul-gateway-9501 项目】定义微服务代理地址与 url 地址。

```
zuul:
  ignored-services: # 禁用微服务列表
    "*"
  routes:
    mydept.path: /dept-proxy/**
    mydept.url: http://dept-8001.com:8001/dept/
```

访问地址为 http://zdmin:mldnjava@gateway-9501.com:9501/dept-proxy/list。

在本书中建议使用传统方式进行配置，这样的配置相对会简洁一些。

3.【mldncloud-zuul-gateway-9501 项目】设置公共前缀。

```
zuul:
  prefix: /mldn-proxy                # 定义公共代理前缀
  ignored-services:                  # 禁用微服务列表
    "*"
  routes:
    mldncloud-dept-service: /dept-proxy/**
```

在进行访问时需要追加 http://zdmin:mldnjava@gateway-9501.com:9501/mldn-proxy/dept-proxy/dept/list 前缀。

提示：消费端通过 Zuul 代理访问。

Zuul 已经成功地配置了代理与部门微服务之间的访问，所有微服务的认证信息都在 Zuul 代理中进行了配置，此时消费端必须通过 Zuul 进行访问。

范例：【mldncloud-api】修改 IDeptService 接口，通过 Zuul 微服务访问。

```
@FeignClient(value = "MLDNCLOUD-ZUUL-GATEWAY", configuration = FeignClientConfig.class,
fallbackFactory=DeptServiceFallbackFactory.class)
public interface IDeptService {
	@PostMapping("/mldn-proxy/dept-proxy/dept/add")
	public DeptDTO add(DeptDTO dto) ;
	@GetMapping("/mldn-proxy/dept-proxy/dept/get/{deptno}")
	public DeptDTO get(@PathVariable("deptno") long deptno) ;
	@GetMapping("/mldn-proxy/dept-proxy/dept/list")
	public List<DeptDTO> list() ;
}
```

范例：【mldncloud-api】修改 FeignClientConfig 配置类，设置 Zuul 访问密码。

```
	@Bean
	public BasicAuthRequestInterceptor getBasicAuthRequestInterceptor() {
		return new BasicAuthRequestInterceptor("zdmin", "mldnjava");
	}
```

此时就可以通过消费端调用 Zuul 代理的部门微服务。

9.4.4　Zuul 服务降级

服务降级指的是当某个微服务不可用时，可以使用默认的信息进行数据返回的处理机制。Zuul 本身是一个微服务的代理网关，本身也提供有服务降级机制。

提示：客户端与 Zuul 网关服务降级。

前面讲解了利用 Hystrix 熔断机制实现的服务降级，该降级机制属于客户端降级。本节所讲解的是针对 Zuul 网关的实现的降级机制，如图 9-17 所示。

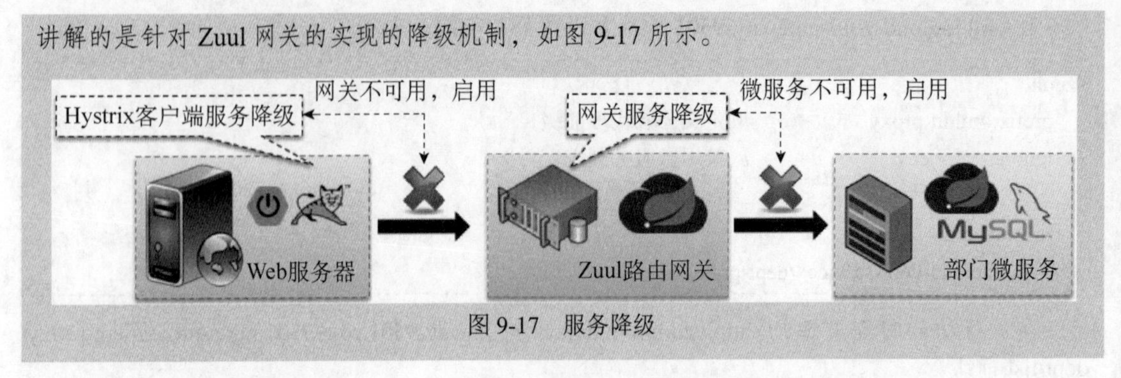

图 9-17　服务降级

范例：【mldncloud-zuul-gateway-9501 项目】建立 Zuul 网关的失败回退类。

```
package cn.mldn.mldncloud.fallback;
import java.io.ByteArrayInputStream;
import java.io.IOException;
import java.io.InputStream;
import org.springframework.cloud.netflix.zuul.filters.route.FallbackProvider;
import org.springframework.http.HttpHeaders;
import org.springframework.http.HttpStatus;
import org.springframework.http.client.ClientHttpResponse;
import org.springframework.stereotype.Component;
@Component
public class DeptServiceFallback implements FallbackProvider {          // 失败回退
    @Override
    public String getRoute() {
        return "mldncloud-dept-service";                                 // 微服务名称
    }
    @Override
    public ClientHttpResponse fallbackResponse() {
        return new ClientHttpResponse() {
            @Override
            public InputStream getBody() throws IOException {
                return new ByteArrayInputStream(
                    ("{\"deptno\":-99,\"dname\":\"【ERROR】Zuul-Fallback\""
                    + ",\"loc\":\"Gateway客户端提供\"}").getBytes());       // 服务降级数据
            }
            @Override
            public HttpHeaders getHeaders() {                            // 返回头信息
                HttpHeaders headers = new HttpHeaders() ;
                headers.set("Content-Type", "text/html; charset=UTF-8");
                return headers;
            }
```

```
            @Override
            public HttpStatus getStatusCode() throws IOException {          // HTTP状态码
                return HttpStatus.BAD_REQUEST;
            }
            @Override
            public int getRawStatusCode() throws IOException {              // HTTP状态码
                return HttpStatus.BAD_REQUEST.value();
            }
            @Override
            public String getStatusText() throws IOException {              // 状态信息
                return HttpStatus.BAD_REQUEST.getReasonPhrase();
            }
            @Override
            public void close() {
            }
        };
    }
    @Override
    public ClientHttpResponse fallbackResponse(Throwable cause) {
        return this.fallbackResponse();
    }
}
```

此时，当部门微服务不可用时，会返回 Zuul 服务降级信息，返回结果如图 9-18 所示。

图 9-18　Zuul 失败回退

9.4.5　上传微服务

微服务除了可以进行业务处理之外，也可以针对上传功能进行创建，所有的上传微服务依然需要向 Eureka 中注册，这样就可以在 Zuul 中进行微服务代理操作。

注意：不建议构建上传微服务。

在实际的开发过程中，利用微服务端实现上传业务并不是合理做法，从实际开发来讲，上传的功能一般都在 Web 消费端完成，最好的做法是直接利用 Web 消费端将上传文件保存到文件服务器中（如 FastDFS）。即使现在使用了微服务做上传，那么一般情况下也会将其保存到文件服务器中。所以本节只针对 Zuul 的功能进行技术性的讨论。

1.【mldncloud-upload-service-8101 项目】修改 application.yml 配置文件，追加上传配置。

```
spring:
  application:
    name: mldncloud-upload-service                    # 定义微服务名称
  http:
    multipart:
      enabled: true                                   # 启用HTTP上传处理
      max-file-size: 100MB                            # 设置单个文件的最大长度
      max-request-size: 100MB                         # 设置最大的请求文件的大小
      file-size-threshold: 1MB                        # 当上传文件达到1MB的时候进行磁盘写入
      location: /                                     # 上传的临时目录
server:
  port: 8101                                          # 服务端口
eureka:
  client:                                             # 客户端进行Eureka注册的配置
    service-url:                                      # 定义Eureka服务地址
      defaultZone: http://edmin:mldnjava@eureka-7001.com:7001/eureka
  instance:
    instance-id: upload-8101.com                      # 主机名称
info:
  app.name: mldncloud-upload-service
  company.name: www.mldn.cn
  build.artifactId: $project.artifactId$
  build.version: $project.version$
```

2.【mldncloud-upload-service-8101 项目】定义上传 Rest 微服务。

```
package cn.mldn.mldncloud.rest;
import java.io.File;
import java.util.HashMap;
import java.util.Map;
import java.util.UUID;
import org.springframework.web.bind.annotation.RequestMapping;
import org.springframework.web.bind.annotation.RequestMethod;
import org.springframework.web.bind.annotation.RequestParam;
import org.springframework.web.bind.annotation.RestController;
import org.springframework.web.context.request.RequestContextHolder;
import org.springframework.web.context.request.ServletRequestAttributes;
import org.springframework.web.multipart.MultipartFile;
import com.netflix.hystrix.contrib.javanica.annotation.HystrixCommand;
@RestController
public class UploadRest {
```

```
    @RequestMapping(value = "/upload", method = RequestMethod.POST)
    @HystrixCommand
    public Object upload(@RequestParam("photo") MultipartFile photo,HttpServletRequest request)
                throws Exception {
        Map<String,Object> map = new HashMap<String,Object>() ;
        String fileName = "mldn-file-nophoto" ;
        if (photo != null) {                                          // 有文件上传
            map.put("photo-name", photo.getName()) ;
            map.put("content-type", photo.getContentType()) ;
            map.put("photo-size", photo.getSize()) ;
            fileName = "mldn-file-" + UUID.randomUUID() + "."
                    + photo.getContentType().substring(
                            photo.getContentType().lastIndexOf("/") + 1);// 创建文件名称
            System.out.println(fileName);
            String filePath = request.getServletContext().getRealPath("/") + fileName;
            photo.transferTo(new File(filePath));                     // 文件保存
            map.put("save-path", filePath) ;
            map.put("save-file", fileName) ;
        }
        return map ;
    }
}
```

3.【操作系统】修改 hosts 配置文件，增加新的主机信息。

```
127.0.0.1   upload-8101.com
```

4.【mldncloud-upload-service-8101 项目】启动上传微服务，然后可以利用 curl 命令进行上传测试，测试成功会返回相应的上传信息。

```
curl -F "photo=@jixianit.png" http://mldnjava:hello@upload-8101.com:8101/upload
```

5.【mldncloud-zuul-gateway-9501 项目】如果需要 Zuul 进行上传微服务代理，还需要修改 application.yml，追加微服务的代理配置项。

```
zuul:
  prefix: /mldn-proxy                          # 定义公共代理前缀
  ignored-services:                            # 禁用微服务列表
    "*"
  routes:
    mldncloud-dept-service: /dept-proxy/**
    mldncloud-upload-service: /upload-proxy/**
```

开启代理之后，就可以使用 curl 命令通过 Zuul 代理进行上传，但是如果 Zuul 要代理上传微服务则必须在代理路径前追加 zuul/**（表示将所有上传操作交由目标微服务控制）才可以访问，测试命令如下：

```
curl -F "photo=@jixianit.png" http://zdmin:mldnjava@gateway-9501.com:9501/zuul/mldn-proxy/upload-proxy/upload
```

6. 【mldncloud-consumer-upload 项目】如果消费端要调用上传微服务，则无法使用 Feign 进行接口转换，而要利用 httpclient 依赖库进行处理。修改 pom.xml 配置文件，追加依赖库配置。

```
<dependency>
    <groupId>org.apache.httpcomponents</groupId>
    <artifactId>httpclient</artifactId>
</dependency>
<dependency>
    <groupId>org.apache.httpcomponents</groupId>
    <artifactId>httpmime</artifactId>
</dependency>
<dependency>
    <groupId>org.springframework.boot</groupId>
    <artifactId>spring-boot-starter-thymeleaf</artifactId>
</dependency>
```

7. 【mldncloud-consumer-upload 项目】建立 src/main/view 目录，将其提升为源文件目录，随后将所需要的 Thymeleaf 页面文件保存到此目录中，并且在里面建立 templates 子目录。

8. 【mldncloud-consumer-upload 项目】建立 src/main/view/templates/upload.html 页面。

```
<!DOCTYPE HTML>
<html xmlns:th="http://www.thymeleaf.org">
<head>
<title>SpringCloud微服务</title>
<link rel="icon" type="image/x-icon" href="/images/mldn.ico" />
<meta http-equiv="Content-Type" content="text/html;charset=UTF-8" />
</head>
<body>
    <form th:action="@{/consumer/upload}" method="post" enctype="multipart/form-data">
        姓名：<input type="text" name="name" id="name" value="MLDN"/><br/>
        照片：<input type="file" name="photo" id="photo"/><br/>
        <input type="submit" value="提交"/>
        <input type="reset" value="重置"/>
    </form>
</body>
</html>
```

9. 【mldncloud-consumer-upload 项目】建立一个 ConsumerUploadController 控制层程序类。

```
package cn.mldn.mldncloud.consumer.controller;
import java.nio.charset.Charset;
import org.apache.http.HttpEntity;
import org.apache.http.HttpResponse;
import org.apache.http.auth.AuthScope;
import org.apache.http.auth.Credentials;
import org.apache.http.auth.UsernamePasswordCredentials;
import org.apache.http.client.CredentialsProvider;
import org.apache.http.client.methods.HttpPost;
import org.apache.http.client.protocol.HttpClientContext;
import org.apache.http.entity.ContentType;
import org.apache.http.entity.mime.MultipartEntityBuilder;
import org.apache.http.impl.client.BasicCredentialsProvider;
import org.apache.http.impl.client.CloseableHttpClient;
import org.apache.http.impl.client.HttpClients;
import org.apache.http.util.EntityUtils;
import org.springframework.stereotype.Controller;
import org.springframework.web.bind.annotation.RequestMapping;
import org.springframework.web.bind.annotation.RequestMethod;
import org.springframework.web.bind.annotation.ResponseBody;
import org.springframework.web.multipart.MultipartFile;
@Controller
public class ConsumerUploadController {
    // 设置要进行远程上传微服务调用的代理地址
    public static final String UPLOAD_URL =
            "http://gateway-9501.com:9501/zuul/mldn-proxy/upload-proxy/upload";
    @RequestMapping(value = "/consumer/uploadPre", method = RequestMethod.GET)
    public String uploadPre() {
        return "upload";
    }
    @RequestMapping(value = "/consumer/upload", method = RequestMethod.POST)
    public @ResponseBody Object upload(String name, MultipartFile photo) throws Exception {
        if (photo != null) {
            CloseableHttpClient httpClient = HttpClients.createDefault(); // 创建HttpClient对象
            CredentialsProvider credsProvider = new BasicCredentialsProvider(); // 创建认证信息
            Credentials credentials = new UsernamePasswordCredentials("zdmin",
                    "mldnjava");                                         // 创建一条认证操作信息
            credsProvider.setCredentials(AuthScope.ANY, credentials);    // 认证请求信息
            HttpClientContext httpContext = HttpClientContext.create();  // HTTP处理操作上下文
            httpContext.setCredentialsProvider(credsProvider);           // 设置认证提供信息
```

```
                HttpPost httpPost = new HttpPost(UPLOAD_URL);                    // 设置访问请求地址
                HttpEntity entity = MultipartEntityBuilder.create()
                        .addBinaryBody("photo", photo.getBytes(),
                            ContentType.create("image/png"), "temp.png").build();
                httpPost.setEntity(entity);    // 将请求的实体信息进行发送
                HttpResponse response = httpClient.execute(httpPost, httpContext) ; // 请求发送
                return EntityUtils.toString(response.getEntity(),Charset.forName("UTF-8"));
            }
            return "nophoto";
        }
}
```

在此程序类中将伪造一个 POST 请求，将上传文件发送到上传微服务，上传成功后就可以看见微服务返回的信息。

9.5 本章小结

1. Ribbon 是一个工作在消费端的负载均衡组件，SpringBoot 消费端可以通过 Ribbon 调用 Eureka 中注册的微服务。

2. SpringCloud 微服务的负载均衡采用的是服务名称的管理，即同一个服务名称的微服务会自动注册到同一组微服务信息中，Ribbon 中可以利用 IRule 接口子类配置负载均衡策略。

3. Feign 是基于 Ribbon 组件的应用，可以利用 Feign 实现远程 Restful 与接口间的映射转换。

4. Hystrix 提供的是熔断机制，可以在某一个微服务出现问题后自动熔断，以防止雪崩效应出现。

5. HystrixDashboard 提供微服务访问监控，利用 Turbine 可以实现一组微服务的监控。但对于认证的微服务，则需要进行安全访问排除。

6. Zuul 提供有网关路由功能，利用 Zuul 可以实现一组微服务的划分。同时利用路由配置，可以使微服务的访问更加安全。

第 10 章

SpringCloudConfig

通过本章学习，可以达到以下目标：

1. 理解 SpringCloudConfig 与集群微服务的配置管理。
2. 掌握 SpringCloudConfig 集成配置与信息抓取处理。
3. 掌握 SpringCloudConfig 与仓库匹配模式。
4. 掌握 SpringCloudConfig 加密访问处理。
5. 掌握 SpringCloudConfig 高可用配置。
6. 掌握 SpringCloudBus 服务总线配置，并且可以利用 RabbitMQ 实现自动配置抓取。

SpringCloudConfig 是专门为微服务提供的统一配置中心，其利用 Git 或 SVN 这样的版本控制工具实现配置文件的分布式存储，而后所有的微服务都可以通过 SpringCloudConfig 进行微服务配置项的动态抓取，也可以方便地实现不同 profile 配置间的切换。

10.1 SpringCloudConfig 简介

在传统单实例的项目环境中，为了方便配置文件的使用，往往将其直接定义在实例主机之内，如图 10-1 所示。随着业务的不断拆分与微服务实例的不断增加，这样的配置方式必然会造成配置文件的维护困难。最好的方法是对配置文件进行统一管理，例如将配置文件保存到 Git 或 SVN 这样的版本控制工具中，然后利用 SpringCloudConfig 实现配置文件的抓取与使用，如图 10-2 所示。

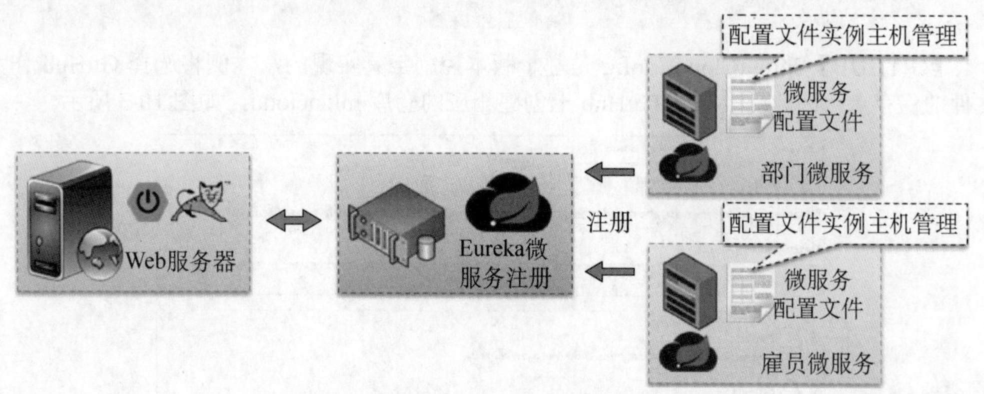

图 10-1　传统微服务每台主机维护配置文件

通过图 10-2 可以发现，所有的 SpringCloudConfig 微服务都可以统一注册到 Eureka 中，这样就可以定义多个 SpringCloudConfig 微服务，从而实现微服务的集群设计。但是，即使在项目中引入 SpringCloudConfig 到每一个微服务中，也依然需要提供核心的配置文件（SpringCloudConfig 微服务地址等信息）。

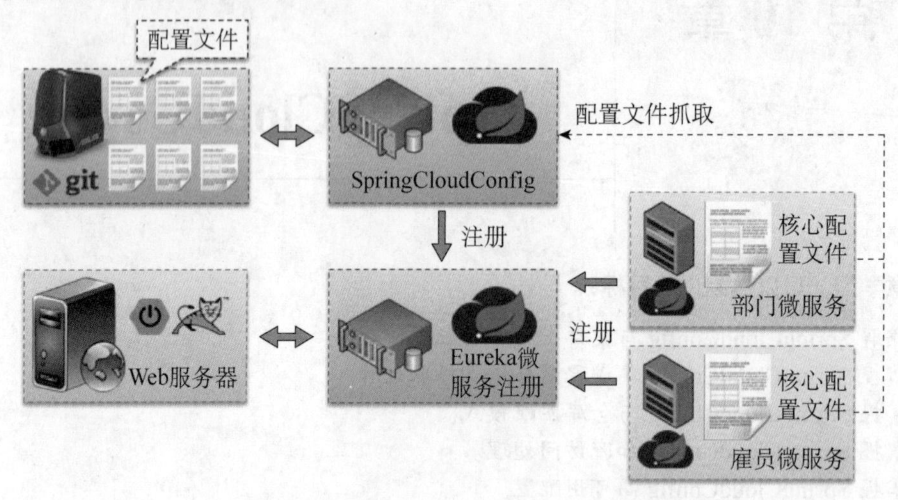

图 10-2 SpringCloudConfig 管理配置文件

10.2 配置 SpringCloudConfig 服务端

SpringCloudConfig 是以微服务的形式存在的，所以需要开发者自己创建配置服务器。此后，所有需要抓取配置文件的微服务都可以通过此服务器实现配置文件抓取。本例讲解的 SpringCloudConfig 将基于 Git 实现配置文件管理，需要把所有的配置文件保存在 GitHub 服务器中。

提示：关于 Config 微服务。

本例程序中已经实现了主机与 GitHub 之间的 SSH 免登录配置，可以直接进行仓库的克隆与推送。同时，定义配置服务器的过程中使用了之前配置的 mldncloud-security 模块，微服务的认证信息为 mldnjava/hello。

1.【GITHUB】SpringCloudConfig 是基于版本控制工具实现的，本例将选择 GitHub 作为配置文件的保存点，因此需要先在 GitHub 上创建相应的仓库 mldncloud，如图 10-3 所示。

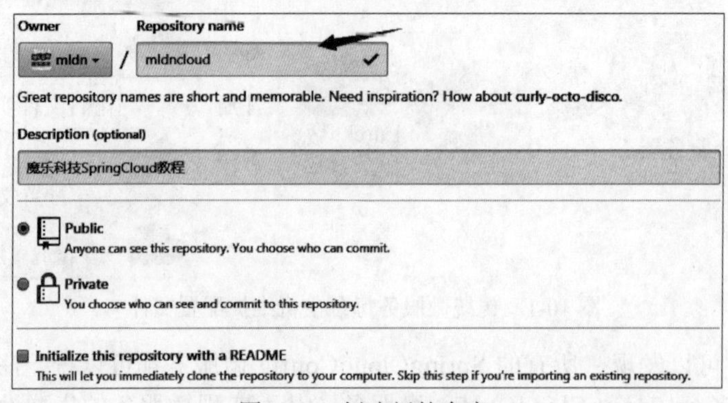

图 10-3 创建新的仓库

仓库创建完成后，会给出仓库的访问地址 git@github.com:mldn/mldncloud.git。

2.【操作系统】通过 git 命令克隆远程仓库 git clone git@github.com:mldn/mldncloud.git，此时会在操作目录中形成 mldncloud 目录（里面包含有.git 仓库文件）。

3.【操作系统】在 mldncloud 仓库目录中创建 application.yml 配置文件，具体内容如下。

```
spring:
  profiles:
    active:
    - dev
---
spring:
  profiles: dev
  application:
    name: mldnconfig-test-dev
---
spring:
  profiles: product
  application:
    name: mldnconfig-test-product
```

在配置文件中定义了两个 profile 环境（dev、product），这样在访问时就可以根据 profile 名称加载不同配置项。

提示：注意文件编码。

通过 GitHub 保存配置文件且文件中存在中文信息时，编码一定要使用 UTF-8 编码，否则在进行加载时会出现属性加载失败的错误提示信息。

4.【操作系统】将本地仓库创建的 application.yml 配置文件上传到远程 GitHub 之中。

将更新提交到暂存库之中	git add
将暂存库中的内容提交到仓库	git commit -m "add application.yml file"
进行远程仓库的同步推送	git push origin master

随后可以在 GitHub 上看见上传的 application.yml 配置文件，如图 10-4 所示。

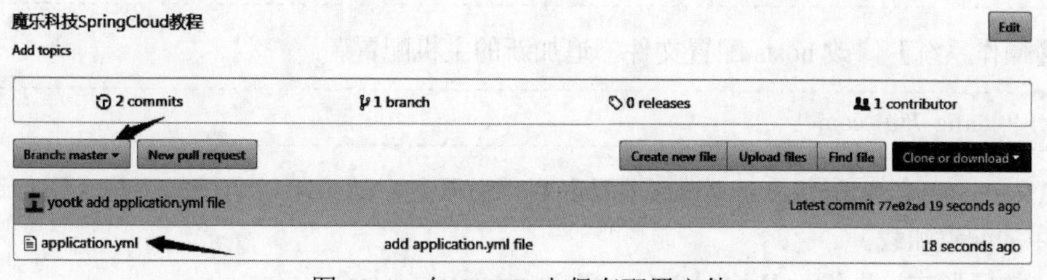

图 10-4　在 GitHub 上保存配置文件

5.【mldncloud-config-server-7501 项目】修改 pom.xml 配置文件，追加 config 相关依赖库。

```
<dependency>
    <groupId>org.springframework.cloud</groupId>
    <artifactId>spring-cloud-config-server</artifactId>
</dependency>
```

6.【mldncloud-config-server-7501 项目】修改 application.yml 配置文件，追加 Git 仓库配置项。

```
spring:
  application:
    name: mldncloud-config-server                    # 定义微服务名称
  cloud:                                             # 进行SpringCloud的相关配置
    config:                                          # 进行SpringCloudConfig的相关配置
      server:                                        # 设置服务的连接地址
        git:                                         # 设置Git连接地址
          uri: git@github.com:mldn/mldncloud.git
server:
  port: 7501                                         # 服务端口
eureka:
  client:                                            # 客户端进行Eureka注册的配置
    service-url:                                     # 定义Eureka服务地址
      defaultZone: http://edmin:mldnjava@eureka-7001.com:7001/eureka
  instance:
    instance-id: config-7501.com                     # 显示主机名称
```

7.【mldncloud-config-server-7501 项目】定义程序启动类。

```
@SpringBootApplication
@EnableEurekaClient                                  // 启用Eureka客户端
@EnableDiscoveryClient
@EnableConfigServer                                  // 配置服务器
public class StartConfigApplication7501 {
    public static void main(String[] args) {
        SpringApplication.run(StartConfigApplication7501.class, args);
    }
}
```

8.【操作系统】修改 hosts 配置文件，追加新的主机配置。

```
127.0.0.1   config-7501.com
```

9.【mldncloud-config-server-7501 项目】启动微服务，随后可以通过如下形式进行远程 Git 服务器中的配置加载。

☑ 访问形式一，**/{application}-{profile}.yml**。

|- 获取 **dev** 环境：http://mldnjava:hello@config-7501.com:7501/application-dev.yml。

|- 获取 **product** 环境：http://mldnjava:hello@config-7501.com:7501/application-product.yml。

☑ 访问形式二，**/{application}/{profile}[/{label}]**。

|- 获取 **dev** 环境：http://mldnjava:hello@config-7501.com:7501/application/dev/master。

|- 获取 **product** 环境：http://mldnjava:hello@config-7501.com:7501/application/product/master。

☑ 访问形式三，**/{label}/{application}-{profile}.yml**。

|- 获取 **dev** 环境：http://mldnjava:hello@config-7501.com:7501/master/application-dev.yml。

|- 获取 **product** 环境：http://mldnjava:hello@config-7501.com:7501/master/application-product.yml。

在实际的项目开发中，master 分支肯定是不能够进行修改更新的，也就是说 master 上的内容一定是可以使用的，所以往往会设置一个 dev 分支处理项，这样如果使用/{label}/{application}-{profile}.yml 格式，将可以更好地进行分支的定位。

提示：使用 properties 实现 SpringCloudConfig 配置。

本程序采用 yml 配置文件实现了配置。如果想采用 properties 文件处理，就必须准备多个 properties，如 application-dev.properties、application- beta.properties 等。

10.3　SpringCloudConfig 客户端抓取配置信息

SpringCloudConfig 服务端搭建完成后，可以与 GitHub 连接，获取相关 profile 的配置信息。而后这些读取的信息应该与具体的微服务结合（SpringCloudConfig 客户端），实现配置信息的统一管理。

1.【GitHub】定义 mldncloud-config-client.yml 配置文件（与要创建的微服务名称相同），并推送到 GitHub 之中。

```
spring:
  profiles:
    active:
    - dev
---
spring:
  profiles: dev
  application:
    name: mldncloud-config-client                    # 定义微服务名称
server:
  port: 8101                                          # 服务端口
eureka:
  client:                                             # 客户端进行Eureka注册的配置
    service-url:                                      # 定义Eureka服务地址
      defaultZone: http://edmin:mldnjava@eureka-7001.com:7001/eureka
```

```
  instance:
    instance-id: config-client.com                    # 显示主机名称
---
spring:
  profiles: product
  application:
    name: mldncloud-config-client                     # 定义微服务名称
server:
  port: 8102                                          # 服务端口
eureka:
  client:                                             # 客户端进行Eureka注册的配置
    service-url:                                      # 定义Eureka服务地址
      defaultZone: http://edmin:mldnjava@eureka-7001.com:7001/eureka
  instance:
    instance-id: config-client.com                    # 显示主机名称
```

2.【mldncloud-config-client 项目】修改 pom.xml 配置文件，追加相关的依赖库。

```
<dependency>
    <groupId>org.springframework.cloud</groupId>
    <artifactId>spring-cloud-starter-config</artifactId>
</dependency>
```

3.【mldncloud-config-client 项目】建立 src/main/resources/bootstrap.yml 配置文件，定义 SpringCloudConfig 服务信息。

```
spring:
  cloud:
    config:                                   # Config服务器
      name: mldncloud-config-client   # 定义要读取的资源文件的名称
      profile: dev                            # 定义profile的名称
      label: master                           # 定义配置文件所在的分支
      uri: http://config-7501.com:7501 # SpringCloudConfig的服务地址
      username: mldnjava                      # 连接的用户名
      password: hello                         # 连接的密码
```

在本程序定义的配置文件中直接设置了 SpringCloudConfig 服务地址，并且定义了要通过配置服务器抓取的配置文件名称为 mldncloud-config-client、profile 为 dev，所在仓库为 master。

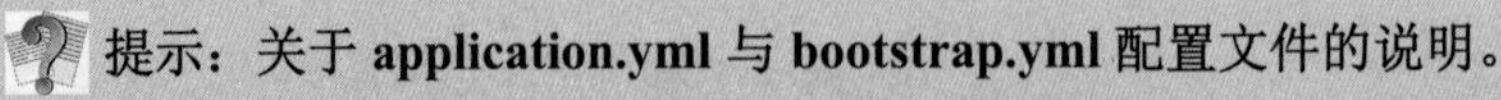

提示：关于 **application.yml** 与 **bootstrap.yml** 配置文件的说明。

☑ application.yml：对应的是用户级的资源配置。

☑ bootstrap.yml：对应的是系统级的资源配置，其优先级会更高。

4.【mldncloud-config-client 项目】建立 ConfigClientRest 程序类，读取配置文件信息。

```
package cn.mldn.mldncloud.rest;
import org.springframework.beans.factory.annotation.Value;
import org.springframework.web.bind.annotation.GetMapping;
import org.springframework.web.bind.annotation.RestController;
@RestController
public class ConfigClientRest {
    @Value("${spring.application.name}")
    private String applicationName;          // 应用的服务名称
    @Value("${eureka.client.service-url.defaultZone}")
    private String eurekaServers;            // 设置所有的Eureka服务信息项
    @GetMapping("/config")
    public String getConfig() {
        return "ApplicationName = " + this.applicationName + "、EurekaServers = "
                + this.eurekaServers;
    }
}
```

5.【操作系统】修改 hosts 配置文件，追加主机配置项。

```
127.0.0.1    config-client.com
```

当程序启动之后会通过 SpringCloudConfig 抓取指定的配置文件信息，同时控制台会输出如下提示信息：

启动时抓取配置	Fetching config from server at: http://config-7501.com:7501
配置抓取文件	Located environment: name=mldncloud-config-client, profiles=[product], label=master, version=cc82b4ce435d95ba43d7a6f52610c8e57436007a, state=null

当配置项抓取成功后，就会利用 GitHub 保存的配置文件对项目进行配置。当开发者需要切换配置时，直接修改 profile 的名称即可。

提示：远程配置优先。

如果本地项目中提供的 application.yml 配置与 SpringCloudConfig 服务端远程抓取的配置重复，则会以远程配置为优先考虑。

10.4　单仓库目录匹配

在实际项目开发过程中，为了方便管理配置文件，往往需要在仓库中创建多个微服务配置的目录，并在目录中提供相应的配置资源文件，而后所有的微服务就可以通过 SpringCloudConfig

服务端进行目录匹配，以实现配置项加载，如图 10-5 所示。

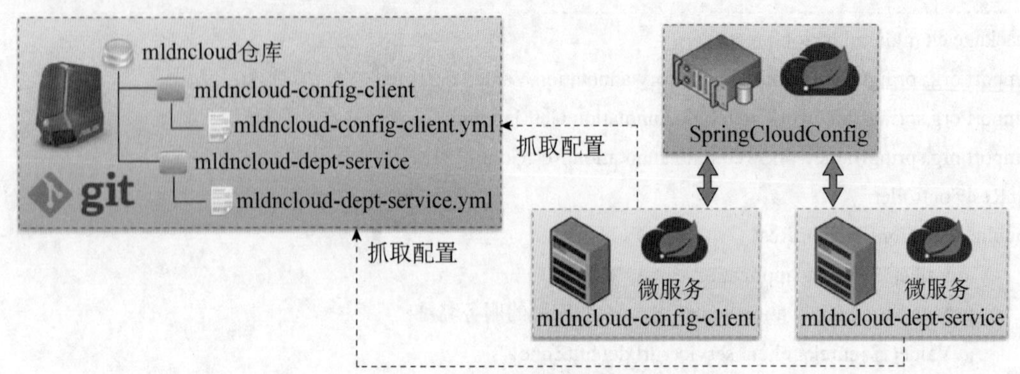

图 10-5　单仓库目录匹配

1.【GITHUB】在 mldncloud 仓库之中建立 mldncloud-config-client 目录，并将 mldncloud-config-client.yml 配置文件保存到此目录中，随后将其提交到 GitHub 中，保存后的路径如图 10-6 所示。

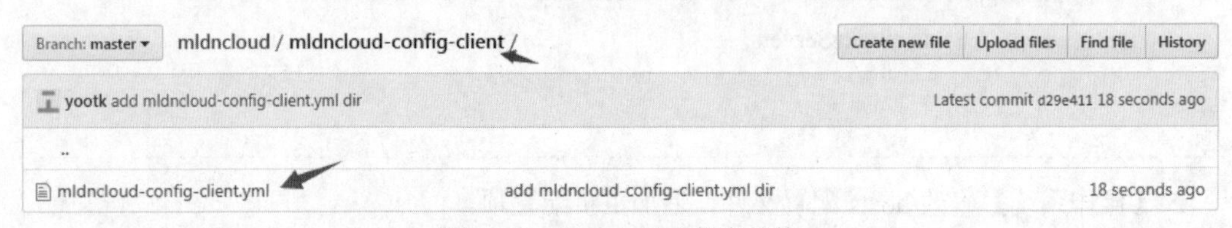

图 10-6　GitHub 目录保存结构

2.【mldncloud-config-server-7501 项目】修改 src/main/resources/application.yml 配置文件。

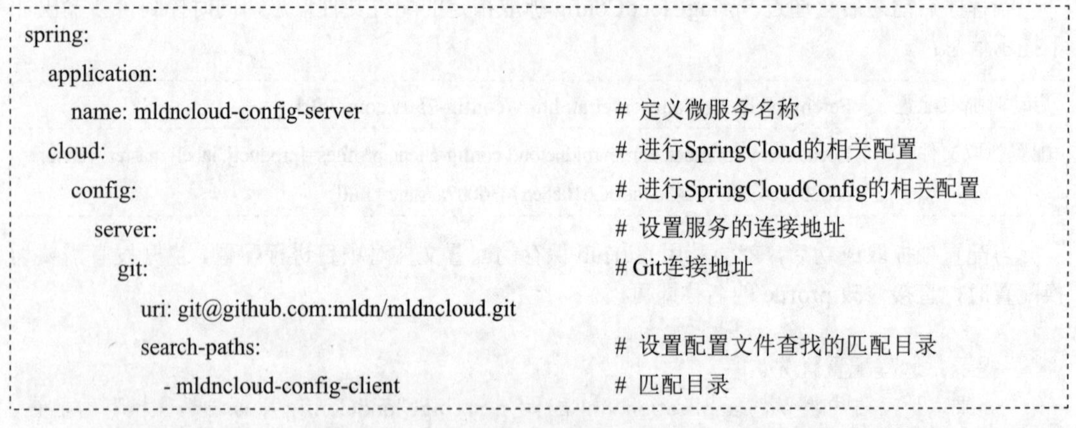

```
spring:
  application:
    name: mldncloud-config-server                    # 定义微服务名称
  cloud:                                             # 进行SpringCloud的相关配置
    config:                                          # 进行SpringCloudConfig的相关配置
      server:                                        # 设置服务的连接地址
        git:                                         # Git连接地址
          uri: git@github.com:mldn/mldncloud.git
          search-paths:                              # 设置配置文件查找的匹配目录
            - mldncloud-config-client                # 匹配目录
```

3.【mldncloud-config-server-7501 项目】启动配置微服务，就可以通过指定 GitHub 加载指定配置目录下的配置目录，访问路径为 http://mldnjava:hello@config-7501.com:7501/master/mldncloud-config-client-dev.yml。

提示：使用*实现目录匹配。

在本程序中使用 search-paths 定义了仓库中的目录名称。如果目录有很多，整体配置就会非常复杂。这个时候如果仓库中的所有目录都是以 mldncloud 开头的，则可以使用 mldncloud-* 的形式匹配所有相关目录。

10.5　多仓库自动匹配

为了方便进行配置文件管理，也可以在 GitHub 上建立多个应用仓库，利用不同的仓库保存不同的微服务配置文件，然后利用微服务的名称进行应用仓库的自动匹配，如图 10-7 所示。

1.【GITHUB】在 GitHub 上创建一个新的仓库 mldncloud-config-client，如图 10-8 所示。

2.【操作系统】将远程仓库克隆到本地（git clone git@github.com:mldn/mldncloud-config-client.git）。

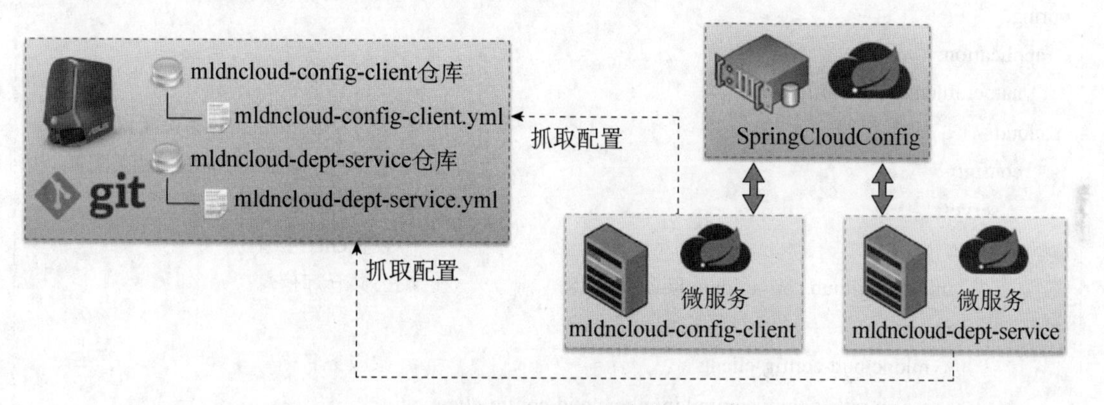

图 10-7　多仓库自动匹配

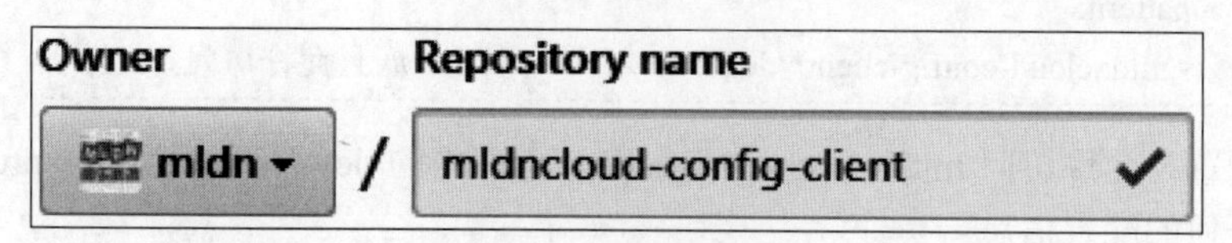

图 10-8　创建新仓库

3.【操作系统】在 mldncloud-config-client 目录中创建 mldncloud-config-client.yml 配置文件，内容与之前相同，并且将配置文件推送到 GitHub 中。

4.【mldncloud-config-server-7501 项目】修改 application.yml 配置文件，追加仓库匹配。

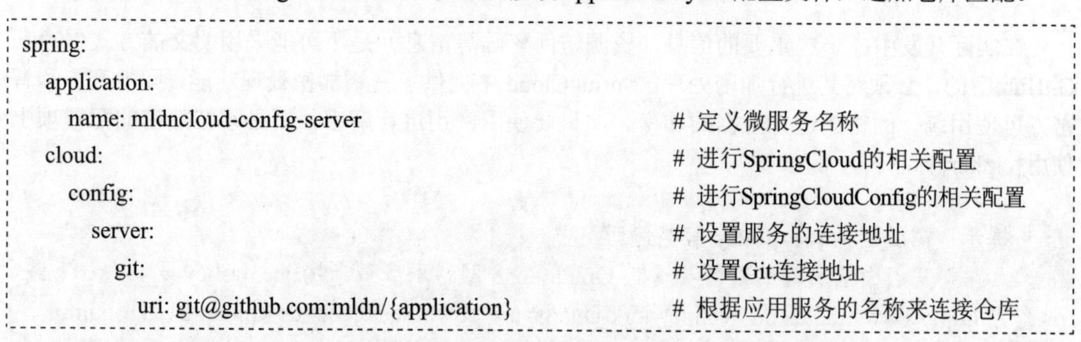

```
spring:
  application:
    name: mldncloud-config-server                # 定义微服务名称
  cloud:                                         # 进行SpringCloud的相关配置
    config:                                      # 进行SpringCloudConfig的相关配置
      server:                                    # 设置服务的连接地址
        git:                                     # 设置Git连接地址
          uri: git@github.com:mldn/{application} # 根据应用服务的名称来连接仓库
```

访问地址为 http://mldnjava:hello@config-7501.com:7501/master/mldncloud-config-client-dev.yml。此时会根据{application}自动匹配要使用的仓库，而在进行访问的时候需要开发者输入仓库名称，而后会自动加载仓库中的 application.yml 文件。

10.6 仓库匹配模式

仓库匹配模式指的是在进行仓库配置资源获得的时候，可以通过一些限制让用户无法获得某些资源（只能够获得指定资源）。例如，10.5 节虽然进行了多仓库的配置，但是在配置的过程中发现用户可以加载所有的 profile 信息。接下来，需要设置一些规则，让用户只能够加载指定的内容。

范例：【mldncloud-config-server-7501 项目】修改 application.yml 配置文件。

```
spring:
  application:
    name: mldncloud-config-server                    # 定义微服务名称
  cloud:                                             # 进行SpringCloud的相关配置
    config:                                          # 进行SpringCloudConfig的相关配置
      server:                                        # 设置服务的连接地址
        git:                                         # 设置Git连接地址
          uri: git@github.com:mldn/mldncloud.git     # 公共仓库信息
          repos:
            mldncloud-config-client:                 # 微服务名称
              uri: git@github.com:mldn/mldncloud-config-client.git
              pattern:
              - mldncloud-config-client*/dev*        # 只允许访问 dev 环境
```

此时，配置只允许访问 mldncloud-config-client 仓库中的 dev 环境。当加载 product 环境时，将不会得到任何的配置信息。

10.7 密钥加密处理

在实际开发中，一些重要的信息（资源访问密码等信息）是不可能采用明文的方式保存在 GitHub 中的，必须对其进行加密处理。SpringCloud 中提供了密钥加密处理功能，即加密方和解密方可使用统一的密钥进行数据的加密。如果要使用密钥进行解密，则必须在加密的配置项上使用{cipher}标注。

提示：需要更换 jce_policy 策略包。

如果要进行加密处理，则需要通过 Oracle 官方网站下载最新的 jce_policy 支持包，下载地址为 http://www.oracle.com/technetwork/java/javase/downloads/jce8- download-2133166.html。而后将下载后的内容保存到 JAVA_HOME\jre\lib\security 目录中进行内容替换（jre 同时替换）。同时还需要考虑 SpringCloud（Dalston.SR1 版本可以正常使用，后续版本无法正常使用）配置的版本，不同的版本即使更换了策略包，可能也会出现问题（{"description":"No key was installed for encryption service","status":"NO_KEY"}）。

1.【mldncloud-config-server-7501 项目】修改 application.yml 配置文件，追加密钥配置。

```
encrypt:
  key: mldnjava.cn      # 设置了一个加密的访问密钥
```

2.【操作系统】启动 mldncloud-config-server-7501 微服务，并且通过 curl 进行信息加密。

加密处理	curl -X POST http://mldnjava:hello@config-7501.com:7501/encrypt -d mysqladmin
加密结果	ec90dd10a225f9873c5e178388de29525dbb778d5c189f84f0b4406220c34e0f

3.【操作系统】如果要对加密后的数据进行解密处理，则可以通过/decrypt 路径访问。

```
curl -X POST http://mldnjava:hello@config-7501.com:7501/decrypt -d
ec90dd10a225f9873c5e178388de29525dbb778d5c189f84f0b4406220c34e0f
```

此时，就可以通过指定的加密解密路径进行密钥加密。

4.【Git 本地仓库】采用加密信息定义要上传的信息，修改 mldncloud-config-encrypt.yml 文件，随后将其上传到 GitHub 服务器的 mldncloud 仓库中。需要进行加密处理的配置前必须要有{cipher}标记。

```
spring:
  profiles:
    active:
    - dev
---
spring:
  profiles: dev
  application:
    name: mldncloud-config-encrypt                          # 定义微服务名称
  datasource:
    type: com.alibaba.druid.pool.DruidDataSource            # 配置当前要使用的数据源的操作类型
    driver-class-name: org.gjt.mm.mysql.Driver              # 配置MySQL的驱动程序类
    url: jdbc:mysql://localhost:3306/mldn8001               # 数据库连接地址
    username: root                                          # 数据库用户名
    password: '{cipher}ec90dd10a225f9873c5e178388de29525dbb778d5c189f84f0b4406220c34e0f'
```

5.【mldncloud-config-server-7501 项目】启动微服务，加载远程 GitHub 中的 mldncloud-config-encrypt.yml 配置文件，地址为 http://mldnjava:hello@config-7501.com:7501/master/mldncloud-config-encrypt-dev.yml。此时加密的信息将以明文的形式进行显示，而未加密的信息将正常显示。

10.8　KeyStore 加密处理

还有一种加密的方式会更加方便，就是直接利用 JKS 操作来完成。通过这种方式实现的加密更加安全，且只需要有一个 JKS 配置文件即可实现加密与解密。

1.【操作系统】生成一个 JKS 配置文件，执行后可以获得 keystore.p12 文件。

```
keytool -genkeypair -alias mytestkey -keyalg RSA -dname "CN=Web Server,OU=Unit,O=Organization,
L=City,S=State,C=US" -keypass changeit -keystore keystore.jks -storepass mldnjava
```

2.【mldncloud-config-server-7501 项目】将生成的 keystore.jks 文件配置到 src/main/resources 目录中，如图 10-9 所示。

```
◢ src/main/resources
    application.yml
    keystore.jks
```

图 10-9　保存 keystore.jks 文件

3.【mldncloud-config-server-7501 项目】修改 application.yml，追加 keystore 相关配置。

```
encrypt:
  keyStore:
    location: classpath:/keystore.jks                # keystore.jks的配置文件的路径
    password: mldnjava                               # keystore的密码
    alias: mytestkey                                 # 别名
    secret: changeit                                 # keypass 密码
```

4.【mldncloud 项目】修改资源配置，允许*.jks 输出。

```
<resources>
    <resource>
        <directory>src/main/resources</directory>
        <includes>
            <include>**/*.properties</include>
            <include>**/*.yml</include>
            <include>**/*.xml</include>
            <include>**/*.tld</include>
            <include>**/*.p12</include>
            <include>**/*.jks</include>
        </includes>
        <filtering>false</filtering>
    </resource>
</resources>
```

5.【mldncloud-config-server-7501 项目】启动微服务，进行加密与解密处理。

加密处理	curl -X POST http://mldnjava:hello@config-7501.com:7501/encrypt -d mysqladmin
解密处理	curl -X POST http://mldnjava:hello@config-7501.com:7501/decrypt -d 加密信息

此时会取得一个很长的 JKS 加密信息，并且将这些保存在 mldncloud-config-encrypt.yml 配

置文件中。与密钥加密相同，JKS 也需要在加密信息处使用{cipher}标记进行声明。

10.9　SpringCloudConfig 高可用

SpringCloudConfig 服务可以实现配置文件的动态加载处理，如果 SpringCloudConfig 服务出现问题，则会造成整体微服务的瘫痪，所以引入 SpringCloudConfig 高可用机制可以避免单主机使用缺陷。SpringCloudConfig 实现高可用机制主要可以依靠 Eureka 注册中心实现，而后所有的微服务客户端就可以通过 Eureka 中提供的 SpringCloudConfig 服务名称进行服务连接，如图 10-10 所示。

1.【mldncloud 项目】建立 3 个微服务项目：mldncloud-config-server-7501、mldncloud-config-server-7502 和 mldncloud-config-server-7503。

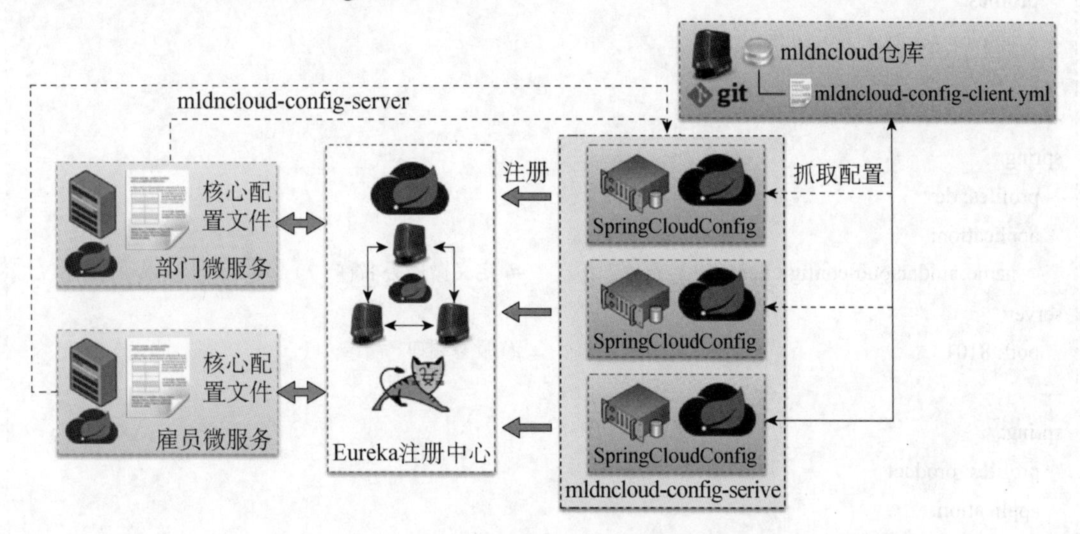

图 10-10　SpringCloudConfig 高可用

2.【mldncloud-config-server-*项目】修改 application.yml 配置文件。3 个 SpringCloudConfig 服务器都与 mldncloud 仓库连接，同时设置的微服务名称均为 mldncloud-config-server，唯一不同的是每个微服务的端口号与 Eureka 中绑定的主机名称。

mldncloud-config-server-7501	mldncloud-config-server-7502	mldncloud-config-server-7503

server:
  port: 7501
	server:
	  port: 7502
		server:
		  port: 7503

instance:
  instance-id: config-7501.com
	instance:
	  instance-id: config-7502.com
		instance:
		  instance-id: config-7503.com

3.【操作系统】修改 hosts 配置文件，追加主机名称。

```
127.0.0.1    config-7501.com
127.0.0.1    config-7502.com
127.0.0.1    config-7503.com
```

4.【mldncloud-config-server-*项目】启动 Eureka 注册服务，同时启动 3 个 SpringCloudConfig 服务，而后可以通过 Eureka 注册中心看到如图 10-11 所示信息。

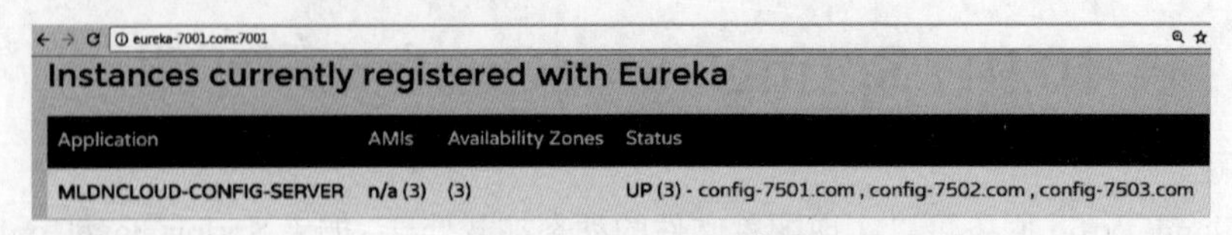

图 10-11　Eureka 注册信息

5.【Git 本地仓库】在本地仓库中建立 mldncloud-config-client.yml 文件，并将其提交到 GitHub 之中。

```
spring:
  profiles:
    active:
    - dev
---
spring:
  profiles: dev
  application:
    name: mldncloud-config-client                    # 定义微服务名称
server:
  port: 8101                                         # 服务端口
---
spring:
  profiles: product
  application:
    name: mldncloud-config-client                    # 定义微服务名称
server:
  port: 8102                                         # 服务端口
```

6.【mldncloud-config-client 项目】修改 application.yml 配置文件，由于此时需要通过 Eureka 获取 SpringCloudConfig 服务信息，因此需要在此配置文件中定义 Eureka 的访问地址。

```
spring:
  cloud:
    config:                                          # Config服务器
      name: mldncloud-config-client                  # 定义要读取的资源文件的名称
      profile: product                               # 定义profile的名称
      label: master                                  # 定义配置文件所在的分支
      username: mldnjava                             # 连接的用户名
      password: hello                                # 连接的密码
      discovery:
        enabled: true                                # 通过配置中心加载配置文件
```

```
        service-id: MLDNCLOUD-CONFIG-SERVER        # 在Eureka之中注册的服务ID
eureka:
  client:                                          # 客户端进行Eureka注册的配置
    service-url:                                   # 定义Eureka服务地址
      defaultZone: http://edmin:mldnjava@eureka-7001.com:7001/eureka
  instance:
    instance-id: config-client.com                 # 显示主机名称
```

此时，SpringCloudConfig 客户端程序将通过 Eureka 中提供的微服务名称 MLDNCLOUD-CONFIG-SERVER 连接 SpringCloudConfig 服务器信息，如果某一台配置服务器出现问题，则其他服务主机会自动更换至其他可用主机提供支持。

10.10　SpringCloudBus 服务总线

SpringCloudBus 是 SpringCloud 消息总线，主要功能是通过消息组件代理各个连接分布点，这样当配置文件发生变更之后就可以通过 SpringCloudBus 实现配置信息的动态抓取，以实现配置项的动态更新，具体操作如图 10-12 所示。

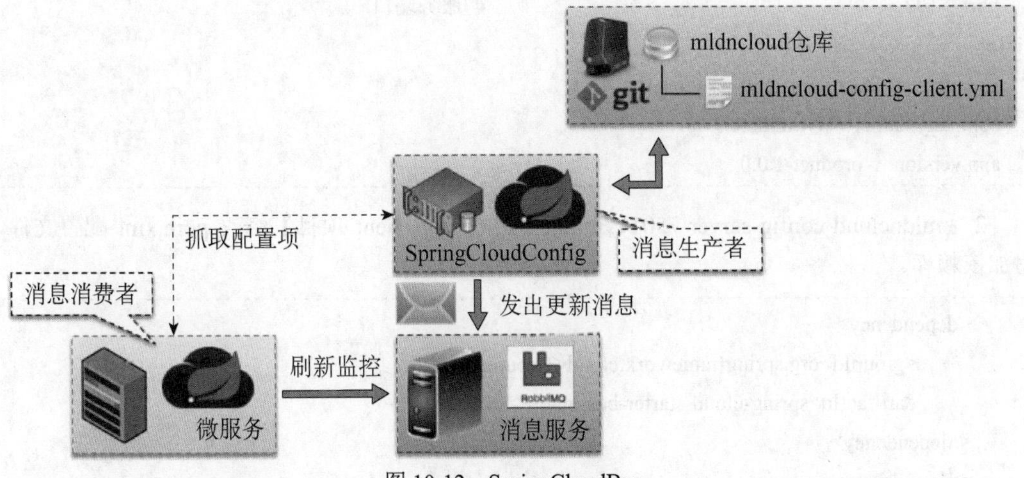

图 10-12　SpringCloudBus

在图 10-12 中可以看到，当 Git 仓库中的配置文件发生信息更新之后，将会通过 SpringCloudConfig 向消息组件中发出一个更新消息。消息消费者（微服务）接收到此消息之后，会通过 SpringCloudConfig 重新进行配置抓取。

1.【GIT 本地仓库】建立 mldncloud-config-client.yml 配置文件，并且将此配置文件上传到 mldncloud 仓库中。

```
spring:
  profiles:
    active:
```

```
    - dev
---
spring:
  profiles: dev
  application:
    name: mldncloud-config-client                    # 定义微服务名称
server:
  port: 8101                                          # 服务端口
info:
  app.name: mldn-microcloud-dev-1
  company.name: www.mldn.cn
  app.version: v-dev-1.0.0
---
spring:
  profiles: product
  application:
    name: mldncloud-config-client                    # 定义微服务名称
server:
  port: 8102                                          # 服务端口
info:
  app.name: mldn-microcloud-product-1
  company.name: www.mldn.cn
  app.version: v-product-1.0.0
```

2.【mldncloud-config-server-7501、mldncloud-config-client 项目】修改 pom.xml 配置文件，追加依赖库。

```
    <dependency>
        <groupId>org.springframework.cloud</groupId>
        <artifactId>spring-cloud-starter-bus-amqp</artifactId>
    </dependency>
    <dependency>
        <groupId>org.springframework.boot</groupId>
        <artifactId>spring-boot-starter-actuator</artifactId>
    </dependency>
```

3.【mldncloud-config-server-7501 项目】修改 application.yml 配置文件，追加 RabbitMQ 连接配置。

```
spring:
  rabbitmq:                          # 集成RabbitMQ作为消息服务总线处理
    host: rabbitmq-single            # RabbitMQ主机服务地址
    port: 5672                       # RabbitMQ的监听端口
```

```
    username: mldn                      # 用户名
    password: java                      # 密码
  application:
    name: mldncloud-config-server       # 定义微服务名称
  cloud:                                # 进行SpringCloud的相关配置
    config:                             # 进行SpringCloudConfig的相关配置
      git:                              # 设置Git连接地址
        uri: git@github.com:mldn/mldncloud.git
server:
  port: 7501                            # 服务端口
eureka:
  client:                               # 客户端进行Eureka注册的配置
    service-url:                        # 定义Eureka服务地址
      defaultZone: http://edmin:mldnjava@eureka-7001.com:7001/eureka
  instance:
    instance-id: config-7501.com        # 显示主机名称
```

当微服务启动时，会在后台发现/bus/refresh、/refresh 两个路径信息。

4.【mldncloud-security 项目】如果要实现 SpringCloudBus 更新，则需要修改安全策略，追加新的角色。

```
@Resource
public void configGlobal(AuthenticationManagerBuilder auth)
        throws Exception {    // 配置用户名与密码
    auth.inMemoryAuthentication().withUser("mldnjava").password("hello")
            .roles("USER","ACTUATOR").and().withUser("admin").password("hello")
            .roles("USER", "ADMIN").and();
}
```

5.【mldncloud-config-client 项目】所有微服务作为 RabbitMQ 消息消费端，修改 application.yml 配置消息组件。

```
spring:
  rabbitmq:                             # 现在将集成RabbitMQ作为消息服务总线处理
    host: rabbitmq-single               # RabbitMQ主机服务地址
    port: 5672                          # RabbitMQ的监听端口
    username: mldn                      # 用户名
    password: java                      # 密码
```

6.【mldncloud-config-client 项目】建立一个 InfoConfig 配置类，该类实现 mldncloud-config-client.yml 配置文件信息。

```
package cn.mldn.mldncloud.config;
import org.springframework.beans.factory.annotation.Value;
```

```
import org.springframework.cloud.context.config.annotation.RefreshScope;
import org.springframework.stereotype.Component;
@Component
@RefreshScope                          // 通过SpringCloudBus获取更新信息
public class InfoConfig {              // 将所有可能动态获取的配置内容写在一个类中，需要处引用
    @Value("${info.app.name}")
    private String appName ;
    @Value("${info.company.name}")
    private String companyName ;
    @Value("${info.app.version}")
    private String appVersion ;
    // setter、getter略
}
```

7.【mldncloud-config-client 项目】建立一个新的控制器，读取 InfoConfig 信息。

```
package cn.mldn.mldncloud.rest;
import org.springframework.beans.factory.annotation.Autowired;
import org.springframework.web.bind.annotation.GetMapping;
import org.springframework.web.bind.annotation.RestController;
import cn.mldn.mldncloud.config.InfoConfig;
@RestController
public class InfoConfigClientRest {
    @Autowired
    private InfoConfig infoConfig;    // 注入InfoConfig对象
    @GetMapping("/config")
    public Object getConfig() {       // 取得配置信息
        return "ApplicationName = " + this.infoConfig.getAppName()
            + "、CompanyName = " + this.infoConfig.getCompanyName()
            + "、ApplicationVersion = " + this.infoConfig.getAppVersion();
    }
}
```

此时程序启动之后，mldncloud-config-client 微服务（消息消费者）会通过 mldncloud-config-server-7501 微服务（消息提供者）抓取配置信息，同时在 RabbitMQ 中也会建立新的交换空间，如图 10-13 所示。

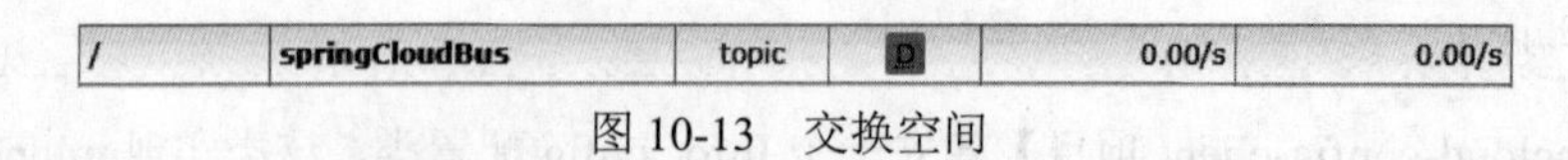

/	springCloudBus	topic	D	0.00/s	0.00/s

图 10-13　交换空间

当用户更改 mldncloud 仓库中的 mldncloud-config-client.yml 配置文件后，默认情况下 InfoConfigClientRest 程序类将无法抓取到最新的配置信息，只有执行以下命令才可以抓取到新的配置。

```
curl -X POST http://mldnjava:hello@config-7501.com:7501/bus/refresh
```

执行完本程序后会在控制台出现重新抓取配置的提示信息，当再次访问 InfoConfigClientRest 程序时就可以实现配置的动态加载，同时 RabbitMQ 中也会出现消息访问记录，如图 10-14 所示。

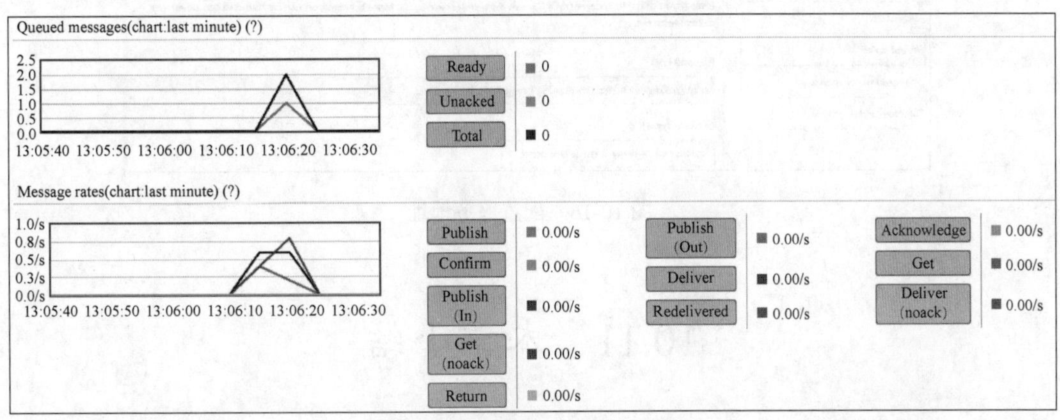

图 10-14　RabbitMQ 执行监控

提示：配置 SpringCloudBus 跟踪。

如果用户有需要，也可以获取一些跟踪轨迹信息，映射路径为/trace。要想实现这样的更新处理，必须在 mldncloud-config-server-7501 项目中追加一个新的配置项。

范例： 修改 application.yml 配置文件，追加 spring.cloud.bus.trace.enabled=true 配置项。

```
spring:
  cloud:  # 进行SpringCloud的相关配置
    bus:
      trace:
        enabled: true
```

这个时候就可以通过 trace 进行配置的更新追踪（http://mldnjava:hello@config-7501.com :7501/trace）。

虽然此时可以通过手动实现配置刷新，但是这种刷新不够智能。如果现在开发者在 GitHub 上保存了配置文件，则可以利用 mldncloud 仓库中的 Webhooks 调用/bus/refresh 路径动态刷新，如图 10-15 与图 10-16 所示。如果想正常配置，则需要开发者提供公网服务地址。

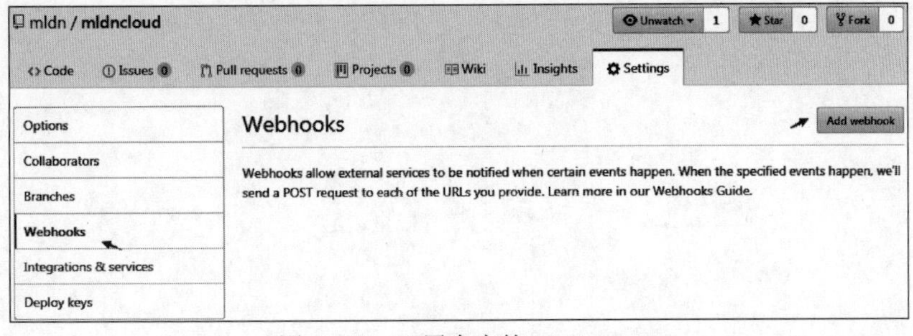

图 10-15　配置仓库的 Webhooks

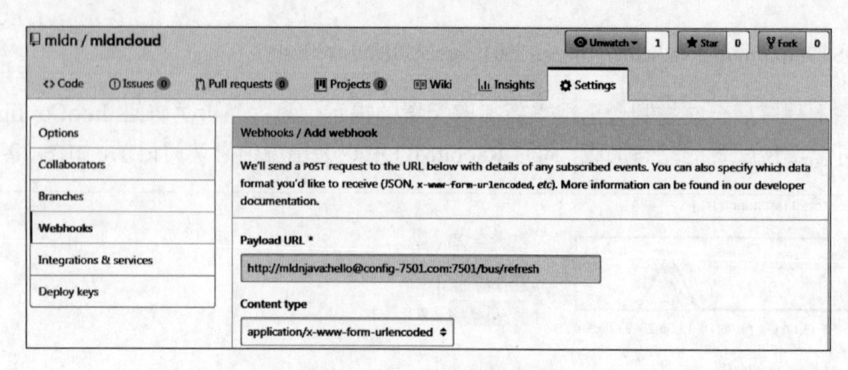

图 10-16　配置服务地址

10.11　本章小结

1. SpringCloudConfig 是提供配置文件统一管理的微服务，可以利用软件版本控制仓库（Git、SVN）实现配置保存。

2. SpringCloudConfig 服务端可以在一个仓库中实现多个配置文件的抓取，也可以通过应用仓库自动选择实现多个仓库配置文件的抓取。

3. SpringCloudConfig 客户端要通过 bootstrap.yml 配置 SpringCloudConfig 服务端地址，这样在客户端启动时就可以自动实现配置文件加载。

4. 在 SpringCloudConfig 中利用密钥与 KeyStore 实现重要信息加密。

5. SpringCloudConfig 服务端作为一个微服务，可以在 Eureka 中注册，以实现配置微服务的高可用。

6. 利用 SpringCloudBus 可以实现配置文件的动态抓取，并且可以结合 GitHub 中的 Webhooks 实现配置自动更新。

第 11 章

SpringCloudStream

通过本章学习，可以达到以下目标：

1. 理解 SpringCloudStream 的主要作用与设计结构。
2. 使用 RabbitMQ 与 SpringCloudStream 整合实现流数据处理。

SpringCloudStream 是构建消息驱动的微服务应用程序的框架。其基于 SpringBoot 建立独立的生产级 Spring 应用程序，并且可以方便地与 RabbitMQ 实现整合应用。

11.1 SpringCloudStream 简介

在企业项目处理中，消息组件是进行业务加强的主要技术手段，利用消息组件不仅可以提升系统操作的吞吐量，也可以避免大并发状态下的用户业务处理问题，SpringCloud 技术作为微服务的一种实现架构，能很好地支持消息组件的整合应用。SpringCloudStream 的整合架构如图 11-1 所示。

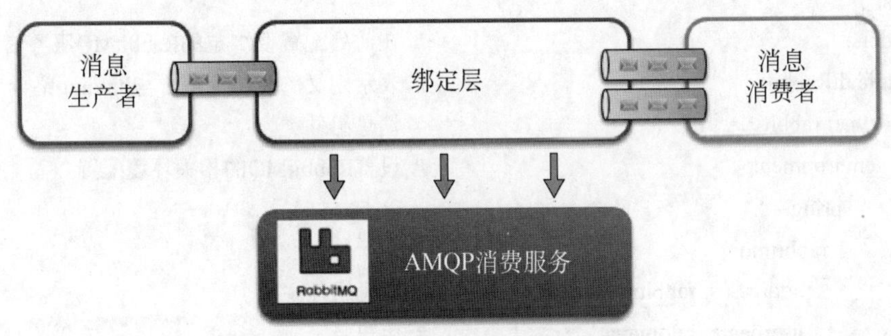

图 11-1　SpringCloudStream 架构

开发者通过定义绑定器作为中间层，实现了应用程序与消息中间件细节之间的隔离。通过向应用程序暴露统一的 Channel 通道，使得应用程序不需要再考虑各种不同消息中间件的实现。当需要升级消息中间件，或者更换其他消息中间件产品时，开发需要做的就是更换对应的 Binder 绑定器，而不需要修改任何应用逻辑。

提问：直接使用消息组件也可以实现，为什么还需要 SpringCloudStream？

如果开发者直接使用 RabbitMQ 也可以实现消息的生产者与消费者处理，为什么还要单独去提供 SpringCloudStream？

 **回答：SpringCloudStream 可发挥出 Restful 特点。**

之所以需要提供 SpringCloudStream，主要原因在于 SpringCloud 与 Restful 之间存在联系。最简单的理解就是：当用户使用自定义类对象进行传输时，SpringCloud 会结合 Jackson，将对象转换为 JSON 处理结构，而 SpringCloudStream 继续发挥此特点，基于消息组件实现了对象与 JSON 间的转换及传输，这样开发者处理起来会更加方便。

11.2 Stream 生产者

在整合消息系统的过程中，一定会有一个消息的生产者，同时为了使用者开发方便，往往也会建立单独的消息发送业务接口。本例将使用 RabbitMQ 实现消息的发送。

1.【mldncloud-stream-provider-8201 项目】修改 pom.xml 配置文件，引入依赖库。

```
<dependency>
    <groupId>org.springframework.cloud</groupId>
    <artifactId>spring-cloud-starter-stream-rabbit</artifactId>
</dependency>
```

2.【mldncloud-stream-provider-8201 项目】修改 application.yml 配置文件，引入相关配置。

```
spring:
  cloud:
    stream:
      binders:                              # 在此处配置要绑定的RabbitMQ服务信息
        defaultRabbit:                      # 表示定义的名称，用于与binding整合
          type: rabbit                      # 消息组件类型
          environment:                      # 设置RabbitMQ的相关环境配置
            spring:
              rabbitmq:
                addresses: rabbitmq-single  # 消息组件主机
                username: mldnjava          # 用户名
                password: hello             # 密码
                virtual-host: /             # 虚拟主机
      bindings:                             # 服务的整合处理
        output:                             # 通道的名称
          destination: MLDNExchange         # Exchange名称定义
          content-type: application/json    # 设置消息类型，本次为对象json
          binder: defaultRabbit             # 设置要绑定的消息服务的具体设置
  application:
    name: mldncloud-stream-provider         # 定义微服务名称
```

3.【mldncloud-stream-provider-8201 项目】建立一个消息发送接口。

```
package cn.mldn.mldncloud.service;
import cn.mldn.mldncloud.dto.DeptDTO;
public interface IMessageProvider {
    /**
     * 实现消息的发送，本次发送的消息是一个对象（自动变为json）
     * @param dto DTO对象，该对象不为null
     */
    public void send(DeptDTO dto) ;
}
```

在本业务接口中要发送的是一个 DTO 对象，这个对象在发送时将自动转为 JSON 数据格式进行传输。

4.【mldncloud-stream-provider-8201 项目】建立消息发送接口实现子类。

```
package cn.mldn.mldncloud.service.impl;
import javax.annotation.Resource;
import org.springframework.cloud.stream.annotation.EnableBinding;
import org.springframework.cloud.stream.messaging.Source;
import org.springframework.integration.support.MessageBuilder;
import org.springframework.messaging.MessageChannel;
import cn.mldn.mldncloud.dto.DeptDTO;
import cn.mldn.mldncloud.service.IMessageProvider;
@EnableBinding(Source.class)                                        // 消息发送管道的定义
public class MessageProviderImpl implements IMessageProvider {
    @Resource
    private MessageChannel output;                                  // 消息的发送管道
    @Override
    public void send(DeptDTO dto) {
        this.output.send(MessageBuilder.withPayload(dto).build());  // 创建并发送消息
    }
}
```

本程序中注入了 MessageChannel 对象，利用该对象可以加载 application.yml 中的配置，进行消息的发送操作。

提示：关于 org.springframework.cloud.stream.messaging.Source 接口。

Source 接口中定义了相应的发送管道信息配置，具体代码定义如下。

范例：Source 接口定义。

```
public interface Source {
    String OUTPUT = "output";    // 之前所设置的消息发送的管道
```

```
    @Output(Source.OUTPUT)
    MessageChannel output();
}
```

本程序定义的发送通道为 output，这与 application.yml 中配置的 bindings.output 名称相同。而在随后定义的消费端程序处也会有一个与之对应的 Sink 接口，里面定义了输入管道配置。

5.【mldncloud-stream-provider-8201 项目】编写测试类，实现消息发送配置。

```
package cn.mldn.mldncloud.test;
import javax.annotation.Resource;
import org.junit.Test;
import org.junit.runner.RunWith;
import org.springframework.boot.test.context.SpringBootTest;
import org.springframework.test.context.junit4.SpringJUnit4ClassRunner;
import org.springframework.test.context.web.WebAppConfiguration;
import cn.mldn.mldncloud.StartStreamApplication8201;
import cn.mldn.mldncloud.dto.DeptDTO;
import cn.mldn.mldncloud.service.IMessageProvider;
@RunWith(SpringJUnit4ClassRunner.class)
@SpringBootTest(classes=StartStreamApplication8201.class)
@WebAppConfiguration
public class TestMessageProvider {
    @Resource
    private IMessageProvider messageProvider ;              // 注入业务接口实例
    @Test
    public void testSend() {
        DeptDTO dto = new DeptDTO() ;                       // 实例化DTO对象
        dto.setDeptno(99L);
        dto.setDname("魔乐科技教学研发部");
        dto.setLoc("北京-天安门");
        this.messageProvider.send(dto);                     // 消息发送
    }
}
```

启动测试程序类后，会在 RabbitMQ 控制台中发现相关的 Exchange 定义，如图 11-2 所示。

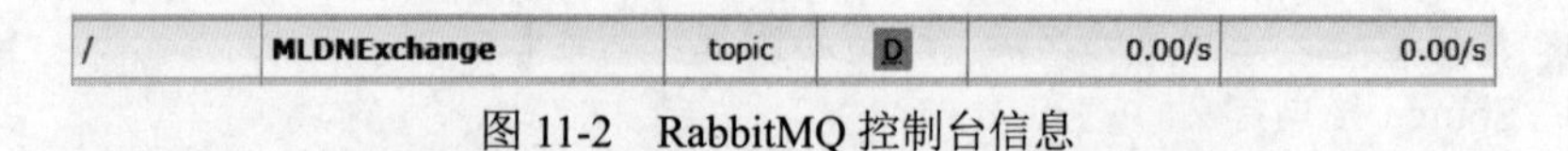

图 11-2　RabbitMQ 控制台信息

当消息成功发送之后，可以通过 RabbitMQ 看见相关的数据访问监控，如图 11-3 所示。

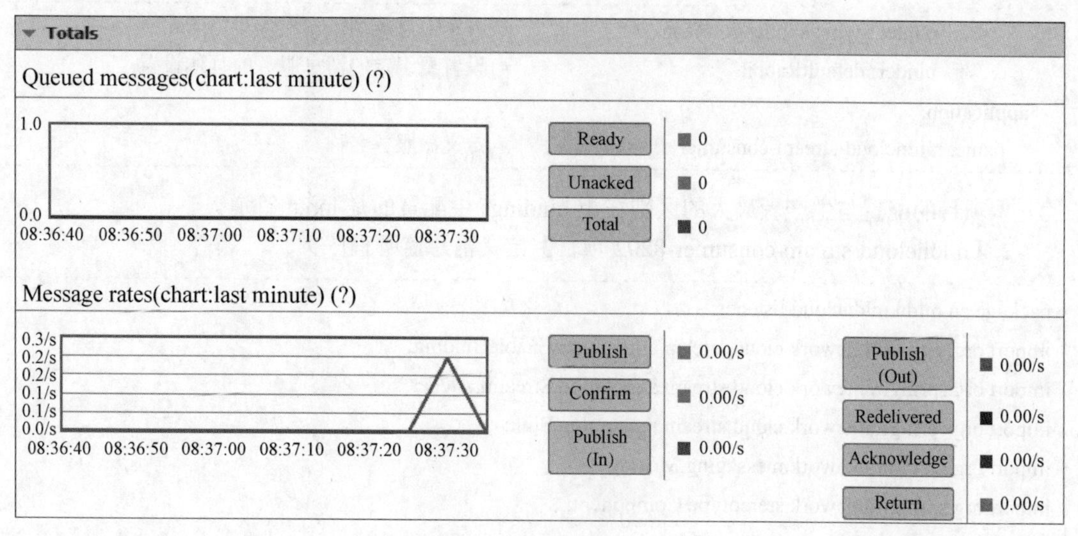

图 11-3　消息监控界面

11.3　Stream 消费者

如果要通过 RabbitMQ 获取消息，则一定要创建消息消费端程序。由于 SpringCloudStream 实现的是消息与 Restful 之间的转换处理，所以消费者可以通过接收的 JSON 数据，实现与类对象间的转换。

1.【mldncloud-stream-consumer-8202 项目】本项目使用的依赖库与 mldncloud-stream-provider-8201 项目一样，区别在于需要针对 application.yml 进行消费配置。

```
spring:
  cloud:
    stream:
      binders:                              # 在此处配置要绑定的RabbitMQ服务信息
        defaultRabbit:                      # 表示定义的名称，用于与binding整合
          type: rabbit                      # 消息组件类型
          environment:                      # 设置RabbitMQ的相关环境配置
            spring:
              rabbitmq:
                addresses: rabbitmq-single  # 消息组件主机
                username: mldnjava          # 用户名
                password: hello             # 密码
                virtual-host: /             # 虚拟主机
      bindings:                             # 服务的整合处理
        input:                              # 通道的名称
          destination: MLDNExchange         # Exchange名称定义
```

```
          content-type: application/json                # 设置消息类型，本次为对象json
          binder: defaultRabbit                         # 设置要绑定的消息服务的具体设置
  application:
    name: mldncloud-stream-consumer                     # 定义微服务名称
```

本项目的配置与生产者最大的区别在于 bindings 中配置的是 input 管道。

2.【mldncloud-stream-consumer-8202 项目】定义消息监听程序类。

```
package cn.mldn.mldncloud.listener;
import org.springframework.cloud.stream.annotation.EnableBinding;
import org.springframework.cloud.stream.annotation.StreamListener;
import org.springframework.cloud.stream.messaging.Sink;
import org.springframework.messaging.Message;
import org.springframework.stereotype.Component;
import cn.mldn.mldncloud.dto.DeptDTO;
@Component
@EnableBinding(Sink.class)
public class MessageListener {
    @StreamListener(Sink.INPUT)
    public void input(Message<DeptDTO> message) {
        System.err.println("【*** 消息接收 ***】" + message.getPayload());
    }
}
```

配置完成后可以启动消费端程序，而后就可以实现 Stream 消息的发送。发送和接收过程中都是以 JSON 为数据交互格式，在消费端接收后可以自动将 JSON 数据转为 DTO 对象。在 RabbitMQ 控制台可以看见如图 11-4 所示的信息。

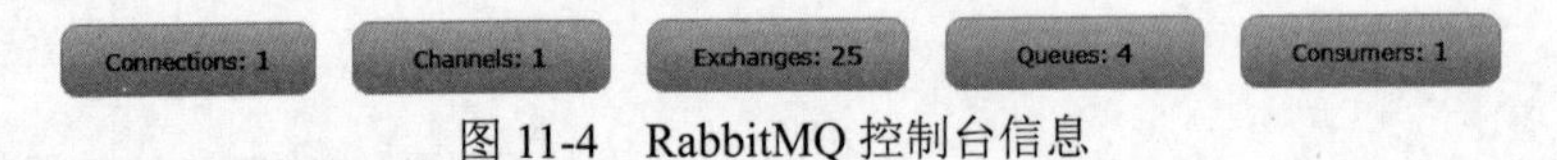

图 11-4 RabbitMQ 控制台信息

11.4 自定义消息通道

一套完整的 SpringCloudStream 中会包含消息的生产者与发送者，生产者要利用 Source 配置消息生产通道，消费者也需要通过 Sink 配置消息消费通道。如果开发者有需要，也可以自定义新的消息通道。

1.【mldncloud-api 项目】由于生产者与消费者需要进行消费通道更改，可以在公共 API 项目中进行通道定义。首先修改 pom.xml 配置文件，引入依赖包。

```
<dependency>
    <groupId>org.springframework.cloud</groupId>
```

```
        <artifactId>spring-cloud-starter-stream-rabbit</artifactId>
    </dependency>
```

2.【mldncloud-api 项目】定义公共通道。

```
package cn.mldn.mldncloud.channel;
import org.springframework.cloud.stream.annotation.Input;
import org.springframework.cloud.stream.annotation.Output;
import org.springframework.messaging.MessageChannel;
import org.springframework.messaging.SubscribableChannel;
public interface DefaultProcess {
    public static final String OUTPUT = "mldn_output";        // 输出通道名称
    public static final String INPUT = "mldn_input";          // 输入通道名称
    @Input(DefaultProcess.INPUT)
    public SubscribableChannel input();
    @Output(DefaultProcess.OUTPUT)
    public MessageChannel output();
}
```

3.【mldncloud-stream-provider-8201 项目】修改 application.yml，使用自定义通道名称。

```
spring:
  cloud:
    stream:
      bindings:                                   # 服务的整合处理
        mldn_output:                              # 通道的名称
          destination: MLDNExchange               # Exchange名称定义
          content-type: application/json          # 设置消息类型，本次为对象json
          binder: defaultRabbit                   # 设置要绑定的消息服务的具体设置
```

4.【mldncloud-stream-provider-8201 项目】修改 MessageProviderImpl 实现子类。

```
package cn.mldn.mldncloud.service.impl;
import javax.annotation.Resource;
import org.springframework.beans.factory.annotation.Qualifier;
import org.springframework.cloud.stream.annotation.EnableBinding;
import org.springframework.integration.support.MessageBuilder;
import org.springframework.messaging.MessageChannel;
import cn.mldn.mldncloud.channel.DefaultProcess;
import cn.mldn.mldncloud.dto.DeptDTO;
import cn.mldn.mldncloud.service.IMessageProvider;
@EnableBinding(DefaultProcess.class)                              // 消息发送管道的定义
public class MessageProviderImpl implements IMessageProvider {
    @Qualifier(DefaultProcess.OUTPUT)                             // 设置一个标记，避免类型重复
```

```
    @Resource
    private MessageChannel output;                                        // 消息的发送管道
    @Override
    public void send(DeptDTO dto) {
        this.output.send(MessageBuilder.withPayload(dto).build());        // 创建并发送消息
    }
}
```

5.【mldncloud-stream-consumer-8202 项目】修改 application.yml，使用自定义通道名称。

```
spring:
  cloud:
    stream:
      bindings:                                    # 服务的整合处理
        mldn_input:                                # 通道的名称
          destination: MLDNExchange                # Exchange名称定义
          content-type: application/json           # 设置消息类型，本次为对象json
          binder: defaultRabbit                    # 设置要绑定的消息服务的具体设置
```

6.【mldncloud-stream-consumer-8202 项目】修改 MessageListener 程序类。

```
package cn.mldn.mldncloud.listener;
import org.springframework.cloud.stream.annotation.EnableBinding;
import org.springframework.cloud.stream.annotation.StreamListener;
import org.springframework.messaging.Message;
import org.springframework.stereotype.Component;
import cn.mldn.mldncloud.channel.DefaultProcess;
import cn.mldn.mldncloud.dto.DeptDTO;
@Component
@EnableBinding(DefaultProcess.class)
public class MessageListener {
    @StreamListener(DefaultProcess.INPUT)
    public void input(Message<DeptDTO> message) {
        System.err.println("【*** 消息接收 ***】" + message.getPayload());
    }
}
```

此时，就可以利用自定义通道实现消息的发送与消费处理。

11.5 分组与持久化

在上面的程序里面成功地实现了消息的发送及接收，但是需要注意一个问题，所发送的消

息在默认情况下属于一种临时消息，也就是说，如果现在没有消费者进行消费处理，那么该消息是不会被保留的。要想实现持久化的消息处理，重点在于消息的消费端配置，同时也需要考虑到一个分组的情况（有分组就表示该消息可以进行持久化）。

范例：【mldncloud-stream-consumer-8202 项目】修改 application.yml 配置文件，追加分组配置。

```
spring:
  cloud:
    stream:
      bindings:                                   # 服务的整合处理
        mldn_input:                               # 通道的名称
          destination: MLDNExchange               # Exchange名称定义
          content-type: application/json          # 设置消息类型，本次为对象json
          binder: defaultRabbit                   # 设置要绑定的消息服务的具体设置
          group: mldn-group                       # 进行操作的分组，实际上就表示持久化
```

此时在消费端配置了 group 分支项，随后运行程序，可以通过 RabbitMQ 控制台看到如图 11-5 所示的信息。

Overview				Messages			Message rates			+/-
Virtual host	Name	Features	State	Ready	Unacked	Total	incoming	deliver / get	ack	
/	MLDNExchange.mldn-group	D	idle	0	0	0				

图 11-5　消息持久化

在 SpringCloudStream 中如果要设置持久化队列，则名称格式为 destination.group。此时关闭消费端的微服务之后，该队列信息依然会被保留在 RabbitMQ 中，只有当消费端取走数据后，该消息才会被删除。

11.6　RoutingKey

默认情况下，之前的程序都是属于广播消息。也就是说，所有的消费者都可以接收发送消息内容。在 RabbitMQ 里面支持有直连消息，而直连消息主要是通过 RoutingKey 来实现，利用直连消息可以实现准确的消息消费端的接收处理。

1.【mldncloud-stream-provider-8201 项目】修改 application.yml 配置文件，追加 RoutingKey 配置。

```
spring:
  cloud:
    stream:
      rabbit:
        bindings:
          mldn_output:
            producer: # 进行生产端配置
              routing-key-expression: '"mldn-key"'
```

2.【mldncloud-stream-consumer-8202 项目】修改 application.yml 配置文件，追加 RoutingKey 配置。

```
spring:
  cloud:
    stream:
      rabbit:                                    # 进行rabbit的相关绑定配置
        bindings:
          mldn_input:
            consumer:                            # 进行消费端配置
              bindingRoutingKey: mldn-key        # 设置一个 RoutingKey 信息
```

此时，生产者与消费者配置了相同的 RoutingKey 信息，所以可以正常实现消息处理。如果消费端配置的 RoutingKey 不同，将无法接收到消息。

11.7 本章小结

1. SpringCloudStream 可以实现消息驱动微服务的搭建。

2. SpringCloudStream 最大的特征是用户采用对象的形式进行程序处理，而在消息传递中可以将对象自动转换为 JSON 结构，同时在消费端也可以实现 JSON 数据与对象之间的转换。

3. SpringCloudStream 支持 RabbitMQ 与 Kafka 两类消息组件，建议采用 RabbitMQ 整合。

4. SpringCloudStream 默认通道使用的是 Source 与 Sink 接口，如果开发者有需要，也可以自定义通道配置。

5. SpringCloudStream 结合 RabbitMQ 时，可以利用消费端的分组配置实现消息持久化存储。

第 12 章 SpringCloudSleuth

通过本章学习，可以达到以下目标：

1. 理解微服务开发问题与环形调用。
2. 理解 Sleuth 跟踪服务的主要作用与调用监控。
3. 理解 Sleuth 数据采集处理。

微服务的开发与调用是一个周期很长并且非常繁杂的处理过程，为了可以监控各个微服务之间的调用情况，在 SpringCloud 里面提供 Sleuth 跟踪技术，可以针对微服务的调用实现信息采集处理。本章将对 SpringCloudSleuth 的使用进行讲解。

12.1 SpringCloudSleuth 简介

微服务是一种子业务的拆分处理机制，在微服务处理架构过程中经常会出现若干个微服务互相调用的情况，如图 12-1 所示。

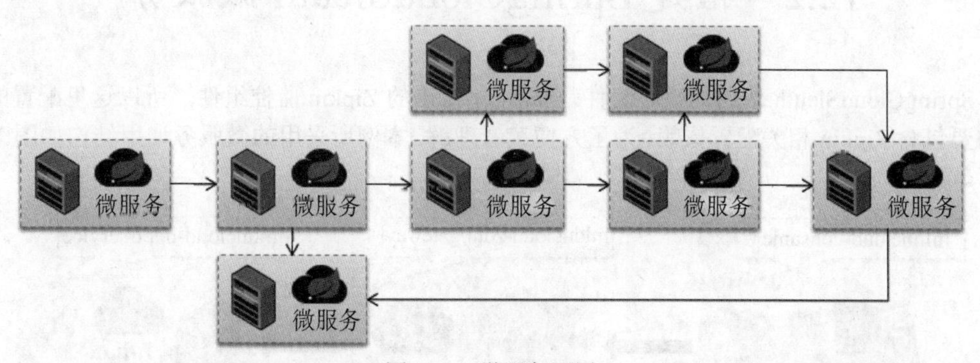

图 12-1 微服务调用

图 12-1 给出的是一种实际开发中可能存在的调用过程，有可能是几个微服务之间互相调用，也有可能为完成某一个大型的业务需要几十个微服务之间互相调用。在这样的场景中，就有可能出现如下几个问题：

☑ 当业务处理执行速度变慢时，有可能是某一个或某几个微服务处理性能不高。该如何去追踪这些处理速度较慢的微服务，从而实现性能的整体提升？

☑ 如果某一个微服务出现问题，应该如何快速找到出现问题的微服务并且加以修复？

☑ 如果现在微服务变为环形调用，那么这些关系该如何描述出来？

所以，一个完善的微服务并不只是简单地进行 RPC 的功能实现，还应该对整体的微服务执行进行监控。SpringCloud 中提供的 Sleuth 技术就可以实现微服务的调用跟踪，它可以自动形成一个调用连接线，通过这个连接线开发者可以轻松找到所有微服务间的关系。所有微服务的调用信息都自动发送到 Sleuth 中，如图 12-2 所示。这样不仅可以采集到微服务调用的关系，也可以获取微服务所耗费的时间，从而进行整体微服务状态的监控以及相应的数据分析。

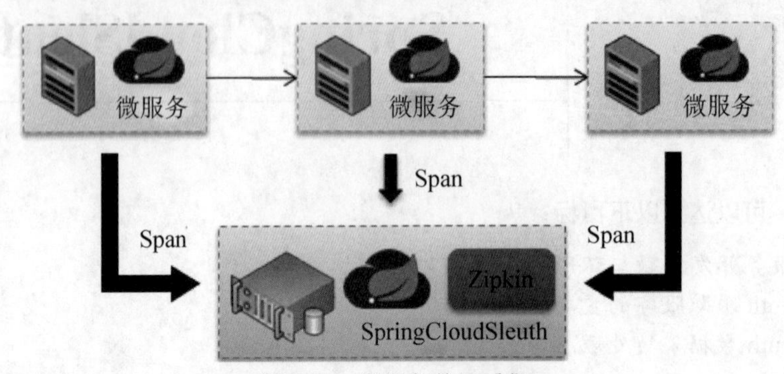

图 12-2　微服务数据采集

在图 12-2 中可以看到，所有微服务发送到 Sleuth 采集微服务上的信息都是以 Span 描述的，每一个 Span 包含 4 个组成部分。

☑　cs-Client Sent：客户端发出一个请求，描述的是一个 Span 开始。

☑　sr-Server Received：服务端接收请求，sr-cs 表示发送的网络延迟。

☑　ss-Server Sent：服务端发送请求（回应处理），ss-sr 表示服务端的消耗时间。

☑　cr-Client Received：客户端接收到服务端数据，cr-ss 表示回复所需要的时间。

12.2　搭建 SpringCloudSleuth 微服务

SpringCloudSleuth 使用的核心组件是 Twitter 推出的 Zipkin 监控组件，所以这里配置的模块需要包含 Zipkin 相关配置依赖。为了方便读者理解，本例所采用的微服务调用结构如图 12-3 所示。

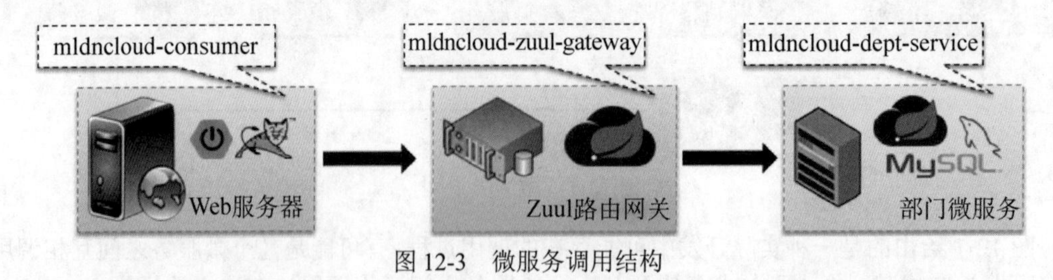

图 12-3　微服务调用结构

1.【mldncloud-sleuth-8601 项目】修改 pom.xml 配置文件，追加如下依赖库。

```
<dependency>
    <groupId>org.springframework.cloud</groupId>
    <artifactId>spring-cloud-starter-sleuth</artifactId>
</dependency>
```

```
<dependency>
    <groupId>org.springframework.cloud</groupId>
    <artifactId>spring-cloud-starter-zipkin</artifactId>
</dependency>
<dependency>
    <groupId>io.zipkin.java</groupId>
    <artifactId>zipkin-server</artifactId>
</dependency>
<dependency>
    <groupId>io.zipkin.java</groupId>
    <artifactId>zipkin-autoconfigure-ui</artifactId>
</dependency>
```

2.【mldncloud-sleuth-8601 项目】修改 application.yml 配置文件。

```
server:
  port: 8601                # 配置监听端口号
spring:
  application:
    name: mldncloud-zipkin-server
```

3.【mldncloud-sleuth-8601 项目】修改程序启动类。

```
package cn.mldn.mldncloud;
import org.springframework.boot.SpringApplication;
import org.springframework.boot.autoconfigure.SpringBootApplication;
import zipkin.server.EnableZipkinServer;
@SpringBootApplication
@EnableZipkinServer                         // 启用Zipkin服务
public class StartSleuthApplication8601 {
    public static void main(String[] args) {
        SpringApplication.run(StartSleuthApplication8601.class, args);
    }
}
```

4.【操作系统】修改 hosts 主机配置文件。

```
127.0.0.1    zipkin.com
```

5.【mldncloud-dept-service-8001 项目、mldncloud-consumer 项目、mldncloud-zuul-gateway-9501 项目】在需要监控的微服务中引入 spring-cloud-starter-zipkin 依赖库。

```
<dependency>
    <groupId>org.springframework.cloud</groupId>
    <artifactId>spring-cloud-starter-zipkin</artifactId>
</dependency>
```

6.【mldncloud-dept-service-8001 项目、mldncloud-consumer 项目、mldncloud-zuul-gateway-9501 项目】修改 application.yml 配置文件，配置 Sleuth 连接信息。

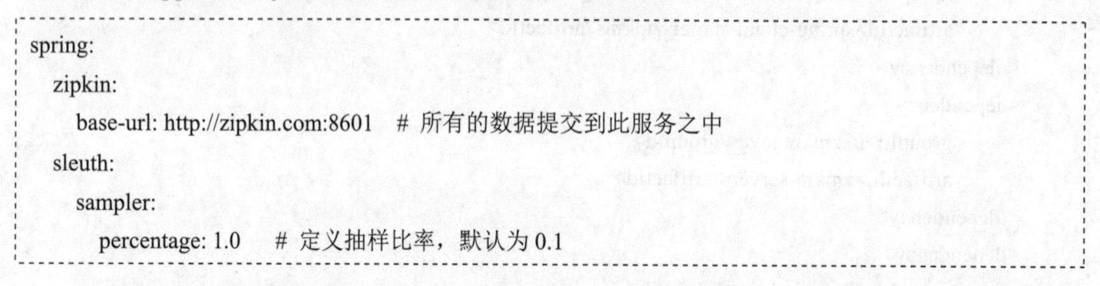

```
spring:
  zipkin:
    base-url: http://zipkin.com:8601   # 所有的数据提交到此服务之中
  sleuth:
    sampler:
      percentage: 1.0      # 定义抽样比率，默认为 0.1
```

在本配置中，抽样比率指的是访问次数百分比，即如果采用默认的 0.1，则表示每 10 次访问进行一次抽样。这里为了让读者观察清晰，将每一次访问都进行了记录。

7. 依次启动所有相关的微服务，并且通过消费端进行部门微服务的调用，随后就可以通过 Zipkin 地址检测到访问信息。图 12-4 显示了所监测到的访问信息列表，图 12-5 显示了其中一次访问监测信息。

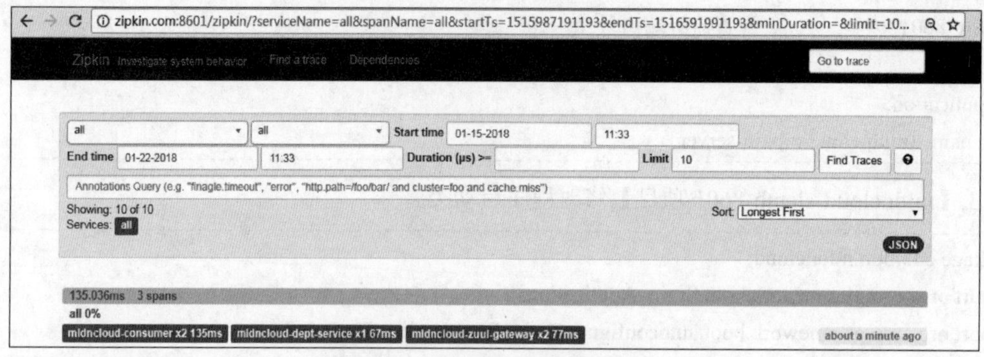

图 12-4　微服务访问监测列表

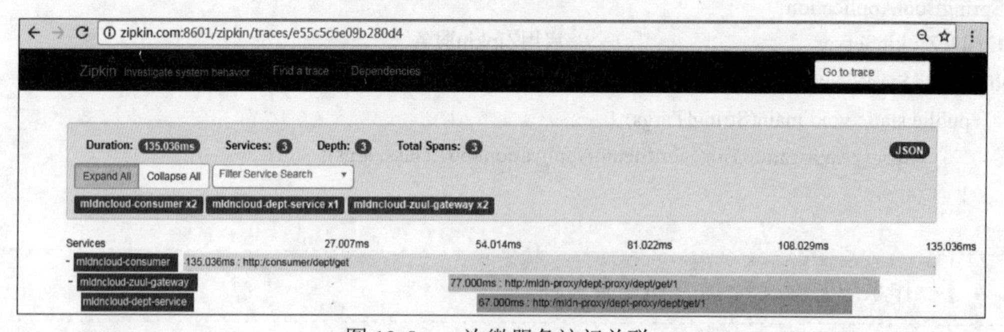

图 12-5　一次微服务访问关联

12.3　Sleuth 数据采集

现在已经成功地实现了一个 SpringCloudSleuth 基础操作。此时需要考虑一个实际的问题：当前所有发送到 Sleuth 服务端的统计汇总操作都是记录在内存中的，也就是说，如果开发者关

闭了 Zipkin 服务端，那么这些统计信息将消失。这样的设计明显是不合理的，应该将这些统计的数据记录保存下来。同时，有可能一个项目中存在许多微服务，这样就需要发送大量的数据信息进入，为了解决这种高并发的问题，可以结合消息组件（Stream）进行缓存处理。本例为了方便，将统计结果保存在数据库之中（MySQL），程序的操作结构如图 12-6 所示。

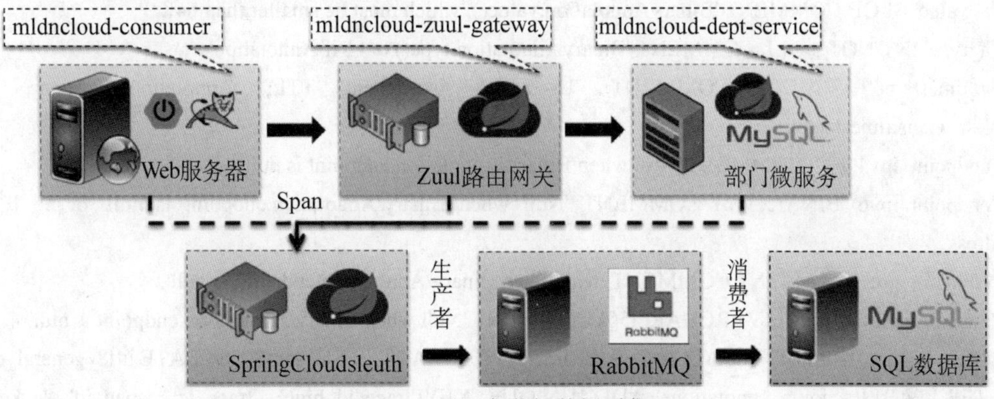

图 12-6　微服务监控数据采集

1.【MySQL 数据库】本次以 MySQL 作为数据采集存储介质，所以执行以下脚本，进行数据库创建。

```
DROP DATABASE IF EXISTS zipkin ;
CREATE DATABASE zipkin CHARACTER SET UTF8 ;
USE zipkin ;
CREATE TABLE IF NOT EXISTS zipkin_spans (
    'trace_id_high' BIGINT NOT NULL DEFAULT 0 COMMENT 'If non zero, this means the trace uses 128
bit traceIds instead of 64 bit',
    'trace_id' BIGINT NOT NULL,
    'id' BIGINT NOT NULL,
    'name' VARCHAR(255) NOT NULL,
    'parent_id' BIGINT,
    'debug' BIT(1),
    'start_ts' BIGINT COMMENT 'Span.timestamp(): epoch micros used for endTs query and to implement
TTL',
    'duration' BIGINT COMMENT 'Span.duration(): micros used for minDuration and maxDuration query'
) ENGINE=InnoDB ROW_FORMAT=COMPRESSED CHARACTER SET=utf8 COLLATE utf8_general_ci;
ALTER TABLE zipkin_spans ADD UNIQUE KEY('trace_id_high', 'trace_id', 'id');
ALTER TABLE zipkin_spans ADD INDEX('trace_id_high', 'trace_id', 'id');
ALTER TABLE zipkin_spans ADD INDEX('trace_id_high', 'trace_id');
ALTER TABLE zipkin_spans ADD INDEX('name');
ALTER TABLE zipkin_spans ADD INDEX('start_ts');
CREATE TABLE IF NOT EXISTS zipkin_annotations (
    'trace_id_high' BIGINT NOT NULL DEFAULT 0 COMMENT 'If non zero, this means the trace uses 128
bit traceIds instead of 64 bit',
```

```
'trace_id' BIGINT NOT NULL COMMENT 'coincides with zipkin_spans.trace_id',
'span_id' BIGINT NOT NULL COMMENT 'coincides with zipkin_spans.id',
'a_key' VARCHAR(255) NOT NULL COMMENT 'BinaryAnnotation.key or Annotation.value if type == -1',
'a_value' BLOB COMMENT 'BinaryAnnotation.value(), which must be smaller than 64KB',
'a_type' INT NOT NULL COMMENT 'BinaryAnnotation.type() or -1 if Annotation',
'a_timestamp' BIGINT COMMENT 'Used to implement TTL; Annotation.timestamp or zipkin_spans.timestamp',
'endpoint_ipv4' INT COMMENT 'Null when Binary/Annotation.endpoint is null',
'endpoint_ipv6' BINARY(16) COMMENT 'Null when Binary/Annotation.endpoint is null, or no IPv6 address',
'endpoint_port' SMALLINT COMMENT 'Null when Binary/Annotation.endpoint is null',
'endpoint_service_name' VARCHAR(255) COMMENT 'Null when Binary/Annotation.endpoint is null'
) ENGINE=InnoDB ROW_FORMAT=COMPRESSED CHARACTER SET=utf8 COLLATE utf8_general_ci;
ALTER TABLE zipkin_annotations ADD UNIQUE KEY('trace_id_high', 'trace_id', 'span_id', 'a_key', 'a_timestamp');
ALTER TABLE zipkin_annotations ADD INDEX('trace_id_high', 'trace_id', 'span_id') ;
ALTER TABLE zipkin_annotations ADD INDEX('trace_id_high', 'trace_id');
ALTER TABLE zipkin_annotations ADD INDEX('endpoint_service_name');
ALTER TABLE zipkin_annotations ADD INDEX('a_type');
ALTER TABLE zipkin_annotations ADD INDEX('a_key');
CREATE TABLE IF NOT EXISTS zipkin_dependencies (
  'day' DATE NOT NULL,
  'parent' VARCHAR(255) NOT NULL,
  'child' VARCHAR(255) NOT NULL,
  'call_count' BIGINT
) ENGINE=InnoDB ROW_FORMAT=COMPRESSED CHARACTER SET=utf8 COLLATE utf8_general_ci;
ALTER TABLE zipkin_dependencies ADD UNIQUE KEY('day', 'parent', 'child');
```

2.【mldncloud-sleuth-8601 项目】修改 pom.xml 配置文件，追加相关依赖库。

```
<dependency>
    <groupId>org.springframework.cloud</groupId>
    <artifactId>spring-cloud-sleuth-zipkin-stream</artifactId>
</dependency>
<dependency>
    <groupId>org.springframework.cloud</groupId>
    <artifactId>spring-cloud-starter-stream-rabbit</artifactId>
</dependency>
<dependency>
    <groupId>org.springframework.boot</groupId>
    <artifactId>spring-boot-starter-jdbc</artifactId>
```

```
    </dependency>
    <dependency>
        <groupId>mysql</groupId>
        <artifactId>mysql-connector-java</artifactId>
    </dependency>
```

需要注意的是，此时需要删除 spring-cloud-starter-zipkin 依赖库。

3.【mldncloud-sleuth-8601 项目】修改 application.yml 配置文件，追加 MySQL 连接与 RabbitMQ 消息组件配置。

```
server:
  port: 8601                                    # 配置监听端口号
spring:
  rabbitmq:
    host: rabbitmq-single                       # 消息主机
    port: 5672                                  # 连接端口
    username: mldn                              # 用户名
    password: java                              # 密码
    virtual-host: /                             # 虚拟主机
  datasource:
    url: jdbc:mysql://localhost:3306/zipkin     # 连接地址
    username: root                              # 用户名
    password: mysqladmin                        # 密码
    driver-class-name: org.gjt.mm.mysql.Driver  # 驱动程序
    initialize: true                            # 初始化
  application:
    name: mldncloud-zipkin-server
zipkin:
  storage:                                      # 设置Zipkin收集的信息通过MySQL进行存储
    type: mysql                                 # 数据库存储
```

4.【mldncloud-dept-service-8001 项目、mldncloud-consumer 项目、mldncloud-zuul-gateway-9501 项目】修改 pom.xml 配置文件，追加依赖配置。

```
    <dependency>
        <groupId>org.springframework.cloud</groupId>
        <artifactId>spring-cloud-sleuth-zipkin-stream</artifactId>
    </dependency>
    <dependency>
        <groupId>org.springframework.cloud</groupId>
        <artifactId>spring-cloud-starter-stream-rabbit</artifactId>
    </dependency>
```

5.【mldncloud-dept-service-8001 项目、mldncloud-consumer 项目、mldncloud-zuul-gateway-9501 项目】修改 application.yml 配置文件，由于此时是通过消息组件进行采集信息的发送，所以删除 zipkin.base-url 配置，追加 RabbitMQ 相关配置，这几个微服务将作为消息生产者存在。

```
spring:
  rabbitmq:
    host: rabbitmq-single                     # 消息主机
    port: 5672                                # 连接端口
    username: mldn                            # 用户名
    password: java                            # 密码
    virtual-host: /                           # 虚拟主机
```

6.【mldncloud-sleuth-8601 项目】修改程序，启动注解。

```
package cn.mldn.mldncloud;
import org.springframework.boot.SpringApplication;
import org.springframework.boot.autoconfigure.SpringBootApplication;
import org.springframework.cloud.sleuth.zipkin.stream.EnableZipkinStreamServer;
@SpringBootApplication
@EnableZipkinStreamServer                                    // 启用Zipkin服务
public class StartSleuthApplication8601 {
    public static void main(String[] args) {
        SpringApplication.run(StartSleuthApplication8601.class, args);
    }
}
```

此时启动各个微服务，这样所有被监听的微服务都将成为消息的生产者，然后 Sleuth 将作为消息消费者，将收到的消息保存到数据库中存储。图 12-7 显示了 RabbitMQ 消息组件监控到的信息。

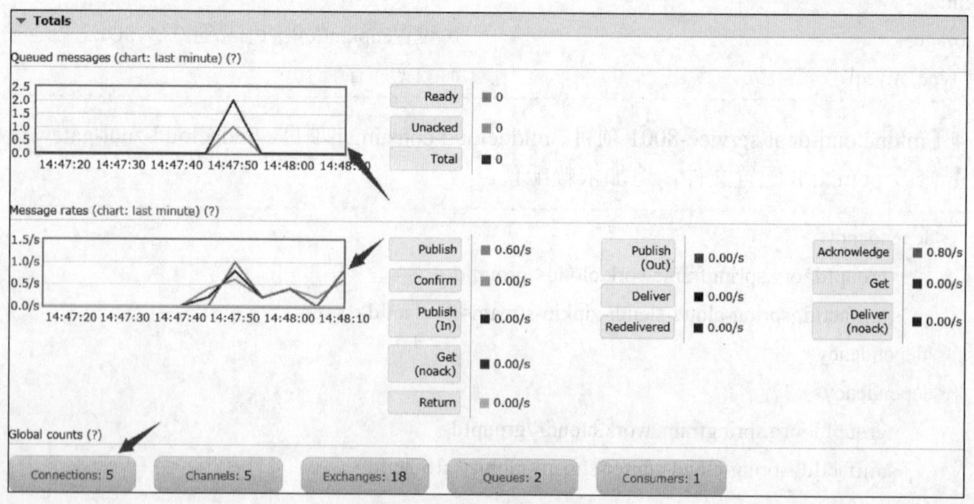

图 12-7 Sleuth 消息监控

12.4　本章小结

1. SpringCloudSleuth 是数据采集微服务，可以与 Zipkin 结合，实现微服务的调用结构观察。

2. SpringCloudSleuth 可以与 RabbitMQ 与 MySQL 结合，实现数据采集。也可以与 ELK 结合，进行数据采集后的分析。

第 13 章

OAuth 认证管理

通过本章学习，可以达到以下目标：

1. 理解传统 RPC 认证的问题以及 OAuth 统一认证的特点。
2. 掌握 SpringCloud 与 OAuth 统一认证的结合使用。
3. 掌握 ClientDetailsService、UserDetailsService 的作用。
4. 掌握授权管理控制。

SpringCloud 作为当下最优秀的 RPC 开发框架，除了其本身的特点之外，最大的优势在于其可以与 OAuth 一起实现统一认证与授权管理架构，这样就为微架构的开发提供了良好的安全机制。本章将为读者讲解如何进行 SpringCloud 与 OAuth 的整合开发。

13.1 SpringCloud 与 OAuth

OAuth 是一个统一的认证标准，其使用范围很广，远不止于 Web 客户登录。在实际的开发过程中，SpringCloud 开发技术越来越普及，其中的各类认证与授权处理问题也日益突出。

SpringCloud 在整个 RPC 技术的实现过程中虽然定义了许多技术，但真正核心的技术只有 Restful、Feign、Ribbon、Hystrix、Zuul 和 Eureka。消费端主要采用的是 OAuth，可以结合 Shiro 实现授权管理，并且利用 Web 端的 OAuth 实现统一认证中心，如图 13-1 所示。

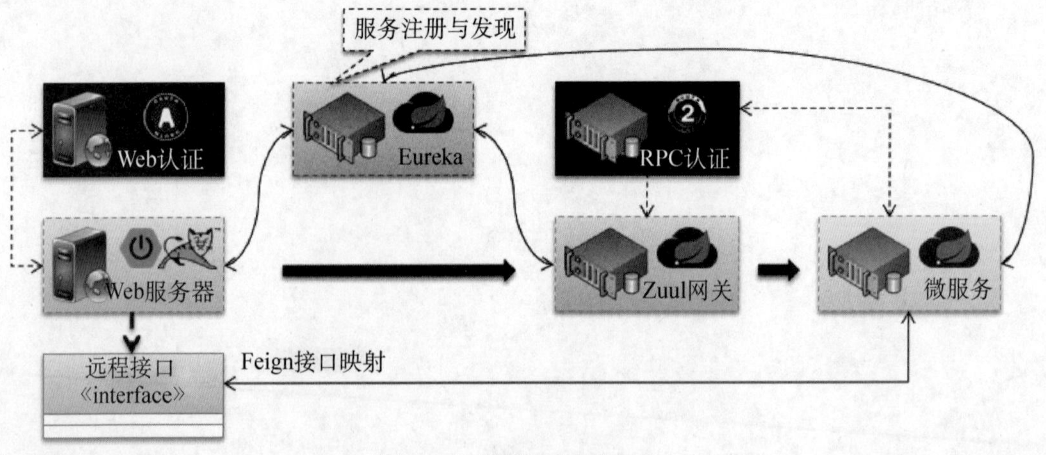

图 13-1　完整的微服务架构

提示：关于本书之前章节中的安全管理。

在之前讲解 SpringCloud 开发内容时，为了帮助读者理解认证问题，采用一个固定用户名和密码进行认证信息的保存，同时为了方便，直接使用一个专门的安全管理配置模块来实现整体的认证信息定义。但没有数据库的支持，这样的认证会显得比较尴尬。例如，如果用户名和密码都不改变，之前的架构没有任何问题；如果要对不同用户进行分别的认证以及授权处理，那么就需要通过数据库来进行认证内容的保存。而且只是简单地采用数据库保存，整体的安全访问依然会存在问题，因为需要配置多个数据源。很明显，这样的设计是不合理的。本章讲解的 OAuth 认证是现在最合理的安全机制。

通过图 13-1 可以看到，在整体微服务设计架构处理中，一般会存在两个 OAuth 认证，一个是针对于前端的认证与授权控制，另一个就是后端 RPC 的 OAuth 认证处理。本章所要讨论的就是 RPC 端的 OAuth 认证。

提示：传统的 SpringSecurity 单机认证。

在之前的 SpringCloud 技术开发项目中，对部门微服务使用了一套认证密码 mldnjava/hello，对网关微服务也使用了一套认证密码 zdmin/mldnjava，因此要管理多组密码才可以访问。可见，单独使用 SpringSecurity 进行认证管理是非常复杂的，而 OAuth 的设计目的正是为了简化这一问题。

如果要结合 OAuth 实现 SpringCloud 统一认证服务，那么 Zuul 除了要实现路由网关功能之外，也需要参与到 OAuth 认证过程中，具体操作如图 13-2 所示。

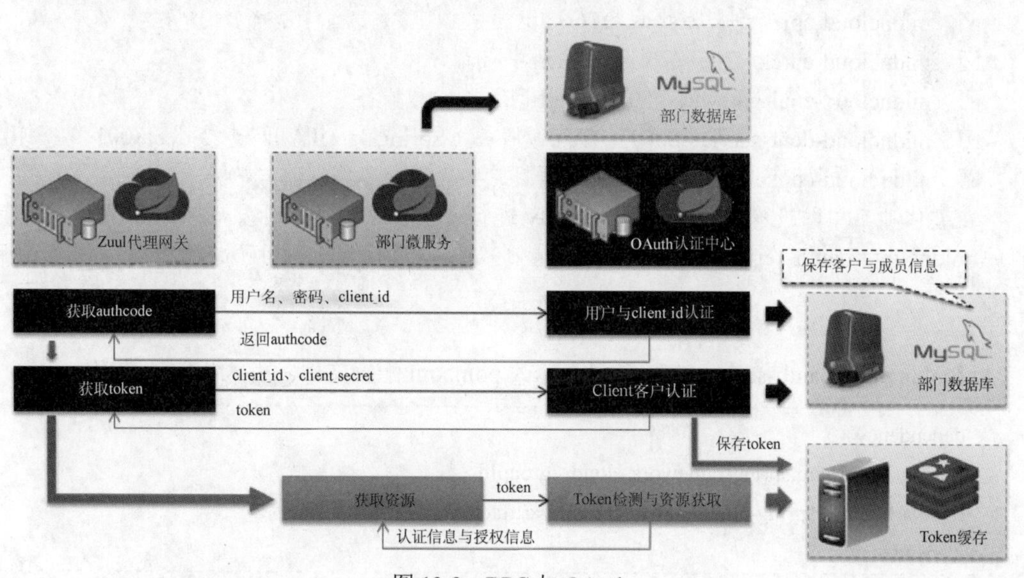

图 13-2　RPC 与 OAuth

在 SpringCloud 开发过程中，Zuul 作为网关代理，只能实现微服务代理转发。但是在 OAuth 认证中，Zuul 网关需要通过 OAuth 获取 authcode 以及 token 认证信息操作。这个认证信息将作为后续微服务的认证标记，可以表示用户的身份，而用户的身份就是整个的微服务之中 OAuth

的核心所在。SpringCloud + OAuth 整合的基本流程如下。

☑ 【Zuul 网关】通过 Zuul 发出用户身份信息（用户名、密码），同时还需要发出 client_id 信息。当 OAuth 接收到此信息之后，首先会检测用户名和密码是否合法，如果合法，则进入数据库中进行 client_id 的合法性检测。如果通过，则返回一个 authcode 信息。

☑ 【ZUUL 网关】根据 authcode 信息再发送 client_id、client_secret 到 Oauth 服务端，并且利用服务端进行客户身份的验证，验证通过之后会将用户信息直接保存在 redis 里面（身份及授权信息）。

☑ 【部门微服务】在部门微服务进行具体操作的时候，要根据 Zuul 传递过来的 token 进行用户信息获取。如果可以获取用户信息，则表示当前用户身份合法，就可以直接调用微服务的方法获取相关操作数据。如果不合法或者没有指定的 token，那么将出现授权错误。

13.2 搭建 OAuth 基础服务

在 Spring 中搭建一个 OAuth 的认证服务是很容易的，只需要引入 spring-cloud- starter-oauth2 依赖库即可实现。同时，Spring 中的 OAuth 服务是在 SpringSecurity 基础上搭建的，当引入此依赖包后会自动引入 SpringSecurity 依赖库。

提示：项目启动环境。

本例讲解的是 OAuth 的整合处理，在整个处理之中将使用如下的项目模块。

☑ mldncloud-api：定义公共的 DTO 与远程接口。

☑ mldncloud-eureka-7001：微服务注册中心。

☑ mldncloud-zuul-gateway-9501：Zuul 网关路由微服务。

☑ mldncloud-dept-service-8001：部门微服务（SpringDataJPA），连接 dept8001 数据库。

☑ mldncloud-consumer：SpringBoot 项目，实现微服务调用。

在整体项目中并没有使用 SpringSecurity 进行安全认证处理，开发者应该首先建立一个新的 mldncloud-oauth-server-8701 项目模块。此外，本项目模块暂时不使用数据库保存认证信息。

1.【mldncloud-oauth-server-8701 项目】修改 pom.xml 配置文件，引入相关依赖库。

```
<dependency>
    <groupId>org.springframework.cloud</groupId>
    <artifactId>spring-cloud-starter-oauth2</artifactId>
</dependency>
```

2.【mldncloud-oauth-server-8701 项目】定义用户认证配置类。

```
package cn.mldn.mldncloud.config;
import org.springframework.context.annotation.Configuration;
import org.springframework.security.config.annotation.authentication.builders
```

```
.AuthenticationManagerBuilder;
import org.springframework.security.config.annotation.web.builders.HttpSecurity;
import org.springframework.security.config.annotation.web.configuration
.WebSecurityConfigurerAdapter;
@Configuration
public class DefaultWebSecurityConfig extends WebSecurityConfigurerAdapter {
    @Override
    protected void configure(AuthenticationManagerBuilder auth) throws Exception {
        auth
            .inMemoryAuthentication()       // 在内存之中定义一个认证信息
            .withUser("mldnjava")           // 用户名
            .password("hello")              // 密码
            .roles("USER");                 // 角色信息
    }
    @Override
    protected void configure(HttpSecurity http) throws Exception {
        http.httpBasic().and().authorizeRequests().anyRequest().fullyAuthenticated();
    }
}
```

3.【mldncloud-oauth-server-8701 项目】要进行 Authcode 的获取，还需要有 client_id、client_secret 信息。此时可以再建立一个单独的配置类，进行授权服务配置。

```
package cn.mldn.mldncloud.config;
import org.springframework.context.annotation.Configuration;
import org.springframework.security.oauth2.config.annotation.configurers
.ClientDetailsServiceConfigurer;
import org.springframework.security.oauth2.config.annotation.web.configuration
.AuthorizationServerConfigurerAdapter;
@Configuration
public class DefaultAuthorizationServerConfig extends AuthorizationServerConfigurerAdapter {
    @Override
    public void configure(ClientDetailsServiceConfigurer clients) throws Exception {
        clients.inMemory()
            .withClient("cmldn")        //client_id信息
            .secret("cjava")            //client_secret信息
            .autoApprove(true)          // 直接进行授权控制
            .authorizedGrantTypes("authorization_code")      // 定义授权类型
            .scopes("webapp") ;         // 授权范围
    }
}
```

本程序类配置有一个 client 认证信息 cmldn/cjava，这样当客户端通过地址进行访问时，只需要配置好相应的客户端认证信息，就可以获取授权码（authcode）。

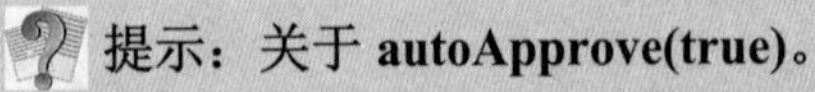

提示：关于 autoApprove(true)。

在本程序中配置 client 认证信息时使用了 autoApprove(true)语句，该语句的主要作用是不等待用户授权，如果没有配置此语句，则在程序执行时会出现如图 13-3 所示的确认界面。很明显，在 RPC 的访问过程中是不应该出现确认界面的。

OAuth Approval

Do you authorize 'cmldn' to access your protected resources?

- scope.webapp: Approve Deny

Authorize

图 13-3　授权界面

4.【mldncloud-oauth-server-8701 项目】修改程序启动类，追加认证服务注解。

```
package cn.mldn.mldncloud;
import org.springframework.boot.SpringApplication;
import org.springframework.boot.autoconfigure.SpringBootApplication;
import org.springframework.security.oauth2.config.annotation.web
.configuration.EnableAuthorizationServer;
@SpringBootApplication
@EnableAuthorizationServer        // 启动授权服务
public class StartOAuthServerApplication8701 {
    public static void main(String[] args) {
        SpringApplication.run(StartOAuthServerApplication8701.class, args);
    }
}
```

5.【操作系统】修改 hosts 配置文件，追加新的主机名称。

```
127.0.0.1        oauth-server.com
```

6.【mldncloud-oauth-server-8701 项目】此时一个基于固定认证信息的 OAuth 服务端搭建完成。当进行微服务启动时，可以在控制台看到如下提示信息。

```
Mapped "{[/oauth/authorize]}"...
Mapped "{[/oauth/authorize],methods=[POST],params=[user_oauth_approval]}" ...
Mapped "{[/oauth/token],methods=[POST]}" ...
Mapped "{[/oauth/token],methods=[GET]}" ...
Mapped "{[/oauth/check_token]}" ...
Mapped "{[/oauth/confirm_access]}" ...
Mapped "{[/oauth/error]}" ...
```

7.【浏览器访问】输入如下访问地址，获取 authcode 信息。

```
http://mldnjava:hello@oauth-server.com:8701/oauth/authorize?
client_id=cmldn&response_type=code&redirect_uri=http://www.mldn.cn
```

该访问地址包含 client_id 信息，由于设置了返回路径 redirect_uri=http://www.mldn.cn，所以当成功获取了 authcode 授权码之后，就可以看见如图 13-4 所示的信息，此时返回的授权码内容为 cN2WeJ（参数名称为 code）。

图 13-4　获取 authcode 信息

8.【操作系统】获取 authcode 之后需要得到 token 信息，此时可以通过 curl 命令来实现。

```
curl -X POST -H "Content-Type:application/x-www-form-urlencoded" -d "grant_type=authorization_code&
code=cN2WeJ&redirect_uri=http://www.mldn.cn" "http://cmldn:cjava@oauth-server.com:8701/oauth/token"
```

命令执行结果：

```
{
    "access_token":"1d7fb0c7-85f1-4e84-99ec-fcb0f9a16e78",
    "token_type":"bearer",
    "expires_in":43199,
    "scope":"webapp"
}
```

其中"access_token":"1d7fb0c7-85f1-4e84-99ec-fcb0f9a16e78"是返回的 token 信息，该配置在 43199 秒内有效，此时一个基础的 OAuth 服务搭建完成。

13.3　使用数据库保存客户信息

在基础 OAuth 微服务里面，为了保存客户信息，专门配置了一个 AuthorizationServerConfigurerAdapter 子类，并且定义了一个固定的客户信息 cmldn/cjava。要针对客户信息进行统一的管理，最好的做法是将这些客户信息保存到数据库中。要想实现这样的处理，需要使用 ClientDetailsService 接口来处理。

提示：实际开发中需要做加密处理。

在实际的开发过程中，客户端信息要采用加密形式来进行处理，为了安全还有可能要进行客户端信息的变更。这些都属于 OAuth 的功能扩展，可以由开发者自行完成。编写本例时考虑到入门读者的学习需要，不采用加密处理形式。

1.【MySQL 数据库】编写数据库脚本，进行客户与授权信息保存（ER 关系见图 13-5）。

```
DROP DATABASE IF EXISTS mldn;
CREATE DATABASE mldn CHARACTER SET UTF8 ;
USE mldn ;
CREATE TABLE client(
    clientid        varchar(50) not null,
    clientsecret    varchar(32),
    scope           varchar(50),
    authorizedgranttypes    varchar(50) ,
    CONSTRAINT pk_mid PRIMARY KEY (clientid)
) engine='innodb';
CREATE TABLE authorities(
     authoid     varchar(50) ,
     title       varchar(50) ,
     CONSTRAINT pk_authoid PRIMARY KEY(authoid)
)engine='innodb' ;
CREATE TABLE client_authorities(
     clientid    varchar(50) ,
     authoid     varchar(50)
) engine='innodb';
INSERT INTO client(clientid,clientsecret,scope,authorizedgranttypes) VALUES
('cadmin','chello','webapp','authorization_code') ;
INSERT INTO client(clientid,clientsecret,scope,authorizedgranttypes) VALUES
('cmldn','cjava','webapp','authorization_code' );
INSERT INTO authorities (authoid,title) VALUES ('CLIENT','访问客户') ;
INSERT INTO client_authorities(clientid,authoid) VALUES ('cadmin','CLIENT') ;
INSERT INTO client_authorities(clientid,authoid) VALUES ('cmldn','CLIENT') ;
```

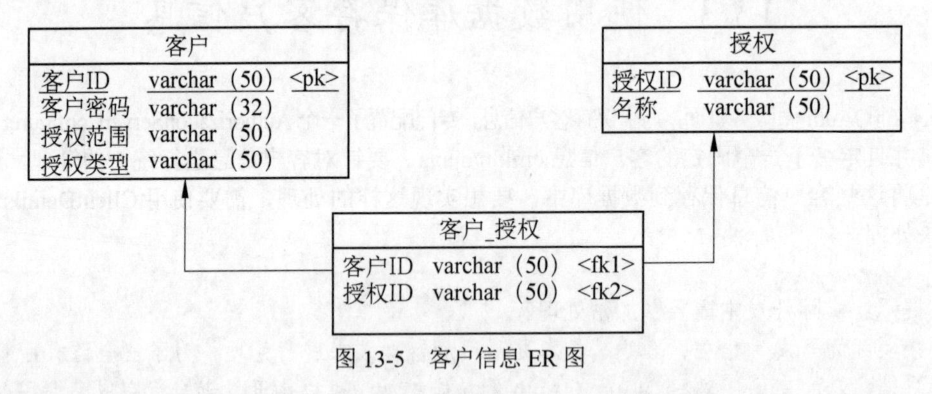

图 13-5　客户信息 ER 图

2.【mldncloud-oauth-server-8701 项目】由于要基于 SpringDataJPA 实现访问，所以建立 Client 持久化类。

```
package cn.mldn.mldncloud.po;
import java.io.Serializable;
```

```
import javax.persistence.Entity;
import javax.persistence.GeneratedValue;
import javax.persistence.GenerationType;
import javax.persistence.Id;
@SuppressWarnings("serial")
@Entity
public class Client implements Serializable {
    @Id
    private String clientid;
    private String authorizedgranttypes;
    private String clientsecret;
    private String scope;
    // setter、getter略
}
```

3.【mldncloud-oauth-server-8701 项目】创建 IClientDAO 接口。

```
package cn.mldn.mldncloud.dao;
import java.util.Set;
import org.springframework.data.jpa.repository.JpaRepository;
import org.springframework.data.jpa.repository.Query;
import org.springframework.data.repository.query.Param;
import cn.mldn.mldncloud.po.Client;
public interface IClientDAO extends JpaRepository<Client,String> {
    @Query(nativeQuery=true,value="SELECT authoid FROM client_authorities WHERE
clientid=:clientid")
    public Set<String> findAllAuthoritiesByClient(@Param(value="clientid") String clientid) ;
}
```

在本程序中，IClientDAO 扩充了一个新的查询方法，同时该方法将使用原生 SQL 查询模式，通过关联表查询用户角色。

4.【mldncloud-oauth-server-8701 项目】定义 IClientService 业务接口。

```
package cn.mldn.mldncloud.service;
import java.util.Map;
public interface IClientService {
    /**
     * 根据客户的编号（client_id）查询出客户的信息以及对应的授权内容
     * @param clientId 客户端ID
     * @return 返回的数据包含如下内容：
     * key = client、value = Client的对象，如果不存在则null;
     * key = allAuthorities、value = 当client信息存在时获取授权内容
```

```
    */
    public Map<String,Object> get(String clientId) ;
}
```

5.【mldncloud-oauth-server-8701 项目】定义 ClientServiceImpl 子类。

```
package cn.mldn.mldncloud.service.impl;
import java.util.HashMap;
import java.util.Map;
import org.springframework.beans.factory.annotation.Autowired;
import org.springframework.stereotype.Service;
import cn.mldn.mldncloud.dao.IClientDAO;
import cn.mldn.mldncloud.po.Client;
import cn.mldn.mldncloud.service.IClientService;
@Service
public class ClientServiceImpl implements IClientService {
    @Autowired
    private IClientDAO clientDAO ;
    @Override
    public Map<String, Object> get(String clientId) {
        Map<String,Object> map = new HashMap<String,Object>() ;
        Client client = this.clientDAO.findOne(clientId) ;          // 查询客户信息
        if (client != null) {                                       // 客户信息存在
            map.put("allAuthorities", this.clientDAO.findAllAuthoritiesByClient (clientId)) ;
        }
        map.put("client", client) ;
        return map;
    }
}
```

6.【mldncloud-oauth-server-8701 项目】定义 ClientDetailsService 接口子类，并与 IClientService 业务接口整合。

```
package cn.mldn.mldncloud.security.util;
import java.util.ArrayList;
import java.util.Arrays;
import java.util.Iterator;
import java.util.List;
import java.util.Map;
import java.util.Set;
import org.springframework.beans.factory.annotation.Autowired;
import org.springframework.security.core.GrantedAuthority;
```

```
import org.springframework.security.core.authority.SimpleGrantedAuthority;
import org.springframework.security.oauth2.provider.ClientDetails;
import org.springframework.security.oauth2.provider.ClientDetailsService;
import org.springframework.security.oauth2.provider.ClientRegistrationException;
import org.springframework.security.oauth2.provider.client.BaseClientDetails;
import cn.mldn.mldncloud.po.Client;
import cn.mldn.mldncloud.service.IClientService;
public class DefaultClientDetailsService implements ClientDetailsService {
    @Autowired
    private IClientService clientService ;                          // 查询客户与授权信息
    @Override
    public ClientDetails loadClientByClientId(String clientId) throws ClientRegistrationException {
        Map<String,Object> map = this.clientService.get(clientId) ;     // 查询信息
        if (map.get("client") == null) {                               // 没有查询到客户信息
            throw new ClientRegistrationException("客户"" +
                    clientId + ""的信息不存在，无法进行OAuth认证处理。");
        }
        Client client = (Client) map.get("client") ;                   // 获取客户信息，将其进行填充
        BaseClientDetails clientDetails = new BaseClientDetails() ;
        clientDetails.setClientId(clientId);
        clientDetails.setClientSecret(client.getClientsecret());       // 需要后续验证
        clientDetails.setAuthorizedGrantTypes(
            Arrays.asList(client.getAuthorizedgranttypes()));
        clientDetails.setScope(Arrays.asList(client.getScope()));
        clientDetails.setAutoApproveScopes(clientDetails.getScope()); // 接收所有授权信息
        Set<String> auth = (Set<String>) map.get("allAuthorities") ;   // 获得所有授权信息
        List<GrantedAuthority> allGrantedAuthority = new ArrayList<GrantedAuthority>() ;
        Iterator<String> iter = auth.iterator() ;
        while(iter.hasNext()) {
            allGrantedAuthority.add(new SimpleGrantedAuthority(iter.next())) ;
        }
        clientDetails.setAuthorities(allGrantedAuthority);
        return clientDetails;
    }
}
```

7.【mldncloud-oauth-server-8701 项目】修改 DefaultAuthorizationServerConfig 程序类，与 ClientDetailsService 子类整合。

```
package cn.mldn.mldncloud.config;
import org.springframework.context.annotation.Configuration;
```

```
import org.springframework.security.oauth2.config.annotation.configurers
.ClientDetailsServiceConfigurer;
import org.springframework.security.oauth2.config.annotation.web.configuration
.AuthorizationServerConfigurerAdapter;
import cn.mldn.mldncloud.security.util.DefaultClientDetailsService;
@Configuration
public class DefaultAuthorizationServerConfig extends AuthorizationServerConfigurerAdapter {
    @Override
    public void configure(ClientDetailsServiceConfigurer clients) throws Exception {
        clients.withClientDetails(new DefaultClientDetailsService()) ;
    }
}
```

此时，客户端信息已经基于数据库实现了保存。如果在用户访问时客户认证信息正确，则可以获取到 authcode 与 token 信息；如果客户信息不正确，则会出现如图 13-6 所示的错误信息提示。

OAuth Error

error="invalid_client", error_description="Bad client credentials"

图 13-6　客户端信息错误

13.4　使用数据库保存微服务认证信息

在 OAuth 整合中，客户信息主要负责实现使用者的身份认证，而用户认证信息才是微服务进行安全检测的重要屏障。开发者可以使用 UserDetailsService 接口，实现用户信息配置。

1.【MySQL 数据库】编写数据库脚本，数据表关系如图 13-7 所示。

```
use mldn ;
CREATE TABLE member(
    mid         varchar(50) not null,
    name        varchar(30),
    password    varchar(32),
    CONSTRAINT pk_mid PRIMARY KEY (mid)
) engine='innodb';
CREATE TABLE role(
    rid         varchar(50) ,
    title       varchar(200) ,
    CONSTRAINT pk_rid PRIMARY KEY(rid)
) engine='innodb' ;
```

```
CREATE TABLE member_role(
    mid     varchar(50) ,
    rid     varchar(50) ,
    CONSTRAINT fk_mid2 FOREIGN KEY(mid) REFERENCES member(mid),
    CONSTRAINT fk_rid2 FOREIGN KEY(rid) REFERENCES role(rid)
) engine='innodb' ;
-- 0表示活跃、1表示锁定
INSERT INTO member(mid,name,password) VALUES ('admin','管理员','hello') ;
INSERT INTO member(mid,name,password) VALUES ('mldnjava','普通人','hello') ;
INSERT INTO member(mid,name,password) VALUES ('mermaid','美人鱼','hello') ;
-- 定义角色信息
INSERT INTO role(rid,title) VALUES ('USER','普通用户') ;
INSERT INTO role(rid,title) VALUES ('ADMIN','管理员') ;
INSERT INTO role(rid,title) VALUES ('GUEST','临时用户') ;
-- 定义用户与角色的关系
INSERT INTO member_role(mid,rid) VALUES ('admin','USER') ;
INSERT INTO member_role(mid,rid) VALUES ('admin','ADMIN') ;
INSERT INTO member_role(mid,rid) VALUES ('admin','GUEST') ;
INSERT INTO member_role(mid,rid) VALUES ('mldnjava','USER') ;
INSERT INTO member_role(mid,rid) VALUES ('mldnjava','GUEST') ;
INSERT INTO member_role(mid,rid) VALUES ('mermaid','GUEST') ;
```

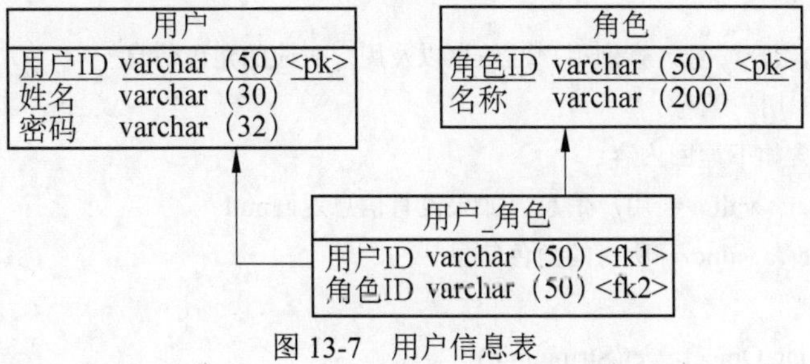

图 13-7　用户信息表

2.【mldncloud-oauth-server-8701 项目】建立 Member 实体类。

```
package cn.mldn.mldncloud.po;
import java.io.Serializable;
import javax.persistence.Entity;
import javax.persistence.Id;
@SuppressWarnings("serial")
@Entity
public class Member implements Serializable {
    @Id
    private String mid;
```

```
    private String name;
    private String password;
    // setter、getter略
}
```

3.【mldncloud-oauth-server-8701 项目】建立 IMemberDAO 接口。

```
package cn.mldn.mldncloud.dao;
import java.util.Set;
import org.springframework.data.jpa.repository.JpaRepository;
import org.springframework.data.jpa.repository.Query;
import org.springframework.data.repository.query.Param;
import cn.mldn.mldncloud.po.Member;
public interface IMemberDAO extends JpaRepository<Member, String> {
    @Query(nativeQuery=true,value="SELECT rid FROM member_role WHERE mid=:mid")
    public Set<String> findAllByMember(@Param("mid") String mid) ;
}
```

4.【mldncloud-oauth-server-8701 项目】建立用户业务接口。

```
package cn.mldn.mldncloud.service;
import java.util.Map;
public interface IMemberService {
    /**
     * 根据指定的用户编号，查询用户的信息以及用户对应的角色信息
     * @param mid 用户编号
     * @return 包含如下返回内容：
     * key = member，value = 用户信息，如果没有信息返回null
     * key = allRoles，value = 所有的角色信息
     */
    public Map<String,Object> get(String mid) ;
}
```

5.【mldncloud-oauth-server-8701 项目】建立 IMemberService 接口子类。

```
package cn.mldn.mldncloud.service.impl;
import java.util.HashMap;
import java.util.Map;
import org.springframework.beans.factory.annotation.Autowired;
import org.springframework.stereotype.Service;
import cn.mldn.mldncloud.dao.IMemberDAO;
import cn.mldn.mldncloud.po.Member;
import cn.mldn.mldncloud.service.IMemberService;
```

```
@Service
public class MemberServiceImpl implements IMemberService {
    @Autowired
    private IMemberDAO memberDAO;
    @Override
    public Map<String, Object> get(String mid) {
        Map<String, Object> map = new HashMap<String, Object>();
        Member member = this.memberDAO.findOne(mid);
        if (member != null) {
            map.put("allRoles", this.memberDAO.findAllByMember(mid));
        }
        map.put("member", member);
        return map;
    }
}
```

6.【mldncloud-oauth-server-8701 项目】建立 UserDetailsService 接口子类，并调用 IMemberService 业务方法。

```
package cn.mldn.mldncloud.security.util;
import java.util.ArrayList;
import java.util.Iterator;
import java.util.List;
import java.util.Map;
import java.util.Set;
import org.springframework.beans.factory.annotation.Autowired;
import org.springframework.security.core.GrantedAuthority;
import org.springframework.security.core.authority.SimpleGrantedAuthority;
import org.springframework.security.core.userdetails.User;
import org.springframework.security.core.userdetails.UserDetails;
import org.springframework.security.core.userdetails.UserDetailsService;
import org.springframework.security.core.userdetails.UsernameNotFoundException;
import cn.mldn.mldncloud.po.Member;
import cn.mldn.mldncloud.service.IMemberService;
public class DefaultUserDetailsService implements UserDetailsService {
    @Autowired
    private IMemberService memberService;
    @Override
    public UserDetails loadUserByUsername(String username) throws UsernameNotFoundException {
        Map<String, Object> map = this.memberService.get(username);
        Member member = (Member) map.get("member");
        if (member == null) {                                    // 用户信息不存在
```

```
            throw new UsernameNotFoundException("用户“" + username + "”信息不存在！");
        }    // 要获取该用户的全部角色信息（授权信息）
        Set<String> allRoles = (Set<String>) map.get("allRoles");
        Iterator<String> iter = allRoles.iterator();
        List<GrantedAuthority> allGrantedAuthority = new ArrayList<GrantedAuthority>();
        while (iter.hasNext()) {
            allGrantedAuthority.add(new SimpleGrantedAuthority(iter.next()));
        }
        User user = new User(username, member.getPassword(), allGrantedAuthority);
        return user;
    }
}
```

7.【mldncloud-oauth-server-8701 项目】修改 DefaultWebSecurityConfig 配置类，整合 DefaultUserDetailsService 类。

```
package cn.mldn.mldncloud.config;
import org.springframework.context.annotation.Bean;
import org.springframework.context.annotation.Configuration;
import org.springframework.security.config.annotation.authentication.
builders.AuthenticationManagerBuilder;
import org.springframework.security.config.annotation.web.builders.HttpSecurity;
import org.springframework.security.config.annotation.web.
configuration.WebSecurityConfigurerAdapter;
import org.springframework.security.core.userdetails.UserDetailsService;
import cn.mldn.mldncloud.security.util.DefaultUserDetailsService;
@Configuration
public class DefaultWebSecurityConfig extends WebSecurityConfigurerAdapter {
    @Bean
    public UserDetailsService getUserDetailsService() {
        return new DefaultUserDetailsService() ;
    }
    @Override
    protected void configure(AuthenticationManagerBuilder auth) throws Exception {
        auth.userDetailsService(this.getUserDetailsService()) ;    // 使用自定义的用户认证
    }
    @Override
    protected void configure(HttpSecurity http) throws Exception {
        http.httpBasic().and().authorizeRequests().anyRequest().fullyAuthenticated() ;
    }
}
```

此时，OAuth 服务就可以利用 member 表中提供的用户名和密码进行访问了，从而对微服务的认证信息实现了统一管理。

13.5　建立访问资源

在 OAuth 认证之后获取和保存了 token，其核心作用在于资源的获得。本例就来进行资源信息的获得以及配置处理操作。

1.【mldncloud-oauth-server-8701 项目】编写一个可以获取资源的 Restful 服务接口。

```
package cn.mldn.mldncloud.rest;
import java.security.Principal;
import org.springframework.web.bind.annotation.RequestMapping;
import org.springframework.web.bind.annotation.RestController;
@RestController
public class ResourceRest {
    @RequestMapping("/user")                                    // 资源路径
    public Principal resource(Principal user) {                 // 用户资源
        return user ;
    }
}
```

2.【mldncloud-oauth-server-8701 项目】如果已经获取了 token，并且发出请求后 token 验证通过了，就可以返回用户资源信息。如果想正常进行信息的获得（token 合法），还需要追加一个资源安全配置类。

```
package cn.mldn.mldncloud.config;
import javax.servlet.http.HttpServletResponse;
import org.springframework.context.annotation.Configuration;
import org.springframework.security.config.annotation.web.builders.HttpSecurity;
import org.springframework.security.oauth2.config.annotation.web
.configuration.EnableResourceServer;
import org.springframework.security.oauth2.config.annotation.web
.configuration.ResourceServerConfigurerAdapter;
@Configuration
@EnableResourceServer      // 资源配置
public class DefaultResourceServerConfigurerAdapter extends ResourceServerConfigurerAdapter {
    @Override
    public void configure(HttpSecurity http) throws Exception {
        http.csrf().disable()
        .exceptionHandling()
```

```
                .authenticationEntryPoint((request,response,authException)->
                        response.sendError(HttpServletResponse.SC_UNAUTHORIZED))
                .and()
                        .authorizeRequests().anyRequest().authenticated()
                .and().httpBasic() ;
        }
}
```

3.【mldncloud-oauth-server-8701 项目】如果要通过 application.yml 配置文件进行用户信息的认证，那么已经可以正常使用了。如果要通过安全配置类来进行处理，那么将无法正常使用，因为安全配置项存在重复。下面修改 application.yml 配置文件，资源访问是在整个 OAuth 认证流程中的第三步。

```
security:
  oauth2:
    resource:
      filter-order: 3            # 对于Resource的过滤执行流程定义顺序
```

如果没有配置此操作项，则@EnableResourceServer 无法正确进行资源控制。配置完成后重新启动微服务，当 token 正确的时候可以进行正常的访问。如果 token 有错误，将出现如图 13-8 所示的错误信息。

```
- <oauth>
  - <error_description>
      Invalid access token: ccf1fcfb-ff55-4eb0-9b53-218faadb5eea
    </error_description>
    <error>invalid_token</error>
  </oauth>
```

图 13-8　token 错误返回信息

13.6　使用 Redis 保存 token 令牌

在整个 OAuth 处理中，除了要进行客户端的认证之外，还有一个重要的问题——就是进行 token 的保存。现在的 token 实际上是直接保存在内存中的，这样的做法不利于程序的扩展与高性能处理。从开发角度来讲，最好的做法是将其直接保存在 Redis 数据库中。

1.【mldncloud-oauth-server-8701 项目】修改 pom.xml 配置文件，引入 Redis 相关依赖。

```
<dependency>
        <groupId>org.springframework.boot</groupId>
        <artifactId>spring-boot-starter-data-redis</artifactId>
</dependency>
```

2.【mldncloud-oauth-server-8701 项目】修改 application.yml 配置文件，追加 Redis 的相关配

置项。

```
spring:
  redis:                                # Redis相关配置
    host: redis-server                  # 主机名称
    port: 6379                          # 端口号
    password: mldnjava                  # 认证密码
    timeout: 1000                       # 连接超时时间
    database: 0                         # 默认数据库
    pool:                               # 连接池配置
      max-active: 10                    # 最大连接数
      max-idle: 8                       # 最大维持连接数
      min-idle: 2                       # 最小维持连接数
      max-wait: 100                     # 最大等待连接超时时间
```

3.【mldncloud-oauth-server-8701 项目】修改 DefaultAuthorizationServerConfig 配置类，追加 Redis 处理。

```
package cn.mldn.mldncloud.config;
import org.springframework.beans.factory.annotation.Autowired;
import org.springframework.context.annotation.Configuration;
import org.springframework.data.redis.connection.RedisConnectionFactory;
import org.springframework.security.oauth2.config.annotation.configurers
.ClientDetailsServiceConfigurer;
import org.springframework.security.oauth2.config.annotation.web.configuration
.AuthorizationServerConfigurerAdapter;
import org.springframework.security.oauth2.config.annotation.web
.configurers.AuthorizationServerEndpointsConfigurer;
import org.springframework.security.oauth2.provider.token.store.redis.RedisTokenStore;
import cn.mldn.mldncloud.security.util.DefaultClientDetailsService;
@Configuration
public class DefaultAuthorizationServerConfig extends AuthorizationServerConfigurerAdapter {
    @Autowired
    private RedisConnectionFactory redisConnectionFactory ; // 自动注入Redis连接对象
    @Override
    public void configure(AuthorizationServerEndpointsConfigurer endpoints) throws Exception {
        // 建立一个token存储的配置项，将token直接保存在Redis之中
        endpoints.tokenStore(new RedisTokenStore(this.redisConnectionFactory)) ;
    }
    @Override
    public void configure(ClientDetailsServiceConfigurer clients) throws Exception {
        clients.withClientDetails(new DefaultClientDetailsService()) ;
```

```
    }
}
```

通过此时的配置，就可以将 token 保存在 Redis 数据库中。执行后，可以在 Redis 中看到如图 13-9 所示的内容。

```
1) "auth:b9089007-3b8a-481a-a5fe-6fad2d33297c"
2) "auth_to_access:64ad48820be15fbae230f07642f543b0"
3) "access:b9089007-3b8a-481a-a5fe-6fad2d33297c"
4) "client_id_to_access:cmldn"
5) "uname_to_access:cmldn:mldn"
```

图 13-9　Redis 保存 token 信息

13.7　SpringCloud 整合 OAuth

OAuth 的统一认证中心搭建完成之后，下面需要将其与微服务进行整合处理，操作流程如图 13-10 所示。在进行微服务整合处理的时候，一定要清楚以下两个核心原则：

☑　所有的具体微服务信息（dept 微服务）只需要获得资源即可。

☑　所有的 authcode 与 token 操作都应该通过网关（Zuul）来完成。

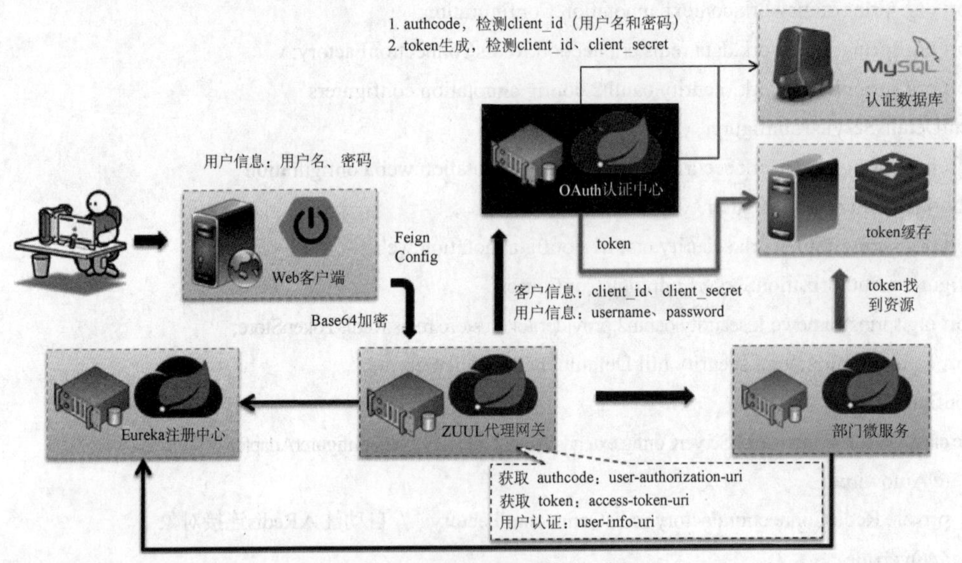

图 13-10　SpringCloud 整合 OAuth

1.【mldncloud-dept-service-8001、mldncloud-zuul-gateway-9501 项目】修改 pom.xml 配置文件，引入 OAuth 相关依赖库。

```
<dependency>
    <groupId>org.springframework.cloud</groupId>
    <artifactId>spring-cloud-starter-oauth2</artifactId>
</dependency>
```

2.【mldncloud-dept-service-8001 项目】修改 application.yml 配置文件，进行 resource 的资源配置。

```
security:
  oauth2:
    resource:
      id: dept-service                                              # 定义标记
      user-info-uri: http://oauth-server.com:8701/user              # 资源路径
      prefer-token-info: false                                      # token 存在的情况下才可以获取资源项
```

3.【mldncloud-dept-service-8001 项目】建立一个资源服务访问的安全配置类。

```
package cn.mldn.mldncloud.config;
import javax.servlet.http.HttpServletResponse;
import org.springframework.context.annotation.Configuration;
import org.springframework.security.config.annotation.web.builders.HttpSecurity;
import org.springframework.security.oauth2.config.annotation.web.configuration
.EnableResourceServer;
import org.springframework.security.oauth2.config.annotation.web.configuration
.ResourceServerConfigurerAdapter;
@Configuration
@EnableResourceServer
public class ClientResourceServerConfig extends ResourceServerConfigurerAdapter {
	@Override
	public void configure(HttpSecurity http) throws Exception {
		http.csrf().disable()
		.exceptionHandling()
		.authenticationEntryPoint((request,response,authException)->
			response.sendError(HttpServletResponse.SC_UNAUTHORIZED))
		.and()
			.authorizeRequests().anyRequest().authenticated()
		.and().httpBasic() ;
	}
}
```

部门微服务配置完以上操作后，将无法直接进行访问。也就是说，此时必须要通过 Zuul 来代理访问。

4.【mldncloud-zuul-gateway-9501 项目】修改 application.yml 配置文件，追加 OAuth 客户端的配置。

```
security:
  oauth2:
    client:                                                  # 配置客户信息
```

```
    user-authorization-uri: http://oauth-server.com:8701/oauth/authorize
    access-token-uri: http://oauth-server.com:8701/oauth/token
    client-id: cmldn                                    # 客户ID
    client-secret: cjava                                # 客户密码
  resource:                                             # 配置资源访问
    user-info-uri: http://oauth-server.com:8701/user
```

5.【mldncloud-zuul-gateway-9501 项目】建立安全配置类。

```
package cn.mldn.mldncloud.config;
import org.springframework.boot.autoconfigure.security.oauth2.client
.EnableOAuth2Sso;
import org.springframework.context.annotation.Configuration;
import org.springframework.security.config.annotation.web.builders.HttpSecurity;
import org.springframework.security.config.annotation.web.configuration
.WebSecurityConfigurerAdapter;
@Configuration
@EnableOAuth2Sso
public class SecurityConfig extends WebSecurityConfigurerAdapter {
    @Override
    protected void configure(HttpSecurity http) throws Exception {
        http
                .antMatcher("/**")                  // 所有的请求都必须经过认证和授权处理
                .authorizeRequests().anyRequest().authenticated()
                .and().authorizeRequests().antMatchers("/","/anon").permitAll()
                .and().csrf().disable() ;
    }
}
```

配置完成之后，将可以通过 Zuul 代理进行微服务访问。同时消费端也可以进行微服务的调用。

6.【mldncloud-consumer 项目】此时微服务的访问密码只需要使用 member 表中定义的信息即可，所以可以修改 FeignClientConfig 配置类中的密码定义。

```
    @Bean
    public BasicAuthRequestInterceptor getBasicAuthRequestInterceptor() {
        return new BasicAuthRequestInterceptor("mldnjava", "hello");
    }
```

这样就可以实现微服务的调用，而整体的用户信息认证都可以通过 OAuth 统一完成。

提示：关于消费端 Feign 映射失败。

在使用消费端进行微服务调用时，有可能出现认证信息无法传输的问题。问题的根源可能

是 feign-httpclient 依赖库问题，所以最好的解决方案是进行该依赖库的单独引入配置。

范例：【mldncloud 项目】修改 pom.xml 配置文件，引入 feign-httpclient 依赖库。

属性配置	<feign-httpclient.version>8.18.0</feign-httpclient.version>
依赖库配置	<dependency> <groupId>com.netflix.feign</groupId> <artifactId>feign-httpclient</artifactId> <version>${feign-httpclient.version}</version> </dependency>

范例：【mldncloud-api】Feign 接口转换在公共 API 项目中完成，修改 pom.xml 配置文件，进行依赖配置。

```
<dependency>
    <groupId>com.netflix.feign</groupId>
    <artifactId>feign-httpclient</artifactId>
</dependency>
```

配置完成就可以解决 Feign 的信息传输问题。

7.【mldncloud-dept-service-8001 项目】当获取了 OAuth 统一认证后，实际上不仅仅是为了进行所谓的认证使用，还有授权检测处理问题，在 OAuth 定义的时候提供有角色信息，而这样的角色信息可以拿来进行授权检测。修改部门的微服务提供的 Res 程序，追加一个授权检测。

```
@GetMapping("/dept/list")
@HystrixCommand                                    // Hystrix监控注解
@PreAuthorize("hasAuthority('ADMIN')")             // 具有ADMIN角色的用户才可以访问
public Object list() {
    return this.deptService.list() ;               // 部门信息列表
}
```

8.【mldncloud-dept-service-8001 项目】授权检测如果要生效，还需要做一个方法的拦截配置。修改微服务启动主类。

```
package cn.mldn.mldncloud;
import org.springframework.boot.SpringApplication;
import org.springframework.boot.autoconfigure.SpringBootApplication;
import org.springframework.cloud.client.circuitbreaker.EnableCircuitBreaker;
import org.springframework.cloud.client.discovery.EnableDiscoveryClient;
import org.springframework.cloud.netflix.eureka.EnableEurekaClient;
import org.springframework.cloud.netflix.hystrix.EnableHystrix;
import org.springframework.data.jpa.repository.config.EnableJpaRepositories;
import org.springframework.security.config.annotation.method
.configuration.EnableGlobalMethodSecurity;
```

```
@SpringBootApplication
@EnableEurekaClient                                  // 启用Eureka客户端
@EnableDiscoveryClient
@EnableCircuitBreaker                                // 启用熔断机制
@EnableHystrix                                       // 启用Hystrix支持
@EnableJpaRepositories(basePackages="cn.mldn.mldncloud.dao")
@EnableGlobalMethodSecurity(prePostEnabled=true)
public class StartDeptServiceApplication8001 {
    public static void main(String[] args) {
        SpringApplication.run(StartDeptServiceApplication8001.class, args);
    }
}
```

只有在程序启动主类配置了@EnableGlobalMethodSecurity 注解之后，DeptRest 类中方法上的@PreAuthorize 注解才会生效，这样的配置就可以实现项目角色的统一管理。

13.8 本章小结

1. OAuth 除了可以在 Web 端实现单点登录整合之外，也可以与 SpringCloud 结合使用。

2. OAuth 在与 SpringCloud 整合时，可以使用 ClientDetailsService 与 UserDetailsService 实现数据库信息的访问。

3. OAuth 访问获得的 token 信息一定要保存在 Redis 中，并且获取 token 认证信息的请求可以通过 token 获取用户完整资源。

4. SpringCloud 在与 OAuth 整合时，一定要修改 application.yml 配置的 security.oauth2.resource.filter-order 选项，否则用户 token 将不会被检测。

5. SpringCloud 整合 OAuth 时，需要在 Zuul 网关中生成 token，而后在具体微服务访问时只需要通过 token 获取用户资源即可，同时也可以针对用户的角色进行统一管理。

第三部分

微服务辅助篇

- **RabbitMQ 消息组件**
- **Docker 虚拟化容器**

第 14 章

RabbitMQ 消息组件

通过本章学习，可以达到以下目标：

1. 理解 JMS 与 AMQP 的区别。
2. 掌握 RabbitMQ 的安装与配置。
3. 掌握 RabbitMQ 与 Java 程序访问。
4. 掌握 Spring 与 RabbitMQ 整合。

RabbitMQ 是 AMQP（Advanced Message Queuing Protocol，高级消息队列协议）的技术实现，也是高性能消息服务组件的代表之作。由于其隶属于 Pivotal 公司，所以与 Spring 的整合效果是最完善的。在 SpringCloud 开发技术中大量使用了 RabbitMQ 消息组件，本章将为读者详细地讲解 RabbitMQ 组件的安装与使用方法（基于 Linux 配置）。

14.1 RabbitMQ 简介

RabbitMQ 是一种消息队列服务，同样也是在进行系统整合时的一种通信手段，其运行模式遵循“生产者—消费者”模型，即会存在若干个消息生产者以及若干个消息消费者。与 JavaEE 提出的 JMS 标准不同之处在于：RabbitMQ 是由 ERLang 开发的基于 AMQP 应用层协议标准的一种消息组件，所以其处理性能要比 JMS 组件更高。RabbitMQ 官方网站的网址为 http://www.rabbitmq.com，其界面如图 14-1 所示。

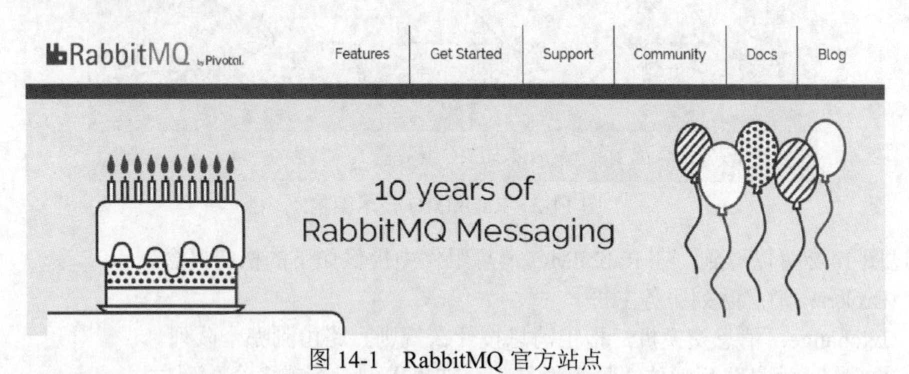

图 14-1　RabbitMQ 官方站点

提示：常见消息组件。

消息组件主要划分为 JMS 组件和 AMQP 组件两类。

☑ JMS（Java Message Service）组件：ActiveMQ 性能较差。

☑ AMQP 组件（协议）：性能是最高的，而 AMQP 有两个主要的开源。

|- RabbitMQ：使用最为广泛，响应速度快。

|- Kafka：是大数据时代作为数据采集的重要组件，处理速度更高。

RabbitMQ 是由 RabbitMQ Technologies Ltd 开发并且提供商业支持的。该公司在 2010 年 4 月被 SpringSource（VMWare 的一个部门）收购，在 2013 年 5 月被并入 Pivotal。

RabbitMQ 最初起源于金融系统，用于在分布式系统中存储转发消息，在易用性、扩展性、高可用性等方面表现不俗。具体特点包括以下几个方面。

☑ **可靠性（Reliability）**：RabbitMQ 使用一些机制来保证可靠性，如持久化、传输确认、发布确认等。

☑ **灵活的路由（Flexible Routing）**：所有的消息要通过 Exchange 来进行路由处理，对于复杂路由也可以将多个 Exchange 绑定在一起使用。

☑ **消息集群（Clustering）**：RabbitMQ 可以采用镜像集群的模式使消息的安全性与处理能力得到提升。

☑ **多种协议（Multi-protocol）**：RabbitMQ 支持多种消息队列协议，如 STOMP、MQTT 等。

☑ **多语言客户端（Many Clients）**：RabbitMQ 几乎支持所有常用语言，如 Java、.NET、Ruby 等。

☑ **管理界面（Management UI）**：RabbitMQ 提供用户管理界面，使开发者可以监控和管理消息。

☑ **跟踪机制（Tracing）**：RabbitMQ 提供消息跟踪机制，方便排查消息异常。

RabbitMQ 除了可实现基本的消息生产与消费之外，还提供了更加完善的消息处理机制。图 14-2 显示了 RabbitMQ 的技术架构。

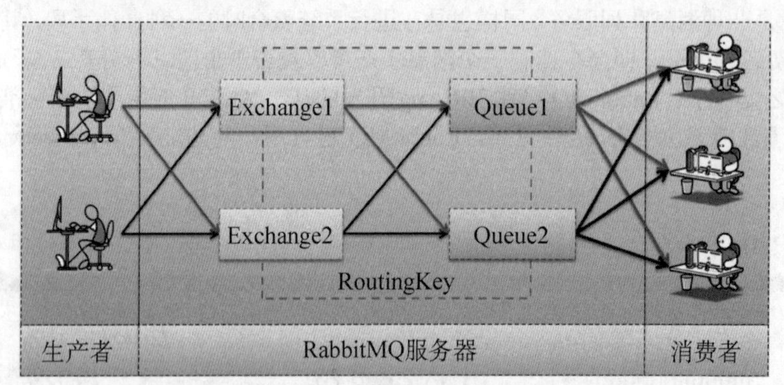

图 14-2　RabbitMQ 技术架构

通过图 14-2 可以看到，在 RabbitMQ 消息组件中提供如下的核心概念。

☑ Broker：消息队列服务主机。

☑ Exchange：消息交换机，指定消息按什么规则、路由到哪个队列。

☑ Queue：消息队列载体，每个消息都会被投入到一个或多个队列中。

☑ Binding：绑定，把 Exchange 和 Queue 按照路由规则绑定起来。

☑ RoutingKey：路由关键字，Exchange 根据这个关键字进行消息投递。

☑ vhost：虚拟主机，一个 broker 里可设多个 vhost，实现用户的权限分离。

☑ producer：消息生产者。

☑ consumer：消息消费者。

☑ Channel：消息通道，在客户端的每个连接里可建立多个 Channel，每个 Channel 代表一个会话任务。

14.2　配置 Erlang 开发环境

RabbitMQ 依靠的开发语言为 Erlang，所以如果想使用 RabbitMQ 服务，首先一定要在系统中进行此开发环境的配置。

1.【rabbitmq-single 主机】将 Erlang 的开发包 otp_src_19.2.tar.gz 上传到系统之中，而后进行解压缩。

```
tar xzvf /srv/ftp/otp_src_19.2.tar.gz -C /usr/local/src/
```

2.【rabbitmq-single 主机】要想进行 Erlang 编译，需要安装一个组件包。

```
apt-get -y install libncurses5-dev
```

3.【rabbitmq-single 主机】将组件解压缩到源代码目录之中。

```
tar xzvf /srv/ftp/otp_src_19.2.tar.gz -C /usr/local/src/
```

4.【rabbitmq-single 主机】进入到 Erlang 源代码所在路径。

```
cd /usr/local/src/otp_src_19.2
```

5.【rabbitmq-single 主机】首先进行安装的配置。

```
./configure --prefix=/usr/local/erlang
```

6.【rabbitmq-single 主机】进行编译与安装。

```
make && make install
```

7.【rabbitmq-single 主机】为了方便使用 Erlang 环境，建议修改一下环境配置文件，追加 ERLANG_HOME 配置项。

打开配置文件	vim /etc/profile
追加配置项	export ERLANG_HOME=/usr/local/erlang export PATH=$PATH:$ERLANG_HOME/bin:
使配置立即生效	source /etc/profile

8.【rabbitmq-single 主机】用 Erlang 语言编写一个 Hello World。

启动 erl 环境	erl
编写程序代码	io:format("Hello World").

此时，如果可以正常执行，则表示 Erlang 安装完成。如果要退出 Erlang 开发环境，则直接输入 halt().命令即可。

14.3 安装并配置 RabbitMQ

通过 RabbitMQ 官方网站下载 RabbitMQ 的开发包，本例使用的开发包版本为 3.6.6，由于是在 Linux 下进行配置的，所以下载的文件名称为 rabbitmq-server-generic-unix-3.6.6.tar.xz。

1.【rabbitmq-single 主机】将 RabbitMQ 开发包 rabbitmq-server-generic-unix-3.6.6.tar.xz 上传到 Linux 系统中。

2.【rabbitmq-single 主机】将 rabbitmq-server-generic-unix-3.6.6.tar.xz 开发包进行解压缩。

☑ 由于上传文件是采用 xz 格式进行压缩的，所以首先需要将 tar.xz 进行解压缩。

```
xz -d /srv/ftp/rabbitmq-server-generic-unix-3.6.6.tar.xz
```

解压完成后可以得到 rabbitmq-server-generic-unix-3.6.6.tar 文件。

☑ 将文件解压缩到指定的目录。

```
tar xvf /srv/ftp/rabbitmq-server-generic-unix-3.6.6.tar -C /usr/local/
```

3.【rabbitmq-single 主机】为了方便进行命令，做一个更名处理。

```
mv /usr/local/rabbitmq_server-3.6.6/ /usr/local/rabbitmq;
```

4.【rabbitmq-single 主机】当一切都准备完毕之后，可以启动 RabbitMQ 组件。

```
/usr/local/rabbitmq/sbin/rabbitmq-server start
```

以上的启动方式为前台启动，用户也可以使用后台启动，代码为/usr/local/rabbitmq/sbin/rabbitmq-server -detached。

5.【rabbitmq-single 主机】配置完成之后需要为 RabbitMQ 配置一个管理员的账户（mldn/java）。

```
/usr/local/rabbitmq/sbin/rabbitmqctl add_user mldn java
```

6.【rabbitmq-single 主机】现在只是创建了一个普通用户，而不是管理员用户，所以需要将其更改为管理员权限。

```
/usr/local/rabbitmq/sbin/rabbitmqctl set_user_tags mldn administrator
```

7.【rabbitmq-single 主机】启动 RabbitMQ 的管理控制台（Web 形式运行，自带监控）。

```
/usr/local/rabbitmq/sbin/rabbitmq-plugins enable rabbitmq_management
```

当管理控制台启动之后，会自动在 15672 的端口上运行一个 Web 监控程序。

8.【rabbitmq-single 主机】修改 mldn 用户可以使用的虚拟主机名称（也可以通过控制台修改）。

```
/usr/local/rabbitmq/sbin/rabbitmqctl set_permissions   -p   / mldn ConfP   WriteP   ReadP
```

9.【操作系统】RabbitMQ 自带管理控制台，但是需要输入完整的主机 IP 地址。为了方便访问，可以修改 hosts 主机配置文件，增加新的主机名称。本例 RabbitMQ 所在主机的 IP 地址为 192.168.68.150。

```
192.168.68.150   rabbitmq-single
```

配置完成后就可以通过浏览器进行访问，地址为 http://rabbitmq-single:15672，随后会出现如图 14-3 所示的登录界面，此时输入之前配置的 mldn/java 账号，就可以进入如图 14-4 所示的管理界面。

图 14-3　RabbitMQ 管理控制台登录界面

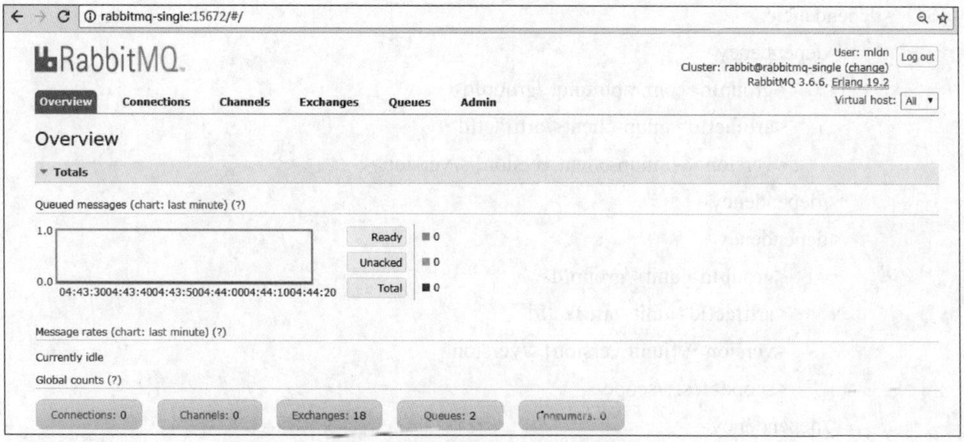

图 14-4　RabbitMQ 管理控制台

14.4　使用 Java 访问 RabbitMQ

要使用 RabbitMQ 组件，首先要获取 RabbitMQ 的相关驱动程序。为了方便管理，本例将利用 Maven 管理工具进行项目依赖包的配置。下面将创建一个 mldnrabbitmq 的父项目，其中对于 pom.xml 文件定义如下。

范例：【mldnrabbitmq 项目】定义统一父 pom.xml 配置文件。

```
<?xml version="1.0" encoding="UTF-8"?>
<project xmlns="http://maven.apache.org/POM/4.0.0"
    xmlns:xsi="http://www.w3.org/2001/XMLSchema-instance"
```

```
xsi:schemaLocation="http://maven.apache.org/POM/4.0.0
    http://maven.apache.org/xsd/maven-4.0.0.xsd">
<modelVersion>4.0.0</modelVersion>
<groupId>cn.mldn</groupId>
<artifactId>mldnrabbitmq</artifactId>
<version>0.0.1</version>
<packaging>pom</packaging>
<name>mldnrabbitmq</name>
<url>http://maven.apache.org</url>
<properties>
    <compiler.version>3.6.1</compiler.version>
    <jdk.version>1.8</jdk.version>
    <junit.version>4.12</junit.version>
    <amqp-client.version>5.0.0</amqp-client.version>
    <project.build.sourceEncoding>UTF-8</project.build.sourceEncoding>
</properties>
<dependencyManagement>
    <dependencies>
        <dependency>
            <groupId>com.rabbitmq</groupId>
            <artifactId>amqp-client</artifactId>
            <version>${amqp-client.version}</version>
        </dependency>
        <dependency>
            <groupId>junit</groupId>
            <artifactId>junit</artifactId>
            <version>${junit.version}</version>
            <scope>test</scope>
        </dependency>
    </dependencies>
</dependencyManagement>
<build>
    <finalName>mldnrabbitmq</finalName>
    <plugins>
        <plugin>
            <groupId>org.apache.maven.plugins</groupId>
            <artifactId>maven-compiler-plugin</artifactId>
            <version>${compiler.version}</version>
            <configuration>
                <source>${jdk.version}</source>
                <target>${jdk.version}</target>
                <encode>${project.build.sourceEncoding}</encode>
```

```
                </configuration>
            </plugin>
        </plugins>
    </build>
</project>
```

本配置文件中重点引入的是 amqp-client 依赖库，随后将依照此父 pom 创建其余的 Maven 子模块。

14.4.1　创建消息生产者

消息生产者的主要作用是向 RabbitMQ 消息组件发送消息。在 RabbitMQ 中所有的消息都需要有主题名称，该主题可以通过 RabbitMQ 控制台创建，也可以通过程序执行时动态创建，本例将采用动态创建的形式完成。

1.【mldnrabbitmq 项目】创建新的项目模块 mldnrabbitmq-queue-provider。

2.【mldnrabbitmq-queue-provider 项目】修改 pom.xml 配置文件，追加 RabbitMQ 相关依赖库。

```
<dependency>
    <groupId>com.rabbitmq</groupId>
    <artifactId>amqp-client</artifactId>
</dependency>
```

3.【mldnrabbitmq-queue-provider 项目】建立消息发送程序类。

```
package cn.mldn.mldnrabbitmq.queue.provider;
import com.rabbitmq.client.Channel;
import com.rabbitmq.client.Connection;
import com.rabbitmq.client.ConnectionFactory;
public class MessageProvider {
    private static final String HOST = "rabbitmq-single";            // 主机名称
    private static final int PORT = 5672;                            // 发送端口
    private static final String USERNAME = "mldn";                   // 用户名
    private static final String PASSWORD = "java";                   // 密码
    private static final String QUEUE_NAME = "mldn.msg.queue";       // 建立一个队列名称
    public static void main(String[] args) throws Exception {
        // 如果要进行RabbitmQ连接，则需要有一个连接工厂类，通过连接工厂创建连接
        ConnectionFactory factory = new ConnectionFactory();         // 创建连接工厂
        factory.setHost(HOST);                                       // 连接主机
        factory.setPort(PORT);                                       // 连接端口
        factory.setUsername(USERNAME);                               // 用户名
        factory.setPassword(PASSWORD);                               // 密码
        // 通过连接工厂可以直接获取一个连接对象
        Connection connection = factory.newConnection();             // 建立新的连接
        // 如果要进行消息的发送，则需要通过连接获取一个Channel
```

```
        Channel channel = connection.createChannel();                    // 创建一个连接通道
        // 对通道中使用的队列进行配置，需要明确地知道队列信息。在创建队列时参数如下
        // 第一个参数（String queue）：队列名称（这个队列可能存在也可能不存在）
        // 第二个参数（boolean durable）：是否为持久存储（未消费不删除）
        // 第三个参数（boolean exclusive）：是否为专用的队列信息，设置为false
        // 第四个参数（boolean autoDelete）：是否允许为自动删除，消费之后删除
        // 第五个参数（Map<String, Object> arguments）：队列处理参数，一般设置为null
        channel.queueDeclare(QUEUE_NAME, true, false, true, null) ;
        for (int x = 0 ; x < 10 ; x ++) {
            String msg = "mldnjava - " + x ;                             // 消息内容
            channel.basicPublish("", QUEUE_NAME, null, msg.getBytes()); // 发送消息
        }
        channel.close();          // 关闭通道
        connection.close();       // 关闭连接
    }
}
```

本程序动态地创建了一个 mldn.msg.queue 队列，由于该队列并不存在，所以会自动进行创建，而后通过控制台可以看到相应的队列信息，如图 14-5 所示。

Overview			Messages			Message rates			+/-
Name	Features	State	Ready	Unacked	Total	incoming	deliver / get	ack	
mldn.msg.queue	D AD	running	519,446	0	519,446	11,569/s	0.00/s	0.00/s	

图 14-5　创建消息队列

14.4.2　创建消息消费者

消息消费者的主要功能是开启对 RabbitMQ 组件的监听，每当有消息发送后都可以及时地进行消息的处理（消费）。如果创建的是非持久化消息（临时消息），则消费端不开启，消息不会保存；如果创建的是持久化队列，则消息会一直等待消费后才会被删除。

1.【mldnrabbitmq 项目】创建新的项目模块 mldnrabbitmq-consumer-a。

2.【mldnrabbitmq-consumer-a 项目】修改 pom.xml 配置文件，配置 amqp-client 依赖库。

3.【mldnrabbitmq-consumer-a 项目】建立消费端程序。

```
package cn.mldn.mldnrabbitmq.consumer.a;
import com.rabbitmq.client.Channel;
import com.rabbitmq.client.Connection;
import com.rabbitmq.client.ConnectionFactory;
import com.rabbitmq.client.Consumer;
import com.rabbitmq.client.DefaultConsumer;
public class MessageConsumerA {
    private static final String HOST = "rabbitmq-single";              // 主机名称
```

```
    private static final int PORT = 5672;                             // 发送端口
    private static final String USERNAME = "mldn";                    // 用户名
    private static final String PASSWORD = "java";                    // 密码
    private static final String QUEUE_NAME = "mldn.msg.queue";        // 建立一个队列名称
    public static void main(String[] args) throws Exception {
        // 如果要进行RabbitmQ连接，则需要有一个连接工厂类，通过连接工厂创建连接
        ConnectionFactory factory = new ConnectionFactory();          // 创建连接工厂
        factory.setHost(HOST);                                        // 连接主机
        factory.setPort(PORT);                                        // 连接端口
        factory.setUsername(USERNAME);                                // 用户名
        factory.setPassword(PASSWORD);                                // 密码
        // 通过连接工厂可以直接获取一个连接对象
        Connection connection = factory.newConnection();              // 建立新的连接
        // 如果要进行消息的发送，则需要通过连接获取一个Channel
        Channel channel = connection.createChannel();                 // 创建一个连接通道
        channel.queueDeclare(QUEUE_NAME, true, false, true, null);
        Consumer consumer = new DefaultConsumer(channel) {
            // 进行消息的处理操作
            public void handleDelivery(String consumerTag, com.rabbitmq.client.Envelope envelope,
com.rabbitmq.client.AMQP.BasicProperties properties, byte[] body) throws java.io.IOException {
                String message = new String(body);                    // 将字节数组变为字符串
                System.err.println("【消息消费者-A】" + message);
            };
        };
        channel.basicConsume(QUEUE_NAME, consumer);                   // 设置消息消费者
    }
}
```

本程序对指定的消息队列 mldn.msg.queue 进行了消费监听，当有消息发送后会自动进行消费处理，同时也可以在 RabbitMQ 中监测到相应信息，如图 14-6 所示。

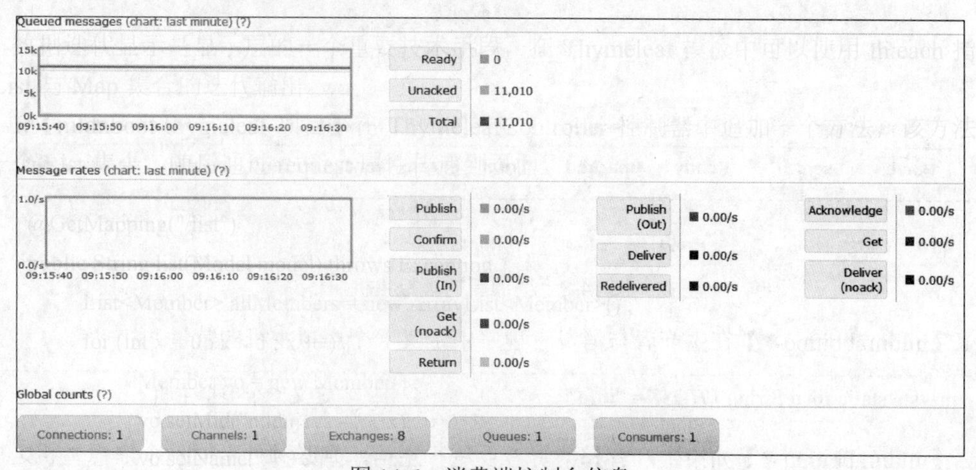

图 14-6　消费端控制台信息

如果消息量很大，那么一个消费者是不够使用的，可以考虑建立多个消费者。一旦建立了多个消费者，那么将自动实现消费者集群。如果有多个消费者，那么 RabbitMQ 会自动指定消息消费的调用轮询。

14.4.3 消息持久化

在进行消息处理的时候会发现存在一个持久化的配置选项。持久化的本质在于：即使 RabbitMQ 停机之后，未消费的消息也可以保存下来，而后等待重新启动之后再进行消费处理。

1.【mldnrabbitmq-queue-provider 项目】将第二个参数（durable）设置为 false，那么就表示该消息队列的信息会被持久化保存；但是如果设置为 false，则在重新启动之后，队列将消失，而其对应的消息也会消失。

```
channel.queueDeclare(QUEUE_NAME, false, false, true, null) ;
```

2.【mldnrabbitmq-queue-provider】定义持久化消息。

```
channel.queueDeclare(QUEUE_NAME, true, false, true, null) ;
channel.basicPublish("", QUEUE_NAME,
        MessageProperties.PERSISTENT_TEXT_PLAIN, msg.getBytes());
```

在进行消息组件设置的时候，一定要使用持久化的消息存储。

14.4.4 虚拟主机

虚拟主机是 RabbitMQ 中提出的消息隔离机制，即不同的消息生产者与消费者可以在不同的虚拟主机中进行消息处理，这样就避免了消息冲突问题。在默认状态下 RabbitMQ 提供一个“/”虚拟主机，如图 14-7 所示。如果开发者有需要，也可以通过控制台进行虚拟主机的配置。例如，本例将配置一个名称为 info 的虚拟主机，同时允许 mldn 用户访问，如图 14-8 所示。

Overview		Messages			Network		Message rates		+/-
Name	Users (?)	Ready	Unacked	Total	From client	To client	publish	deliver / get	
/	guest, mldn	0	10	10	0B/s	0B/s	0.00/s	0.00/s	

图 14-7　默认虚拟主机

Overview		Messages			Network		Message rates	
Name	Users (?)	Ready	Unacked	Total	From client	To client	publish	deliver / get
/	guest, mldn	0	0	0	0B/s	0B/s	0.00/s	0.00/s
info	mldn	NaN	NaN	NaN				

图 14-8　增加新的虚拟主机

1.【mldnrabbitmq-*】在生产者与消费者程序中定义一个虚拟主机的常量。

```
private static final String VHOST = "info" ;                    // 定义一个虚拟主机
```

2.【mldnrabbitmq-*】如果生产者和消费者要进行通信，则必须处于同一个虚拟主机之中，

修改程序的连接配置。

```
// 如果要进行RabbitMQ连接，则一定需要有一个连接工厂类，通过连接工厂创建连接
ConnectionFactory factory = new ConnectionFactory();          // 创建连接工厂
factory.setHost(HOST);                                        // 连接主机
factory.setPort(PORT);                                        // 连接端口
factory.setUsername(USERNAME);                                // 用户名
factory.setPassword(PASSWORD);                                // 密码
factory.setVirtualHost(VHOST);                                // 虚拟主机
```

此时，生产者与消费者将不再使用默认虚拟主机，而使用 info 虚拟主机进行队列消息通信。

14.5 发布订阅模式

在 RabbitMQ 进行消息生产与消费处理中除了可以实现基本的队列消息外，最为重要的就是 Exchange（交换空间）概念，在 RabbitMQ 中对于交换空间有 topic（主题订阅）、direct（直连）、fanout（广播）3 种类型。

提示：请使用不同交换空间。

在 RabbitMQ 中交换空间是不可以混用的，即 fanout 的交换空间只能留给广播模式，如果在此交换空间上使用 direct 模式，则程序无法执行。

14.5.1 广播模式

广播模式指的是同一个消息可以被若干个消费者共同消费。广播模式这种操作需要利用 Exchange 来完成，它的处理结构如图 14-9 所示。

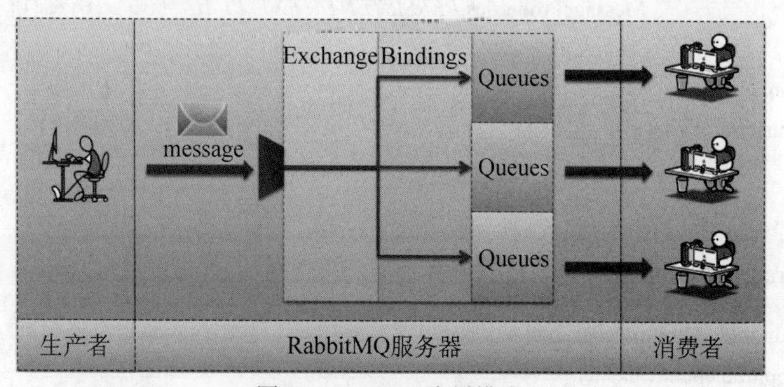

图 14-9　Fanout 广播模式

1.【mldnrabbitmq-queue-provider 项目】修改消息发送程序类。

```
package cn.mldn.mldnrabbitmq.queue.provider;
import com.rabbitmq.client.Channel;
```

```
import com.rabbitmq.client.Connection;
import com.rabbitmq.client.ConnectionFactory;
import com.rabbitmq.client.MessageProperties;
public class MessageProvider {
    private static final String HOST = "rabbitmq-single";              // 主机名称
    private static final int PORT = 5672;                               // 发送端口
    private static final String USERNAME = "mldn";                      // 用户名
    private static final String PASSWORD = "java";                      // 密码
    private static final String EXCHANGE_NAME = "mldn.exchange.fanout"; // 定义exchange名称
    private static final String VHOST = "info";                         // 定义一个虚拟主机
    public static void main(String[] args) throws Exception {
        // 如果要进行RabbitmQ连接，则需要有一个连接工厂类，通过该类创建连接
        ConnectionFactory factory = new ConnectionFactory();            // 创建连接工厂
        factory.setHost(HOST);                                          // 连接主机
        factory.setPort(PORT);                                          // 连接端口
        factory.setUsername(USERNAME);                                  // 用户名
        factory.setPassword(PASSWORD);                                  // 密码
        factory.setVirtualHost(VHOST);                                  // 虚拟主机
        // 通过连接工厂可以直接获取一个连接对象
        Connection connection = factory.newConnection();                // 建立新的连接
        // 如果要进行消息的发送，则需要通过连接获取一个Channel
        Channel channel = connection.createChannel();                   // 创建连接通道
        channel.exchangeDeclare(EXCHANGE_NAME, "fanout");               // 设置exchange类型
        for (int x = 0; x < 10; x++) {
            String msg = "mldnjava - " + x;                             // 消息内容
            channel.basicPublish(EXCHANGE_NAME, "",
                    MessageProperties.PERSISTENT_TEXT_PLAIN, msg.getBytes());
        }
        channel.close();                                                // 关闭通道
        connection.close();                                             // 关闭连接
    }
}
```

2.【mldnrabbitmq-consumer-*项目】消费者也配置相同的 Exchange 信息。

```
package cn.mldn.mldnrabbitmq.consumer.a;
import com.rabbitmq.client.Channel;
import com.rabbitmq.client.Connection;
import com.rabbitmq.client.ConnectionFactory;
import com.rabbitmq.client.Consumer;
import com.rabbitmq.client.DefaultConsumer;
```

```
public class MessageConsumerA {
    private static final String HOST = "rabbitmq-single";                          // 主机名称
    private static final String EXCHANGE_NAME = "mldn.exchange.fanout";           // 定义exchange名称
    private static final int PORT = 5672;                                          // 发送端口
    private static final String USERNAME = "mldn";                                 // 用户名
    private static final String PASSWORD = "java";                                 // 密码
    private static final String QUEUE_NAME = "mldn.msg.queue";                     // 建立一个队列名称
    private static final String VHOST = "info" ;                                   // 定义一个虚拟主机
    public static void main(String[] args) throws Exception {
        // 如果要进行RabbitmQ连接，则需要有一个连接工厂类，通过该类创建连接
        ConnectionFactory factory = new ConnectionFactory();                       // 创建连接工厂
        factory.setHost(HOST);                                                     // 连接主机
        factory.setPort(PORT);                                                     // 连接端口
        factory.setUsername(USERNAME);                                             // 用户名
        factory.setPassword(PASSWORD);                                             // 密码
        factory.setVirtualHost(VHOST);                                             // 虚拟主机
        // 通过连接工厂可以直接获取一个连接对象
        Connection connection = factory.newConnection();                           // 建立新的连接
        // 如果要进行消息的发送，则需要通过连接获取一个Channel
        Channel channel = connection.createChannel();                              // 创建一个连接通道
        channel.queueDeclare(QUEUE_NAME, true, false, true, null) ;                // 定义队列
        channel.exchangeDeclare(EXCHANGE_NAME, "fanout");                          // 设置Exchange类型
        channel.queueBind(QUEUE_NAME, EXCHANGE_NAME, "") ;
        Consumer consumer = new DefaultConsumer(channel) {                         // 进行消息的处理操作
            public void handleDelivery(String consumerTag,
                    com.rabbitmq.client.Envelope envelope,
                    com.rabbitmq.client.AMQP.BasicProperties properties,
                    byte[] body) throws java.io.IOException {
                String message = new String(body) ;                                // 将字节数组变为字符串
                System.err.println("【消息消费者-A】" + message);
            };
        };
        channel.basicConsume(QUEUE_NAME, consumer) ;                               // 设置消息消费者
    }
}
```

此时实现了广播模式，这样所有的消费者启动之后，将会收到相同的消息内容。

14.5.2 直连模式

直连模式的主要特点是需要一个点对点的连接处理，通过直连模式需要定义一个 RoutingKey 信息，而后具有相同 RoutingKey 信息的消费端才能够进行此消息的消费，如图 14-10 所示。

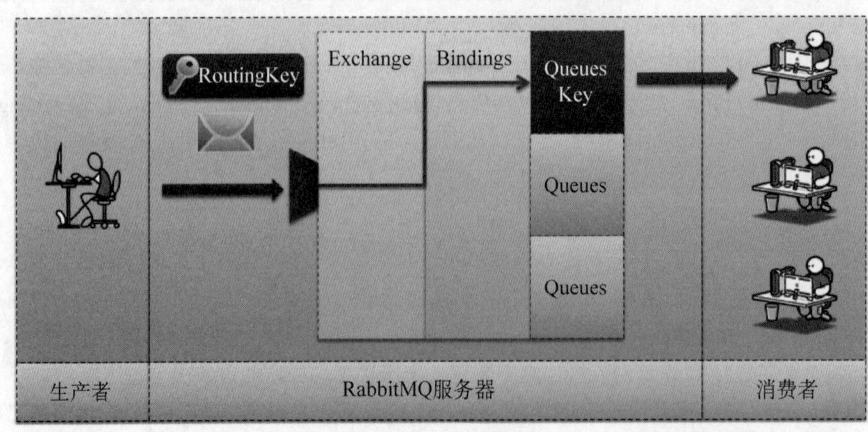

图 14-10　直连模式

1.【mldnrabbitmq-queue-provider 项目】修改消息生产者，在消息发送时追加 RoutingKey 配置项。

创建直连通道	Channel channel = connection.createChannel(); channel.exchangeDeclare(*EXCHANGE_NAME*, "direct");
发送消息	channel.basicPublish(*EXCHANGE_NAME*, "mldn-key", MessageProperties.*PERSISTENT_TEXT_PLAIN*, msg.getBytes());

2.【mldnrabbitmq-consumer-*项目】修改消费端，使用直连模式并设置 RoutingKey 信息。

```
Channel channel = connection.createChannel();                // 创建一个连接通道
String queueName = channel.queueDeclare().getQueue() ;       // 自己创建队列
channel.exchangeDeclare(EXCHANGE_NAME, "direct");            // 设置Exchange类型
channel.queueBind(queueName, EXCHANGE_NAME, "mldn-key") ;
Consumer consumer = new DefaultConsumer(channel) {           // 进行消息的处理操作
    public void handleDelivery(String consumerTag,
            com.rabbitmq.client.Envelope envelope,
            com.rabbitmq.client.AMQP.BasicProperties properties,
            byte[] body) throws java.io.IOException {
        String message = new String(body) ;                  // 将字节数组变为字符串
        System.err.println("【消息消费者-A】" + message);
    };
};
channel.basicConsume(queueName, consumer) ;                  // 设置消息消费者
```

此时，具有相同类型的 Exchange，而后当 RoutingKey 的内容完全一致的时候才可以实现消息的接收处理。在整体的设计中 Exchange 可以管理多个队列信息，所以消费端并没有创建自定义队列名称，而是通过 Channel 自动定义了一个通道。

14.5.3　主题模式

主题模式可以理解为多个直连模式的应用，即消费端可以使用 RoutingKey 匹配模式来实现指定主题的消息获取，如图 14-11 所示。

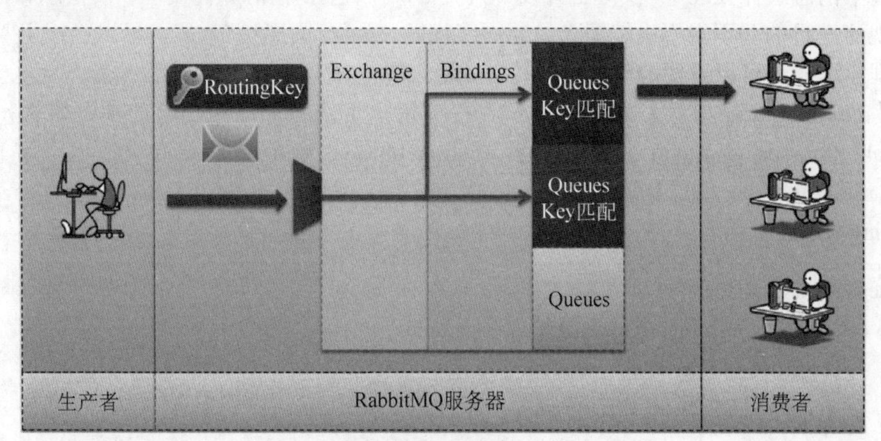

图 14-11　主题消息

在进行主题消息匹配时给出了如下两种匹配标记。

☑ "主题名称.*"：可以匹配一组任意副标题名称。例如，mldn-key.*可以匹配 mldn-key.message 或 mldn-key.java 的 RoutingKey 信息。

☑ "主题名称.#"：可以匹配多组任意副标题名称。例如，mldn-key.#可以匹配 mldn-key.message.java 的 RoutingKey 信息。

1.【mldnrabbitmq-queue-provider 项目】修改消息生产者类，采用主题模式。

```
Connection connection = factory.newConnection();                // 建立新的连接
// 如果要进行消息的发送，则一定需要通过连接获取一个Channel
Channel channel = connection.createChannel();                   // 创建连接通道
channel.exchangeDeclare(EXCHANGE_NAME, "topic");                // 设置Exchange类型
for (int x = 0; x < 10; x++) {
    String msg = "mldnjava - " + x;                             // 消息内容
    channel.basicPublish(EXCHANGE_NAME, "mldn-key.message.java",
            MessageProperties.PERSISTENT_TEXT_PLAIN, msg.getBytes());
}
```

2.【mldnrabbitmq-consumer-*项目】修改消费者类，匹配任意多级副标题。

```
channel.exchangeDeclare(EXCHANGE_NAME, "topic");                // 设置Exchange类型
channel.queueBind(queueName, EXCHANGE_NAME, "mldn-key.#") ;
```

本程序在消费端使用 mldn-key.#配置了 RoutingKey 信息，这样就可以匹配以 mldn-key 开头的任意消息。

14.6 Spring 整合 RabbitMQ

在整个消息组件处理中，实际上开发者只关心两个问题：消息如何发送以及消息如何接收。为了实现这一目的，不得不去编写大量的程序代码。为了简化这些代码，可以直接引入 Spring 进行管理，这样也可以方便地与其他项目进行整合。

1.【mldnrabbitmq 项目】为了让读者观察清楚，本例将创建两个新的项目模块：生产者（mldnrabbitmq-spring-provider）和消费者（mldnrabbitmq-spring-consumer）。

2.【mldnrabbitmq 项目】修改 pom.xml 配置文件，除了引入 Spring 相关依赖包外，还需要引入 spring-rabbit 依赖库。

```
<dependency>
    <groupId>org.springframework.amqp</groupId>
    <artifactId>spring-rabbit</artifactId>
    <version>2.0.0.RELEASE</version>
</dependency>
```

3.【mldnrabbitmq-spring-provider、mldnrabbitmq-spring-consumer 项目】在子模块中引入依赖库。

```
<dependency>
    <groupId>org.springframework.amqp</groupId>
    <artifactId>spring-rabbit</artifactId>
</dependency>
```

4.【mldnrabbitmq-spring-provider 项目】建立 rabbitmq.properties 配置文件，保存 RabbitMQ 核心配置项。

```
# RabbitMQ的连接主机地址
mq.rabbit.host=rabbitmq-single
# RabbitMQ的连接端口号
mq.rabbit.port=5672
# RabbitMQ的虚拟主机名称
mq.rabbit.vhost=info
# RabbitMQ的exchange名称
mq.rabbit.exchange=mldn.msg.spring
# 用户名
mq.rabbit.username=mldn
# 密码
mq.rabbit.password=java
```

5.【mldnrabbitmq-spring-provider 项目】建立 src/main/resources/spring-base.xml 配置文件。

```
<context:component-scan base-package="cn.mldn"/>
<context:property-placeholder location="classpath:config/*.properties"/>
```

6.【mldnrabbitmq-spring-provider 项目】建立 src/main/resources/spring-rabbitmq.xml 配置文件。

```
<!-- 如果要想进行RabbiMQ的操作管理，则首先一定要准备一个连接工厂类 -->
<rabbit:connection-factory id="connectionFactory"
    host="${mq.rabbit.host}" port="${mq.rabbit.port}" username="${mq.rabbit.username}"
    password="${mq.rabbit.password}" virtual-host="${mq.rabbit.vhost}" />
<!-- 所有的连接工厂要求被RabbitMQ所管理 -->
<rabbit:admin connection-factory="connectionFactory" />
<!-- 创建一个队列信息 -->
<rabbit:queue id="myQueue" durable="true" auto-delete="true"
    exclusive="false" name="mldn.queue.title" />
<!-- 下面实现一个直连的操作模式 -->
<rabbit:direct-exchange id="mq-direct"
    name="${mq.rabbit.exchange}" durable="true" auto-delete="true">
    <rabbit:bindings>
        <!-- 现在要求绑定到指定的队列之中 -->
        <rabbit:binding queue="myQueue" key="mldn-key" />
    </rabbit:bindings>
</rabbit:direct-exchange>
<!-- 所有整合的消息系统都会有一个模板 -->
<rabbit:template id="amqpTemplate" exchange="${mq.rabbit.exchange}"
    connection-factory="connectionFactory" />
```

7.【mldnrabbitmq-spring-provider 项目】消息发送应该以一个服务接口的形式出现。建立 IMessageService 业务接口。

```
package cn.mldn.mldnrabbitmq.spring.provider.service;
public interface IMessageService {
    public void send(String msg) ;
}
```

8.【mldnrabbitmq-spring-provider 项目】实现消息发送服务处理。

```
package cn.mldn.mldnrabbitmq.spring.provider.service.impl;
import org.springframework.amqp.core.AmqpTemplate;
import org.springframework.beans.factory.annotation.Autowired;
import org.springframework.stereotype.Service;
import cn.mldn.mldnrabbitmq.spring.provider.service.IMessageService;
@Service
public class MessageServiceImpl implements IMessageService {
```

```
    private static final String ROUTING_KEY = "mldn-key" ;
    @Autowired
    private AmqpTemplate amqpTemplate ;             // 注入消息组件工具类
    @Override
    public void send(String msg) {
        this.amqpTemplate.convertAndSend(ROUTING_KEY, msg);
    }
}
```

9.【mldnrabbitmq-spring-provider 项目】编写测试类，实现消息的发送。

```
package cn.mldn.mldnrabbitmq.spring.provider;
import javax.annotation.Resource;
import org.junit.Test;
import org.junit.runner.RunWith;
import org.springframework.test.context.ContextConfiguration;
import org.springframework.test.context.junit4.SpringJUnit4ClassRunner;
import cn.mldn.mldnrabbitmq.spring.provider.service.IMessageService;
@ContextConfiguration(locations = { "classpath:spring/spring-*.xml" })
@RunWith(SpringJUnit4ClassRunner.class)
public class TestRabbitmqProvider {
    @Resource
    private IMessageService messageService;
    @Test
    public void testSend() {
        this.messageService.send("www.mldn.cn");
    }
}
```

10.【mldnrabbitmq-spring-consumer 项目】建立一个消息监听类（消费者）。

```
package cn.mldn.mldnrabbitmq.spring.consumer;
import org.springframework.amqp.core.Message;
import org.springframework.amqp.core.MessageListener;
// Spring可以整合所有消息系统，但是考虑到操作标准化问题，几乎都会提供一个MessageListener接口
public class MessageConsumer implements MessageListener {
    @Override
    public void onMessage(Message msg) {
        System.out.println("【消费者接收消息】" + new String(msg.getBody()));
    }
}
```

11.【mldnrabbitmq-spring-consumer 项目】创建 Spring 配置文件，此时可以将生产者中的 spring-base.xml 配置文件与 rabbitmq.properties 配置文件复制过来，然后还需要创建 spring-rabbitmq-consumer.xml，在该配置文件中除了要配置连接之外，最重要的就是需要配置消费端的监听程序。

```
<!-- 消息的处理需要有一个消息的监控程序，并且将此类配置到Spring中 -->
<bean id="messageConsumer" class="cn.mldn.mldnrabbitmq.spring.consumer.MessageConsumer"/>
<rabbit:listener-container connection-factory="connectionFactory"><!-- 启动消费监听程序 -->
    <rabbit:listener ref="messageConsumer" queues="myQueue"/>
</rabbit:listener-container>
```

12.【mldnrabbitmq-spring-consumer 项目】编写测试类启动消费者，由于消费者监听类需要一直存在，所以可以直接设置一个长时间修改。

```
package cn.mldn.mldnrabbitmq.spring.consumer;
import java.util.concurrent.TimeUnit;
import org.junit.Test;
import org.junit.runner.RunWith;
import org.springframework.test.context.ContextConfiguration;
import org.springframework.test.context.junit4.SpringJUnit4ClassRunner;
@ContextConfiguration(locations = { "classpath:spring/spring-*.xml" })
@RunWith(SpringJUnit4ClassRunner.class)
public class TestRabbitmqConsumer {
    @Test
    public void testConsumer() throws Exception {
        TimeUnit.DAYS.sleep(Long.MAX_VALUE);
    }
}
```

此时，项目已经实现了基于 Spring 管理的 RabbitMQ 消息生产与消费处理。可以发现，传统框架的整合过程与 SpringBoot 相比，配置项过于烦琐与重复，这也是 SpringBoot 可以迅速流行的主要原因。

14.7　镜像队列

在实际的开发过程中，单主机即便配置再高，也会存在性能瓶颈问题。同样，在整个架构设计之中，消息组件是一个重要的缓冲组件，如果消息组件的性能不足，整个项目就会出现问题，所以就需要使用多个消息组件进行整体设计。本例将为读者讲解 RabbitMQ 集群配置，使用的注解列表如表 14-1 所示。

表 14-1 RabbitMQ 集群主机

No.	主机名称	IP 地址	描述
1	rabbitmq-cluster-a	192.168.68.151	RabbitMQ
2	rabbitmq-cluster-b	192.168.68.152	RabbitMQ
3	rabbitmq-cluster-c	192.168.68.153	RabbitMQ

1.【rabbitmq-cluster-*】3 台主机已经在各自的/etc/hosts 文件里面设置好了主机名称。

```
192.168.68.151    rabbitmq-cluster-a
192.168.68.152    rabbitmq-cluster-b
192.168.68.153    rabbitmq-cluster-c
```

2.【rabbitmq-cluster-*】要想实现 3 台 RabbitMQ 同步处理，必须保证.erlang.cookie 的文件内容是同步的，默认路径为 ~/.erlang.cookie。

3.【rabbitmq-cluster-*】启动 RabbitMQ 服务：/usr/local/rabbitmq/sbin/rabbitmq-server start。

4.【rabbitmq-cluster-a】进行集群搭建前，其他主机要以 rabbitmq-cluster-a 主机为主要的 master，因此需要查看一下 rabbitmq-cluster-a 主机当前的集群状态：/usr/local/rabbitmq/sbin/rabbitmqctlcluster_status。

```
Cluster status of node 'rabbit@rabbitmq-cluster-a' ...
[{nodes,[{disc,['rabbit@rabbitmq-cluster-a']}]},
 {running_nodes,['rabbit@rabbitmq-cluster-a']},
 {cluster_name,<<"rabbit@rabbitmq-cluster-a">>},
 {partitions,[]},
 {alarms,[{'rabbit@rabbitmq-cluster-a',[]}]}]
```

5.【rabbitmq-cluster-b、rabbitmq-cluster-c】配置 rabbitmq-cluster-b 主机与 rabbitmq-cluster-c 主机加入到 rabbitmq-cluster-a 主机的集群，此时需要停止 RabbitMQ 的主服务。

```
/usr/local/rabbitmq/sbin/rabbitmqctl stop_app
```

6.【rabbitmq-cluster-b、rabbitmq-cluster-c】将主机加入到 rabbitmq-cluster-a 主机的服务之中。

```
/usr/local/rabbitmq/sbin/rabbitmqctl join_cluster rabbit@rabbitmq-cluster-a
```

7.【rabbitmq-cluster-b、rabbitmq-cluster-c】恢复节点执行。

```
/usr/local/rabbitmq/sbin/rabbitmqctl start_app
```

8.【rabbitmq-cluster-a】查看当前的集群状态：/usr/local/rabbitmq/sbin/rabbitmqctl cluster_status。

```
Cluster status of node 'rabbit@rabbitmq-cluster-a' ...
[{nodes,[{disc,['rabbit@rabbitmq-cluster-a','rabbit@rabbitmq-cluster-b',
                'rabbit@rabbitmq-cluster-c']}]},
 {running_nodes,['rabbit@rabbitmq-cluster-c','rabbit@rabbitmq-cluster-b',
```

```
                'rabbit@rabbitmq-cluster-a']},
{cluster_name,<<"rabbit@rabbitmq-cluster-a">>},
{partitions,[]},
{alarms,[{'rabbit@rabbitmq-cluster-c',[]},
         {'rabbit@rabbitmq-cluster-b',[]},
         {'rabbit@rabbitmq-cluster-a',[]}]}]
```

9.【rabbitmq-cluster-a】追加一个新的管理员账户 mldn/java。

☑ 追加新的用户：/usr/local/rabbitmq/sbin/rabbitmqctl add_user mldn java。

☑ 追加到管理组中：/usr/local/rabbitmq/sbin/rabbitmqctl set_user_tags mldn administrator。

☑ 设置虚拟主机：/usr/local/rabbitmq/sbin/rabbitmqctl set_permissions -p / mldn ConfP WriteP ReadP。

10.【操作系统】修改 hosts 配置文件，追加主机配置。

```
192.168.68.151  rabbitmq-cluster-a
192.168.68.152  rabbitmq-cluster-b
192.168.68.153  rabbitmq-cluster-c
```

11.【操作系统】输入地址 http://rabbitmq-cluster-a:15672 进行访问，可以看到如图 14-12 所示的提示信息。

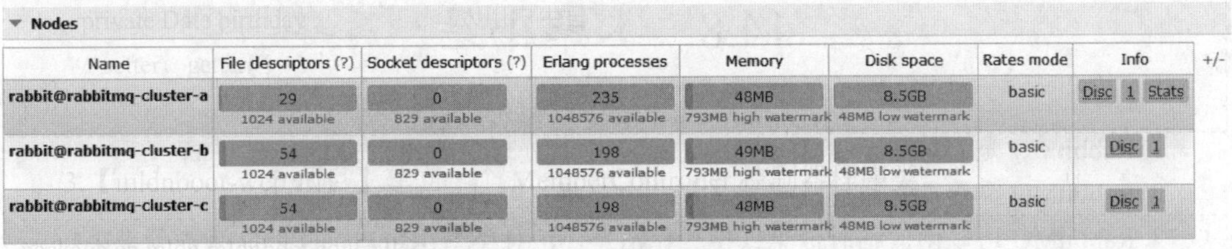

▼ Nodes

Name	File descriptors (?)	Socket descriptors (?)	Erlang processes	Memory	Disk space	Rates mode	Info
rabbit@rabbitmq-cluster-a	29 1024 available	0 829 available	235 1048576 available	48MB 793MB high watermark	8.5GB 48MB low watermark	basic	Disc 1 Stats
rabbit@rabbitmq-cluster-b	54 1024 available	0 829 available	198 1048576 available	49MB 793MB high watermark	8.5GB 48MB low watermark	basic	Disc 1
rabbit@rabbitmq-cluster-c	54 1024 available	0 829 available	198 1048576 available	48MB 793MB high watermark	8.5GB 48MB low watermark	basic	Disc 1

图 14-12　集群主机环境

12.【客户端】集群控制必须满足一个前提：当前用户操作的 rabbitmq-cluster-a 主机宕机之后，其他主机（rabbitmq-cluster-b 或 rabbitmq-cluster-c）可以继续提供这些未消费的消息。这个配置称为镜像队列，需要用户自己手动开启，通过控制台的管理中心进行配置，如图 14-13 所示。

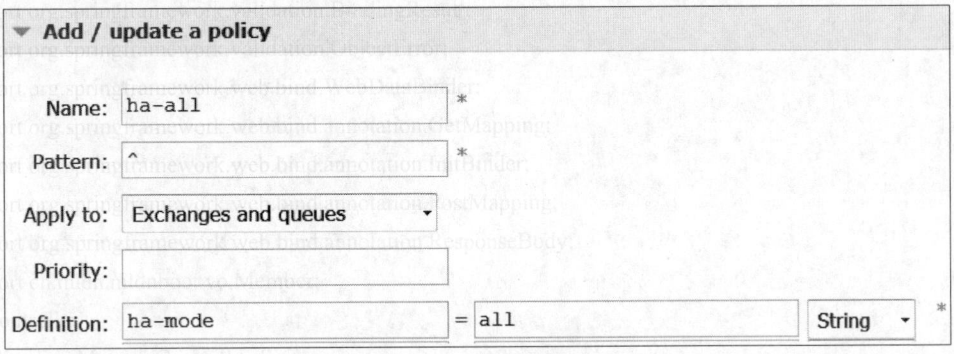

▼ Add / update a policy

Name: ha-all *

Pattern: ^ *

Apply to: Exchanges and queues

Priority:

Definition: ha-mode = all String *

图 14-13　配置镜像队列

**提示：此时并非 HA 机制。**

在 RabbitMQ 中并不支持 HA 机制，只是镜像配置。因为 Java 客户端（不使用 Spring）直接连接 RabbitMQ 时，只能够连接一台 RabbitMQ 主机，这样当客户端连接 rabbitmq-cluster-a 主机并且 rabbitmq-cluster-a 主机宕机后，实际上这个客户端就无法进行消息的发送或消费了。

Spring 整合后的 RabbitMQ 消除了此限制（提升了 RabbitMQ 的 HA 能力），可以配置多台主机。即使有一台主机出现了问题，也可以使用其他正常主机实现操作。

13.【mldnrabbitmq-spring-*项目】修改 rabbitmq.properties 配置文件，配置多台主机地址。

```
# RabbitMQ的连接主机地址
mq.rabbit.address=rabbitmq-cluster-a:5672,rabbitmq-cluster-b:5672,rabbitmq-cluster-c:5672
```

14.【mldnrabbitmq-spring-*项目】修改 spring-rabbitmq-*.xml 配置文件。

```
<rabbit:connection-factory id="connectionFactory"
    addresses="${mq.rabbit.address}" username="${mq.rabbit.username}"
    password="${mq.rabbit.password}" virtual-host="${mq.rabbit.vhost}" />
```

此时，基于 Spring 框架之后，就可以结合 RabbitMQ 镜像队列实现 RabbitMQ 的 HA 机制。

14.8 本章小结

1. RabbitMQ 是一款基于 AMQP 的消息组件，其处理性能要比 JMS 组件更高。
2. RabbitMQ 需要 Erlang 语言环境支持，并且提供完善的管理控制中心。
3. RabbitMQ 可以创建临时消息或持久化消息，消息类型由用户创建的队列类型来决定。
4. RabbitMQ 中支持虚拟主机，以实现不同用户与不同队列之间的隔离。
5. RabbitMQ 中交换空间有 topic（主题订阅）、direct（直连）和 fanout（广播）3 种类型。
6. RabbitMQ 集群需要配置镜像队列后才可以实现消息在多个主机里的保存，但 HA 机制可以依靠 Spring 框架补充完成。

第 15 章

Docker 虚拟化容器

通过本章学习，可以达到以下目标：

1. 掌握 Docker 的主要作用。
2. 掌握 Docker 的安装与配置方法。
3. 掌握 DockerHub 的使用。
4. 掌握 Docker 镜像文件管理。
5. 掌握 SpringBoot（SpringCloud）与 Docker 整合开发。

Docker 是虚拟化云平台技术，利用 Docker 可以极大地简化开发人员与运维人员的工作压力，并且可以在一台主机中运行多个 Docker 容器。最为重要的是，Spring 微架构也是云服务的一种实现，与 Docker 的联系也是极为紧密的。本章将为读者讲解 Docker 技术的使用。

15.1 Docker 简介

Docker 是一种虚拟化容器技术，开发者可以将应用服务封装到 Docker 容器中，随后再根据需要进行 Docker 容器的部署，这样就可以快速地实现项目的发布，也降低了运维人员的环境部署困难程度。

提示：Docker 产生背景。

在传统项目开发过程中经常会出现这样一些问题。项目开发者根据业务需求开发项目代码，而后将成品代码交由运维人员进行生产环境的服务搭建。一位资深的运维人员可以快速、清晰地明白开发者的运行环境，但如果是一些初级运维人员则可能弄不清楚项目的运行环境，甚至有一些冷门的服务根本不会部署，这样就需要不断与开发人员沟通。一旦沟通出现问题，开发人员将不得不自己去进行项目部署。正是在这样的背景下，Docker 虚拟化技术产生了。

Docker 的 Logo 采用了一种集装箱的设计风格，而其应用效果也正如集装箱一样，每一个 Docker 容器都可以运行各自的服务，而后被封装起来不会互相影响（每一个集装箱都是独立的）。

Docker 是一个开源的应用容器引擎，开发者可以打包他们的应用及依赖包到一个可移植的容器中，然后发布到任何流行的 Linux 机器上，从而实现虚拟化。容器完全使用沙箱机制，相互之间不会有任何接口。Docker 官方网站为 https://www.docker.com/，如图 15-1 所示。

Docker 采用的是 C/S 的处理结构，需要有客户端与服务端。一个完整的 Docker 架构（见

图 15-2）由以下 5 个部分组成。

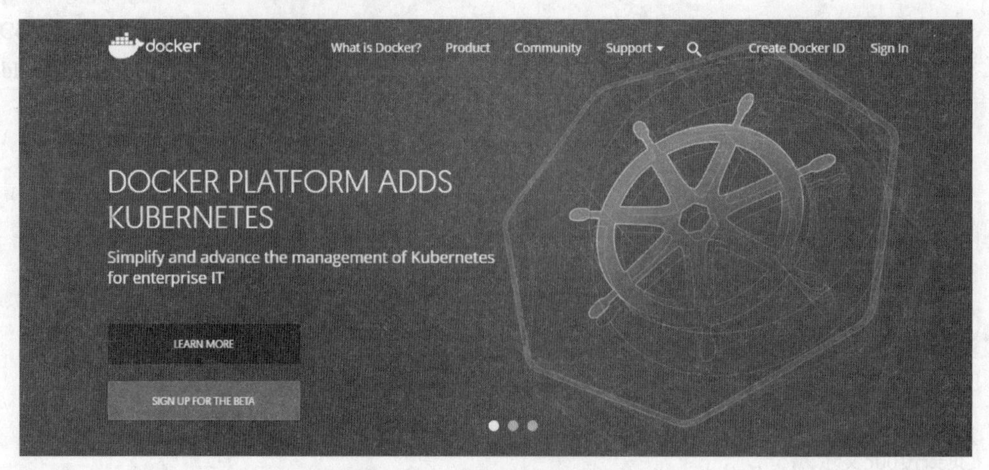

图 15-1　Docker 官方网站

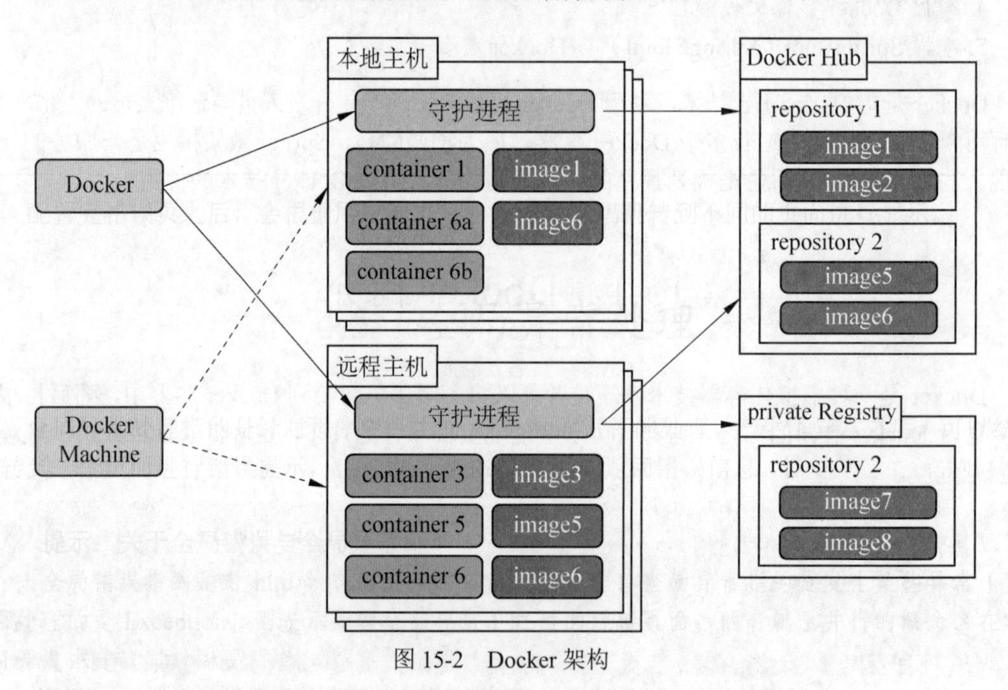

图 15-2　Docker 架构

- ☑ Docker Client 客户端：Docker 的开发环境，可以有多种操作系统支持。
- ☑ Docker Daemon 守护进程：Docker 的服务端进程。
- ☑ Docker Image 镜像：一台主机中可以存放多个 Docker 镜像，每一个 Docker 镜像都是一组服务。
- ☑ Docker Container 容器：提供一个独立的运行系统，可以实现组件部署。
- ☑ Docker Registry 仓库：Docker 提供有 DockerHub 公共仓库，进行镜像统一管理。

提示：**Docker 容器与镜像的关系。**

Docker 容器通过 Docker 镜像来创建，容器与镜像的关系类似于面向对象编程中对象与类之间的关系。

☑ Docker 镜像 = 类。

☑ Docker 容器 = 对象。

15.2　Docker 安装

Docker 可以在各个主流操作系统上进行安装与配置，并且生成的镜像对于所有的系统都是适用的。本例为了方便，将直接使用 Ubuntu 系统进行 Docker 的安装与使用。

1.【Linux 系统】Docker 要求 Ubuntu 系统的内核版本高于 3.10，可以通过如下命令查看当前的内核版本。

```
uname -r
```

此时，系统之中返回的信息为 4.4.0-109-generic，则表示具备 Docker 安装环境要求。

2.【Linux 系统】如果要安装 Docker，建议先更新一下本地软件库。

```
apt-get update
```

3.【Linux 系统】更新成功后，可以使用如下命令进行 Docker 的安装。

```
apt-get -y install docker.io
```

4.【Linux 系统】启动 Docker 服务。

```
systemctl start docker
```

5.【Linux 系统】也可以通过如下命令将 Docker 设置为自动运行。

```
systemctl enable docker
```

6.【Linux 系统】核对当前 Docker 版本。

docker version	
命令执行结果	Client: Version:　1.13.1 API version:　1.26 Go version:　go1.6.2 Git commit:　092cba3 Built:　Thu Nov 2 20:40:23 2017 OS/Arch:　linux/amd64 Server: Version:　1.13.1

```
API version:    1.26 (minimum version 1.12)
Go version:     go1.6.2
Git commit:     092cba3
Built:          Thu Nov  2 20:40:23 2017
OS/Arch:        linux/amd64
Experimental:   false
```

如果可以看到以上的版本信息，则表示已经成功安装了 Docker。

15.3 Docker 配置与使用

安装了 Docker 工具后，开发者就可以直接在本地进行 Docker 镜像的使用了。Docker 镜像可以直接通过 DockerHub 进行抓取，随后启动 Docker 镜像即可实现虚拟化处理。

15.3.1 获取并使用 Docker 镜像

Docker 中所有可以使用的虚拟操作系统都是通过 Docker 镜像定义的。刚刚完成安装的 Docker 主机不存在镜像，可以直接通过 DockerHub 抓取镜像，已经抓取到的镜像则可以直接运行，如图 15-3 所示。

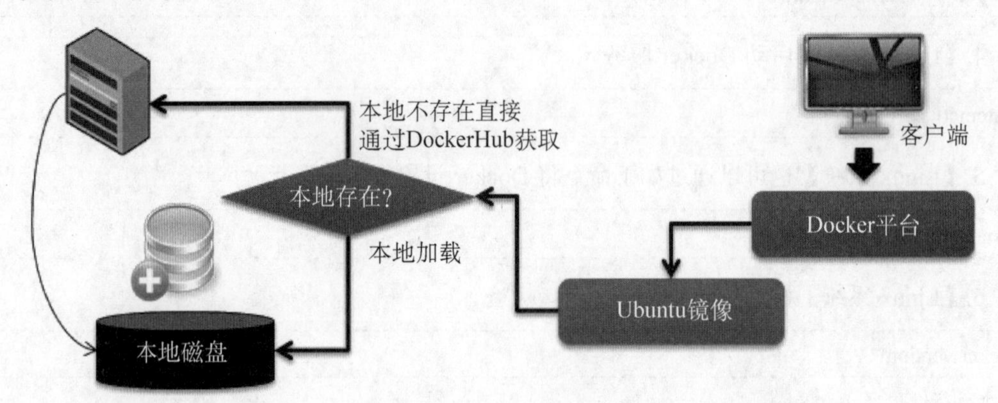

图 15-3 Docker 镜像使用

在 Linux 系统中有一个/bin/echo 命令，该命令的主要功能是对输入的数据进行回显处理。本例将利用 Docker 容器实现这一命令。

1.【Linux 系统】Docker 的所有命令都是以 docker 开头的，要想运行 Docker 镜像（会产生 Docker 容器），需要执行如下命令。

```
docker run ubuntu /bin/echo "Hello World"
```

本程序将直接运行名称为 ubuntu 的镜像文件，由于此镜像文件并不存在于本地，所以会通过 DockerHub 进行镜像抓取，而后再执行镜像中的/bin/echo 命令。执行后，会出现如图 15-4 所示的提示信息。

```
Unable to find image 'ubuntu:latest' locally
latest: Pulling from library/ubuntu
8f7c85c2269a: Pull complete
9e72e494a6dd: Pull complete
3009ec50c887: Pull complete
9d5ffccbec91: Pull complete
e872a2642ce1: Pull complete
Digest: sha256:d3fdf5b1f8e8a155c17d5786280af1f5a04c10e95145a515279cf17abdf0191f
Status: Downloaded newer image for ubuntu:latest
Hello World
```

图 15-4　下载 Docker 镜像

2.【Linux 系统】除了可以在外部直接执行 Docker 镜像之外，也可以进入镜像中进行执行。

```
docker run -i -t ubuntu
```

以上命令也可以简写为 docker run -it ubuntu，其中两个参数的作用如下。

☑　i：在新容器之中指定一个终端。

☑　t：允许直接对容器内的终端进行交互操作处理。

镜像启动之后，会自动分配一个镜像启动容器的 ID，此编号是唯一的，并且一个镜像可以启动多个容器。

3.【Linux 系统】如果要退出使用的终端，可以按 Ctrl + D 快捷键。退出之后，还可以通过 Docker 给出的命令查看当前正在运行的容器。

```
docker ps
```

如果要想查询详细的 Docker 进程信息，则可以使用 docker ps -a 的参数来运行。

15.3.2　Docker 镜像

镜像是整个 Docker 的核心，而只有镜像存在才可以产生 Docker 容器。同时，通过镜像也可以方便地管理各个服务信息。

1.【Linux 系统】查看本机镜像信息。

```
docker images
```

由于此时系统中只有一个 ubuntu 镜像，所以此处只会返回一个镜像信息。

2.【Linux 系统】在 Docker 中提供有镜像信息的查询功能。如果要想查询镜像，可以使用 search 指令完成。

```
docker search ubuntu
```

执行之后会通过 DockerHub 搜索具有关键字 ubuntu 的镜像信息，并将所有版本都进行列表。如果开发者需要指定版本，也可以直接采用如下命令指定镜像版本标记（使用 15 版本）。

```
docker search ubuntu:15
```

3.【Linux 系统】通过 DockerHub 抓取某一个镜像文件。

```
docker pull bharath23/ubuntu-core-15.04-amd64
```

本命令执行之后将进入镜像下载处理，下载完成后可以查看所有的镜像信息，然后得到如图 15-5 所示的信息。

```
REPOSITORY                         TAG       IMAGE ID       CREATED       SIZE
ubuntu                             latest    2a4cca5ac898   8 days ago    111 MB
bharath23/ubuntu-core-15.04-amd64  latest    65dc425bdbd5   2 years ago   131 MB
```

图 15-5　Docker 镜像列表

4.【Linux 系统】对于已经下载的 Docker 镜像，开发者随时都可以启动容器进行镜像的配置。

☑ 进入 ubuntu 镜像 docker run -it ubuntu，此时会分配一个 b4ba10b146df 容器 ID。

☑ 在 ID 为 b4ba10b146df 的容器中进行更新：apt-get update。

☑ 更新完成后，该容器的内容会比原始镜像中多，在不退出的情况下可以进行镜像的保存。

5.【Linux 系统】当用户在容器中进行内容修改后，实际上并不会影响到原始镜像文件，所以可以单独启动一个系统连接，将指定容器 ID 的信息保存到新的镜像之中。

```
docker commit -m="mldn-docker" -a="Lee" b4ba10b146df mldn/ubuntu:base
```

本程序中两个参数的作用如下。

☑ "-m"：表示进行说明的定义，本处定义为-m="mldn-docker"。

☑ "-a"：表示作者，本处定义为-a="Lee"。

6.【Linux 系统】镜像保存完成后可以退出容器，而后查看全部镜像，可以得到如图 15-6 所示的信息。

```
REPOSITORY                         TAG       IMAGE ID       CREATED              SIZE
mldn/ubuntu                        base      78745e02bfbc   About a minute ago   151 MB
ubuntu                             latest    2a4cca5ac898   8 days ago           111 MB
bharath23/ubuntu-core-15.04-amd64  latest    65dc425bdbd5   2 years ago          131 MB
```

图 15-6　Docker 镜像列表

15.3.3　Docker 容器

本例将利用 mldn/ubuntu:base 镜像进行容器配置，主要是在镜像中配置 JDK 与 Tomcat 访问环境。

1.【Linux 系统】将 JDK 与 Tomcat 的开发包上传到/srv/ftp 目录中。

2.【Linux 系统】启动 Docker 容器（容器 ID 为 0598185621f1，启动后不要退出容器）。

```
docker run -it mldn/ubuntu:base
```

3.【Linux 系统】将系统中保存的 JDK 复制到 Docker 容器之中。

```
docker cp /srv/ftp/jdk-8u73-linux-x64.tar.gz 0598185621f1:/usr/
```

4.【Linux 系统】将系统中保存的 Tomcat 复制到 Docker 容器之中。

```
docker cp /srv/ftp/apache-tomcat-9.0.4.tar.gz 0598185621f1:/usr/
```

5.【Docker-0598185621f1 容器】开发包复制完毕之后，可以直接返回已经启动的 Docker 镜

像中查看文件是否存在。如果存在，则表示复制成功。随后对这两个开发包进行解压缩，解压缩到/usr/local 目录中。

解压缩 JDK	tar xzvf /usr/jdk-8u73-linux-x64.tar.gz -C /usr/local/
解压缩 Tomcat	tar xzvf /usr/apache-tomcat-9.0.4.tar.gz -C /usr/local/

6.【Docker-0598185621f1 容器】为方便管理，进行 JDK 与 Tomcat 目录更名配置。

JDK 目录更名	mv /usr/local/jdk1.8.0_73/ /usr/local/jdk
Tomcat 目录更名	mv /usr/local/apache-tomcat-9.0.4/ /usr/local/tomcat

7.【Docker-0598185621f1 容器】为方便配置，可以下载 vim 程序：apt-get -y install vim。

8.【Docker-0598185621f1 容器】修改 Tomcat 配置文件，追加 JAVA_HOME、JRE_HOME。

☑　打开配置文件 vim /usr/local/tomcat/bin/setclasspath.sh。

☑　追加环境配置项。

```
export JAVA_HOME=/usr/local/jdk
export JRE_HOME=/usr/local/jdk/jre
```

9.【Linux 系统】环境配置完成之后，为方便后续使用，可以在外部系统中保存当前配置好的容器为新的镜像。

```
docker commit -m="mldn-tomcat" -a="Lee" 0598185621f1 mldn/ubuntu-tomcat:base
```

保存之后可以通过 docker images 命令查看镜像，得到如图 15-7 所示的列表信息。

```
REPOSITORY                          TAG       IMAGE ID       CREATED          SIZE
mldn/ubuntu-tomcat                  base      ab6a49d5fb0f   19 seconds ago   778 MB
mldn/ubuntu                         base      78745e02bfbc   22 minutes ago   151 MB
ubuntu                              latest    2a4cca5ac898   8 days ago       111 MB
bharath23/ubuntu-core-15.04-amd64   latest    65dc425bdbd5   2 years ago      131 MB
```

图 15-7　镜像列表

10.【Linux 系统】如果在 Docker 容器内部启动 Tomcat 使用的是 Docker 容器中的 8080 端口，会发现这个端口无法被外部主机用户所访问，需要进行一个端口映射。先退出 Docker-0598185621f1 容器（Ctrl + D），而后采用端口映射的模式启动 mldn/ubuntu-tomcat:base 镜像。

```
docker run -p 80:8080 -it mldn/ubuntu-tomcat:base
```

这里将外部的 80 端口与 Docker 容器内部的 8080 端口进行映射，随后会分配一个新的容器编号 616a352c93aa。

11.【Docker- 616a352c93aa 容器】启动 Tomcat 进程：/usr/local/tomcat/bin/catalina.sh start，此时就可以在外部通过 80 端口访问 Docker 中 8080 端口运行的 Tomcat 服务了。

如果要退出当前 Docker 容器，不能再使用 Ctrl + D 快捷键，此命令表示一旦退出，服务就停止了。运行在 Docker 中的 Tomcat 肯定是希望可以在后台继续执行，这时可以使用 Ctrl + P +Q 组合键让其在后台继续执行。

15.4 Docker 镜像管理

配置完成 Docker 之后，该配置的镜像需要提供给运维人员进行部署处理操作。为了能够部署，需要将 Docker 镜像保存下来。Docker 镜像的保存方式有两种：直接利用文件保存和通过 DockerHub 保存。

15.4.1 通过文件保存 Docker 镜像

在 Docker 工具安装完成后会提供一个 save 命令，利用此命令即可实现镜像文件的保存。

1.【Linux 系统】将 mldn/ubuntu:base 镜像保存到/srv/ftp 目录之中，名称为 mldn.base。

```
docker save -o /srv/ftp/mldn.base mldn/ubuntu:base
```

2.【Linux 系统】为了观察镜像导入处理，将原始的 mldn/ubuntu:base 镜像删除。

```
docker rmi mldn/ubuntu:base
```

执行之后再次查询全部镜像，可以发现该镜像已经不在镜像列表之中。

提示：关于后台运行容器冲突。

如果要删除指定的镜像，则必须保证该镜像已经没有运行的容器存在。如果存在，则会出现如下所示的错误提示信息：

```
Error response from daemon: conflict: unable to remove repository reference "mldn/ubuntu:base" (must force) - container 0598185621f1 is using its referenced image 78745e02bfbc
```

这个时候需要查看一下是否有运行的容器，结束容器运行后再进行删除。使用 docker ps -a 可查看所有后台运行的容器信息，使用如下的命令可结束容器的运行。

☑ 停止正在运行的容器：docker stop 容器 ID。

☑ 结束一个容器运行：docker rm 容器 ID。

☑ 结束全部的后台运行的 Docker 容器：docker rm -f 'docker ps -a -q'。

3.【Linux 系统】将保存的镜像文件导入 Docker 中。

```
docker load --input /srv/ftp/mldn.base
```

此时再次查询镜像信息，可以发现 mldn/ubuntu:base 镜像已经恢复，如图 15-8 所示。

REPOSITORY	TAG	IMAGE ID	CREATED	SIZE
mldn/ubuntu-tomcat	base	ab6a49d5fb0f	19 minutes ago	778 MB
mldn/ubuntu	base	78745e02bfbc	41 minutes ago	151 MB
ubuntu	latest	2a4cca5ac898	8 days ago	111 MB
bharath23/ubuntu-core-15.04-amd64	latest	65dc425bdbd5	2 years ago	131 MB

图 15-8 镜像列表

15.4.2　DockerHub

虽然镜像管理提供了镜像文件的导出与导入支持，但是这种环境只适合于小范围的应用。为了对所有的镜像进行管理，也为了帮助开发者进行个人镜像的分享，在 Docker 中提供有 DockerHub 的站点，开发者可以进行免费注册，并且可以将个人镜像提交到公共仓库中以实现分享。DockerHub 官方网站地址为 https://hub.docker.com，界面如图 15-9 所示。

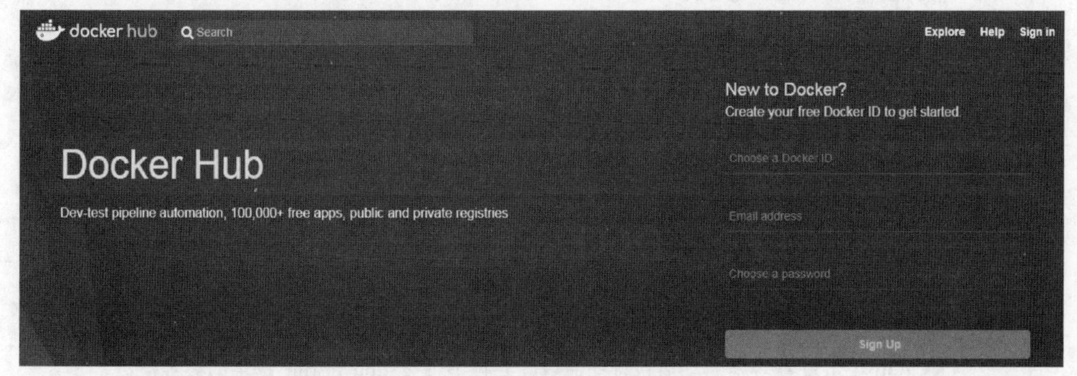

图 15-9　DockerHub 官方站点

1.【Linux 系统】要进行镜像上传，应该使用 dokcer login 命令在本地客户端进行登录。

```
docker login -u 用户名 -p 密码
```

2.【Linux 系统】将 mldn/ubuntu:base 镜像提交到 DockerHub 中。

```
docker push mldn/ubuntu:base
```

镜像发送完成之后，可以通过 DockerHub 看到如下镜像信息，如图 15-10 所示。

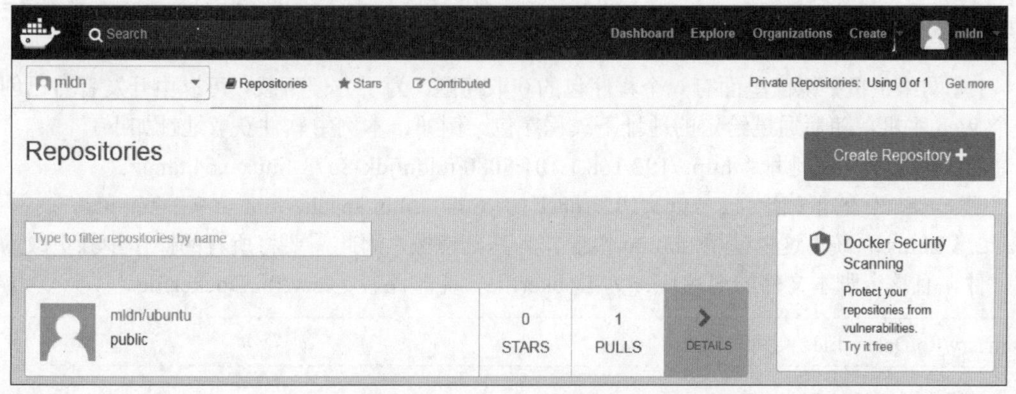

图 15-10　DockerHub 镜像列表

提示：需要有用户名做前缀。

在本程序中创建的镜像名称为 mldn/ubuntu:base，该名称的前缀为当前 DockerHub 的用户名，如果此处配置的用户名不正确，则执行会出现 requested access to the resource is denied 错误信息。

3.【Linux 系统】有需要的时候可以进行 Docker 下载。

```
docker pull mldn/ubuntu:base
```

随着开发者项目的深入，一定会有越来越多的镜像出现。通过 DockerHub 不仅可以方便地进行远程镜像管理，也可以实现镜像的分享。

15.4.3 构建 Docker 镜像

构建 Docker 镜像时，除了可下载后启动容器配置之外，也可以直接通过 Docker 的脚本文件进行创建，但是不管采用何种创建模式，都必须有一个基础镜像进行操作。创建 Dockerfile 脚本文件时需要考虑文件的命令格式，常用的命令如表 15-1 所示。

表 15-1 创建 Dockerfile 文件常用命令

No.	命令	描述
01	ADD	从源系统的文件系统上复制文件到目标容器的文件系统
02	CMD	CMD 可以用于执行特定的命令，镜像构建容器后调用
03	ENTRYPOINT	配置容器启动后执行的命令，并且不可被 docker run 提供的参数覆盖
04	ENV	设置环境变量，这些变量以 key=value 的形式存在，并可以在容器内被脚本或者程序调用
05	EXPOSE	访问端口映射
06	FROM	基础镜像选择
07	MAINTAINER	用于声明作者，放在 FROM 之后
08	RUN	创建镜像时执行命令
09	USER	设置运行容器的 UID
10	VOLUME	Docker 容器访问宿主机上的目录
11	WORKDIR	设置 CMD 指明的命令的运行目录

1.【网络主机】镜像里面有一个程序包的获取问题，为了方便下载，可以由开发者自己创建一个 Web 主机，随后通过给定的地址下载程序包。例如，本例中软件获取地址如下。

☑ JDK 的下载地址：http://192.168.1.101:8080/mldn/jdk-8u73-linux-x64.tar.gz。

☑ Tomcat 的下载地址：http://192.168.1.101:8080/mldn/apache-tomcat-9.0.4.tar.gz。

2.【Linux 系统】这里使用 ubuntu 镜像作为基础镜像，因此需要将所有的操作步骤更改为脚本文件，且这个脚本文件的名称应该为 Dockerfile，完整路径为/srv/ftp/Dockerfile。

```
vim /srv/ftp/Dockerfile
```

3.【Linux 系统】编写 Dockerfile 文件。

```
# 找到一个基础的镜像
FROM ubuntu
# 需要设置一个镜像的说明文件，以及设置作者的信息
MAINTAINER mldn "base@mldn.cn"
```

```
# 镜像下载完成之后需要进行更新处理
RUN apt-get update
# 下载 wget 组件，这样可以通过本地服务器进行指定软件的下载
RUN apt-get -y install wget
# 进行 JDK 的配置，为其设置一个下载目录
# 启动 JDK 的下载操作指令
RUN cd /tmp && wget http://192.168.1.101:8080/mldn/jdk-8u73-linux-x64.tar.gz
# 下载完成之后需要对开发包进行解压缩操作，解压缩到/usr/local 目录
RUN tar xzvf /tmp/jdk-8u73-linux-x64.tar.gz -C /usr/local
# 需要为解压缩后的目录进行更名处理
RUN mv /usr/local/jdk1.8.0_73/ /usr/local/jdk
# 需要在环境的属性之中追加 JDK 配置
ENV JAVA_HOME /usr/local/jdk
ENV PATH $PATH:$JAVA_HOME/jdk/bin
# 进行 Tomcat 的安装与配置
RUN cd /tmp && wget http://192.168.1.101:8080/mldn/apache-tomcat-9.0.4.tar.gz
RUN tar xzvf /tmp/apache-tomcat-9.0.4.tar.gz -C /usr/local/
RUN mv /usr/local/apache-tomcat-9.0.4/ /usr/local/tomcat
# 映射端口
EXPOSE 8080
```

4.【Linux 系统】运行脚本，构建 Docker 镜像。

```
docker build -t mldn/ubuntu-test:dev /srv/ftp/
```

5.【Linux 系统】配置完成后可以直接进行端口映射启动。

```
docker run -p 80:8080 -it mldn/ubuntu-test:dev
```

此时，利用脚本就可以轻松实现镜像创建，使用脚本也可以让镜像的传输更加方便。

15.5　微服务与 Docker

SpringCloud 作为微服务开发技术，其主要特征是进行业务拆分。但如果业务拆分得过细，则会造成服务器的成本攀高。为了解决这样的问题，往往会在项目中整合 Docker 进行微服务部署，这样一台服务器上就可以同时运行多个 SpringCloud 微服务，从而降低服务器运行成本，维护起来也更加方便。

15.5.1　使用 Docker 发布微服务

要将 SpringCloud 微服务程序打包到 Docker 镜像中，需要开启 Docker 的 API 进程，同时还需要在项目中追加相应的打包插件，这样就可以直接将打包后的应用以镜像文件的形式发送到

指定主机上。配置后，也可以将制作的镜像文件直接上传到 DockerHub 中。本例将为读者演示如何将 Eureka 注册中心打包为 Docker 镜像文件。

1.【Linux 系统】编辑 Docker 服务配置文件 docker.service。

```
vim /lib/systemd/system/docker.service
```

2.【Linux 系统】在 docker.service 文件中进行配置，修改 ExecStart 配置项，追加 API 进程。

```
ExecStart=/usr/bin/dockerd -H fd:// -H tcp://0.0.0.0:2375
```

此时的 API 进程运行在本地主机的 2375 端口。

3.【Linux 系统】重新加载 Docker 守护配置项。

```
systemctl daemon-reload
```

4.【Linux 系统】重新启动 Docker 服务进程。

```
service docker restart
```

5.【操作系统】Docker 重新启动完成后可以通过 curl 命令进行测试，如果可以看到版本信息返回，则表示配置成功。

curl 192.168.79.130:2375/version	
返回数据	{"Version":"1.13.1","ApiVersion":"1.26","MinAPIVersion":"1.12","GitCommit":"092cba3","GoVersion":"go1.6.2","Os":"linux","Arch":"amd64","KernelVersion":"4.4.0-109-generic","BuildTime":"2017-11-02T20:40:23.484070968+00:00"}

6.【操作系统】如果要将微服务打包到 DockerHub 中，则还需要在 Maven 的配置文件（conf/settings.xml）中追加 DockerHub 的相关信息。

```
<server>
    <id>服务引用 ID，项目中使用</id>
    <username>DockerHub 用户名</username>
    <password>DockerHub 密码</password>
    <configuration>
        <email>DockerHub 注册邮箱</email>
    </configuration>
</server>
```

7.【mldncloud-eureka-profile 项目】修改 pom.xml 配置文件，配置 docker-maven-plugin 插件。

```
<build>
    <finalName>eureka-server</finalName>
    <plugins>
        <plugin>                <!--该插件的主要功能是进行项目的打包发布处理-->
```

```
            <groupId>org.springframework.boot</groupId>
            <artifactId>spring-boot-maven-plugin</artifactId>
            <configuration>          <!--设置程序执行的主类-->
                <mainClass>
                    cn.mldn.mldncloud.EurekaServerStartApplication7001
                </mainClass>
            </configuration>
            <executions>
                <execution>
                    <goals>
                        <goal>repackage</goal>
                    </goals>
                </execution>
            </executions>
        </plugin>
        <plugin>
            <groupId>com.spotify</groupId>
            <artifactId>docker-maven-plugin</artifactId>
            <version>0.4.12</version>
            <serverId>docker-hub</serverId>       <!--服务id，与settings.xml对应-->
            <configuration>
                <dockerHost>http://192.168.79.130:2375</dockerHost>
                <imageName>mldn/mldncloud-eureka</imageName>     <!--镜像名称-->
                <imageTags>                                      <!--镜像标签-->
                    <imageTag>dev</imageTag>                     <!--可以指定多个标签-->
                    <imageTag>latest</imageTag>                  <!--可以指定多个标签-->
                </imagcTags>
                <baseImage>java</baseImage>                      <!--基础镜像-->
                <forceTags>true</forceTags>                      <!--覆盖已存在的镜像-->
                <entryPoint>
                    ["java", "-jar", "/${project.build.finalName}.jar"]
                </entryPoint>                                    <!--镜像启动命令-->
                <resources>
                    <resource>                                   <!--输出资源-->
                        <targetPath>/</targetPath>
                        <directory>${project.build.directory}</directory>
                        <include>${project.build.finalName}.jar</include>
                    </resource>
                </resources>
            </configuration>
```

```
        </plugin>
    </plugins>
</build>
```

本配置之中有如下几个重要的配置项。

☑ **serverId**：Maven 配置文件（settings.xml）中定义的服务连接 ID。

☑ **dockerHost**：配置 Docker 主机，并且要求此主机开启 API 进程。

☑ **imageTags**：定义镜像的标签，默认会使用 latest。

☑ **imageName**：镜像名称，该镜像名称需要满足"正则[a-z0-9-_.]"，必须有用户名做前缀。

☑ **baseImage**：要使用的基础镜像名称，如果本地 Docker 不存在，则会自动抓取镜像信息。

☑ **entryPoint**：Docker 镜像执行时的命令为 java -jar xxx.jar。

8.【mldncloud-eureka-profile 项目】程序打包编译 clean package docker:build -DpushImage，如图 15-11 所示。

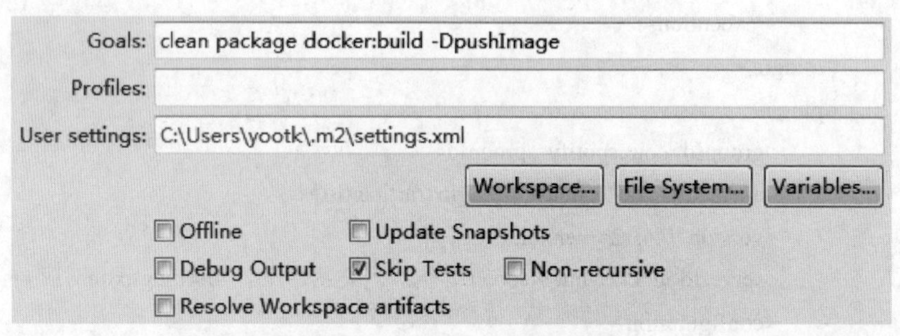

图 15-11 自动打包 Docker 镜像

提示：打包后会自动生成 Dockerfile 文件。

在程序执行打包之后，会自动根据插件配置在 target/docker 目录中生成的 Dockerfile 文件。此文件内容如下：

```
FROM java
ADD /eureka-server.jar //
ENTRYPOINT ["java", "-jar", "/eureka-server.jar"]
```

通过配置文件描述可以发现，此时设置了基础镜像来源，同时在镜像中加入 eureka-server.jar 文件，以及 Docker 镜像启动后自动执行命令。

此时，如果本地没有 java 镜像，文件会自动下载，而后会利用此镜像对生成的 eureka-server.jar 进行 Docker 创建。随后再通过 docker images 查看所有镜像，可以发现提供有 mldncloud-eureka 的镜像信息，如图 15-12 所示。同时在 DockerHub 中也会出现有镜像信息，如图 15-13 所示。

REPOSITORY	TAG	IMAGE ID	CREATED	SIZE
mldn/mldncloud-eureka	dev	ee0ec6fdec60	30 minutes ago	685 MB
mldn/mldncloud-eureka	latest	ee0ec6fdec60	30 minutes ago	685 MB
mldn/ubuntu-test	dev	e98006759b6f	14 hours ago	1.11 GB
mldn/ubuntu-tomcat	base	ab6a49d5fb0f	15 hours ago	778 MB
mldn/ubuntu	base	78745e02bfbc	16 hours ago	151 MB
ubuntu	latest	2a4cca5ac898	8 days ago	111 MB
java	latest	d23bdf5b1b1b	12 months ago	643 MB

图 15-12 Docker 本地镜像

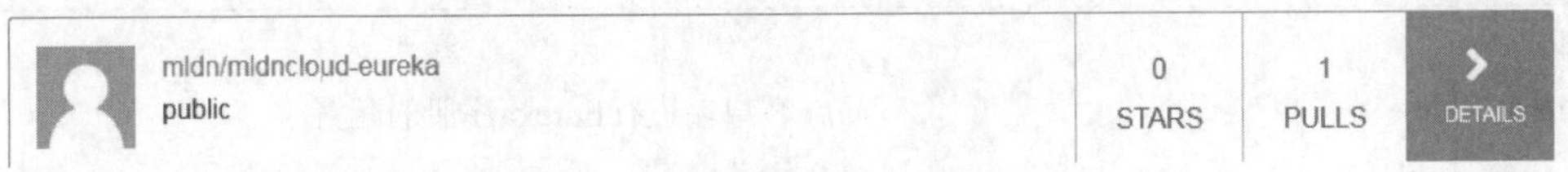

图 15-13　DockerHub 中的镜像信息

9.【Linux 系统】运行指定 Docker 镜像。

```
docker run -p 7001:7001 -itd mldncloud-eureka
```

在启动指定镜像文件时，需要设置映射端口，同时为了保护微服务执行，应该使用-d 参数将其设置为后台启动。

15.5.2　使用 DockerCompose 编排顺序

在一个系统中通常存在大量的微服务，即使具有 Docker 容器支持，也需要解决启动顺序问题。为了方便项目部署，以及更好地应用 Docker 镜像，往往会结合 DockerCompose 工具来对 Docker 镜像的启动顺序进行编排处理。

下面将为读者讲解如何在一个 Docker 主机中实现 Eureka-HA 处理机制。为了方便读者理解，本例将通过 Docker 镜像与 profile 共同实现集群部署。运行的 Docker 容器关系如图 15-14 所示。

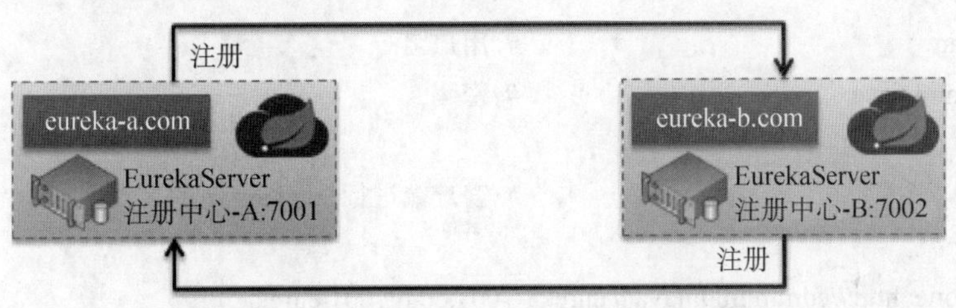

图 15-14　基于 Docker 实现 Eureka-HA 机制

1.【mldncloud-eureka-profile 项目】为了方便处理，本例将通过 profile 配置实现多个 Eureka 微服务。修改 application.yml，追加多个 profile 环境。

```
spring:
  profiles:
    active: dev                          # 定义默认生效的环境
---
spring:
  profiles: dev
server:
  port: 7001                             # 定义运行端口
security:
  basic:
    enabled: true                        # 启用安全认证处理
  user:
    name: edmin                          # 用户名
    password: mldnjava                   # 密码
```

```
eureka:
  client:                                          # 客户端进行Eureka注册的配置
    service-url:
      defaultZone: http://edmin:mldnjava@eureka-7002.com:7002/eureka
    register-with-eureka: false                    # 当前的微服务不注册到Eureka之中
    fetch-registry: false                          # 不通过Eureka获取注册信息
  instance:                                        # Eureak实例定义
    hostname: eureka-7001.com                      # 定义Eureka实例所在的主机名称
---
spring:
  profiles: product
server:
  port: 7002                                       # 定义运行端口
security:
  basic:
    enabled: true                                  # 启用安全认证处理
  user:
    name: edmin                                    # 用户名
    password: mldnjava                             # 密码
eureka:
  client:                                          # 客户端进行Eureka注册的配置
    service-url:
      defaultZone: http://edmin:mldnjava@eureka-7001.com:7001/eureka
    register-with-eureka: false                    # 当前的微服务不注册到Eureka之中
    fetch-registry: false                          # 不通过Eureka获取注册信息
  instance:                                        # Eureak实例定义
    hostname: eureka-7002.com                      # 定义 Eureka 实例所在的主机名称
```

2.【Linux 系统】docker-compose 软件包需要单独安装，利用 apt-get 命令获取。

```
apt-get -y install docker-compose
```

3.【Linux 系统】下载完成后查看 docker-compose 版本。

```
docker-compose --version
```

此时，返回版本信息为 docker-compose version 1.8.0。

4.【Linux 系统】在/usr/local 目录下编写 docker-compose.yml（或 docker-compose.yaml）文件。

```
version: '2'                           # 定义版本
services:                              # 定义要启动的服务
  eurekaserver1:                       # 定义服务名称
    image: mldn/mldncloud-eureka:dev   # 要使用的镜像
```

```
    ports:                               # 映射端口
      - "7001:7001"
    environment:                         # 定义环境属性
      - spring.profiles.active=dev       # 定义 profile 名称
  eurekaserver2:                         # 定义服务名称
    image: mldn/mldncloud-eureka:dev
    ports:                               # 映射端口
      - "7002:7002"
    environment:                         # 定义环境属性
      - spring.profiles.active=product   # 定义 profile 名称
```

5.【Linux 系统】在 docker-compose.yml 所在的目录下执行启动命令。

```
docker-compose up -d
```

本程序采用-d 参数进行镜像服务的后台启动，而后 docker-compose 会按照定义的顺序进行镜像启动，可以通过 environment 属性配置不同的 profile 信息，容器启动后打开任意一个 Eureka 控制台，就可以看到 HA 信息，如图 15-15 所示。

图 15-15　Eureka-HA 正常启动

15.6　本章小结

1. Docker 是基于云服务的一种应用，可以在一台主机上实现若干服务的部署。

2. 一个 Docker 镜像可以创建出若干个 Docker 容器，每一个 Docker 容器独立存在。

3. Docker 镜像可以通过文件保存或加载，也可以提交到 DockerHub 中进行统一管理。

4. 微服务可以利用 Maven 插件实现 Docker 镜像的创建，同时也可以将 Docker 镜像直接提交到 DockerHub 中。

5. 当需要限定微服务启动顺序时，可以利用 DockerCompose 编排服务启动顺序，简化项目部署流程。